Springer Transactions in Civil and Environmental Engineering

More information about this series at http://www.springer.com/series/13593

N. A. Siddiqui · S. M. Tauseef
S. A. Abbasi · Ali S. Rangwala
Editors

Advances in Fire and Process Safety

Select Proceedings of HSFEA 2016

Springer

Editors
N. A. Siddiqui
College of Engineering Studies
University of Petroleum and Energy Studies
Dehradun, Uttarakhand
India

S. A. Abbasi
Centre for Pollution Control and Environmental Engineering
Pondicherry
India

S. M. Tauseef
Environmental Research Institute
University of Petroleum and Energy Studies
Dehradun, Uttarakhand
India

Ali S. Rangwala
Department of Fire Protection Engineering
Worcester Polytechnic Institute
Worcester, MA
USA

ISSN 2363-7633 ISSN 2363-7641 (electronic)
Springer Transactions in Civil and Environmental Engineering
ISBN 978-981-10-7280-2 ISBN 978-981-10-7281-9 (eBook)
https://doi.org/10.1007/978-981-10-7281-9

Library of Congress Control Number: 2017958625

Printed on acid-free paper

This Springer imprint is published by Springer Nature
The registered company is Springer Nature Singapore Pte Ltd.
The registered company address is: 152 Beach Road, #21-01/04 Gateway East, Singapore 189721, Singapore

Preface

The discovery of fire enabled mankind to conquer other life forms and subsequently dominate the planet. As mankind learned various benefits of fire, it also learned that, if not controlled, a fire can destroy its initiator. This realization made humankind strive to find ways and means to control fire.

But it was easier wished than done and the struggle to master fire is still on. Despite a better understanding of fire and technological advancements toward taming it, mankind hasn't been able to achieve total control over fire. Be it forest fires or accidental fires at chemical process industries, fires still leave even the most technologically sophisticated nations helpless in dealing with it.

Similar helplessness is experienced when accidents in chemical process industries continue to occur despite the deeper knowledge of and technological advancements in this field. While some accidents cause temporary harm, like the gas leak from a container depot in New Delhi on May 7, 2017, which resulted in the hospitalization of more than 450 residents including school children, the effect of other accidents can linger for generations and poison several square miles, as happened due to the Bhopal gas tragedy in 1984. The oil spill that followed the accident at British oil rig in Gulf of Mexico resulted in a huge environmental loss estimated at USD 20 billion. Thus, there is a lot more that needs to be understood about fires and process accidents in order to gain total mastery over them.

The key objective of this publication is to update the reader with the latest developments in the field of fire and process safety and inform on related opportunities and challenges. This volume presents select papers that were presented at the international conference on advances in the field of health, safety, fire, environment, allied sciences and engineering (HSFEA 2016), held during November 18–19, 2016, at the University of Petroleum and Energy Studies (UPES), Dehradun. The conference was attended by leading academicians, technocrats, captains of industry, policy makers, budding scholars, and graduate students. The contribution from the authors covers topics ranging from technology that assists in ensuring a safe working environment free from process hazards that could translate into accidents causing loss of life and/or damage to property and the environment. Topics on methods that can be used for hazard identification and risk analysis are also included. Additionally, the importance of ensuring safety and healthy work environment, free from fire and process hazards, is stressed upon.

Dehradun, India N. A. Siddiqui
Dehradun, India S. M. Tauseef
Pondicherry, India S. A. Abbasi
Worcester, USA Ali S. Rangwala

Acknowledgements

This volume has benefitted from contributions from participants at the international conference on advances in the field of health, safety, fire, environment, allied sciences and engineering (HSFEA 2016) from November 18–19, 2016, University of Petroleum and Energy Studies (UPES), Dehradun, and the support and input of a number of individuals and institutions.

We thank Dr. S. J. Chopra (Honorable Chancellor, UPES), Prof. Shrihari Honwad (Vice-Chancellor, UPES), and Prof. Utpal Ghosh (CEO and President, UPES) for their support and encouragement. We are grateful to the Chief Guest of HSFEA 2016—Prof. V. K. Jain (Vice-Chancellor, Doon University)—for gracing the event with his presence; distinguished speakers—Senior Prof. S. A. Abbasi (Professor Emeritus UGC, Pondicherry University), Mr. Devendra Gill (Sr. Additional General Manager, Delhi Metro Railway Corporation), Dr. Tasneem Abbasi (Assistant Professor, Pondicherry University, and concurrently visiting Associate Professor, Worcester Polytechnic Institute, USA), and Dr. R. K. Sharma (General Manager, India Glycols Ltd)—for their talks.

The organizers of HSFEA 2016 wish to thank all the reviewers for their valuable time and comments on the quality of the papers.

We acknowledge the support of our sponsors TATA Projects, Uttarakhand Jal Sansthan, Action for Sustainable, Efficacious Development and Awareness (ASEA), Akbar HSE, PARAM Environmental Solutions, SWAN Enviro, Uttarakhand Environment Protection and Pollution Control Board (UEPPCB).

We also thank the chairs and members of various committees as under:

Steering Committee
Chief Patron
Dr. S. J. Chopra, Chancellor, UPES, Dehradun (India)

Patron(s)
Mr. Utpal Ghosh, President and CEO, UPES, Dehradun (India)
Dr. Shrihari Honwad, Vice-Chancellor, UPES, Dehradun (India)
Dr. S. R. Das, Vice President, Academic Affairs, UPES, Dehradun (India)

General Chair
Dr. Kamal Bansal, Dean, College of Engineering Studies, UPES, Dehradun (India)

Program Chair
Prof. S. A. Abbasi, Emeritus Professor (UGC), Centre for Pollution Control and Environmental Engineering, Pondicherry University, Pondicherry (India)

Organizing Chair
Prof. S. C. Gupta, Associate Dean (COES), UPES, Dehradun (India)
Dr. Piyush Kuchhal, Associate Dean (COES), UPES, Dehradun (India)
Dr. Manish Parteek, Associate Dean (COES), UPES, Dehradun (India)
Dr. S. K. Banerjee, Associate Dean (COES), UPES, Dehradun (India)
Dr. Suresh Kumar, Associate Dean (COES), UPES, Dehradun (India)
Dr. Rajneesh Garg, Associate Dean (COES), UPES, Dehradun (India)

Organizing Co-chair
Dr. N. A. Siddiqui, HOD (HSE), UPES
Dr. S. M. Tauseef, A.P. (SG) and Head Environmental Research Institute (R&D), UPES

Publicity Chair
Dr. N. Siddiqui, HOD (HSE), UPES
Dr. Adesh Kumar, A.P. (SS), EIC, UPES
Dr. Vikas Garg, HOD, Civil Engineering, UPES

Public Relation Chair
Dr. Jitendra Kumar Pandey, Head (R&D), UPES
Dr. Paawan Sharma, A.P. (SS), EIC, UPES
Dr. Mukul Gupta, A.P. (SS), EIC, UPES

Session Management Chair
Dr. N. Parsanthi, A.P. (SG), EIC, UPES
Mr. Ranjan Mishra, A.P. (SG), EIC, UPES

Organizing Committee
Convener
Dr. N. A. Siddiqui, Sr. Associate Professor and HOD (HSE), UPES

Co-convener
Mr. B. P. Yadav, A.P. (SG), HSE, UPES

Conference Secretary
Dr. S. M. Tauseef, A.P. (SG) and Head Environment Research Institute (R&D), UPES

Finance Committee
Mr. Prasenjit Mondal, A.P. (SS), EIC, UPES
Mr. K. Dharani Kumar, A.P., HSE, UPES

Technical Support
Dr. S. M. Tauseef, A.P. (SG) and Head Environment Research Institute (R&D), UPES
Dr. Kanchan Deoli Bahukhandi, A.P. (SS), UPES
Dr. Rajesh Singh, A.P. (SG) and Head Institute of Robotics Tech. (R&D), UPES
Ms. Anita Gehlot, A.P., UPES
Dr. Bhawna Yadav Lamba, A.P., UPES
Dr. Shiley Singhal, A.P. (SG), UPES
Dr. Neeraj Anand, UPES
Dr. Rajnish Garg, UPES
Dr. D. K. Gupta, HOD, Petroleum Engineering, UPES
Dr. Manish Prateek, UPES
Dr. Vikash Garg, HOD, Civil Engineering, UPES
Dr. Suresh Kumar, UPES
Dr. OM Parkash, HOD, Aerospace Engineering, UPES
Dr. Pankaj, Sharma, Sr. Associate Professor, Mechanical Engineering, UPES
Dr. Rajeev Gupta, Associate Professor, Department of Physics, UPES
Dr. Tarun Dhingra, Sr. Associate Professor, College of Management, UPES
Dr. Prasoom Dwidi, Sr. Associate Professor, College of Management, UPES

Sponsorship Committee
Dr. Nihal Siddiqui, HOD (HSE), UPES
Dr. Manish Prateek, Associate Dean (CIT), UPES
Dr. Ajay Kumar, Director PG Program, UPES
Mr. B. P. Yadav, A.P. (SG), UPES

Registration Committee
Mr. Abhishek Nandan, A.P., HSE, UPES
Mr. K. Dharani Kumar, A.P., HSE, UPES
Dr. Kanchan Deoli Bahukhandi, A.P. (SS), UPES
Mr. Prasenjit Mondal, A.P. (SS), EIC, UPES
Mrs. Madhuben Sharma, A.P., UPES

Session Management
Dr. R. K. Tripathi, UPES
Mr. Prasenjit Mondal, A.P. (SS), EIC, UPES
Mrs. Madhuben Sharma, A.P., UPES
Mr. Abhinav Srivastava, UPES
Mr. Ritesh Saini, UPES

Hospitality Committee
Mr. Manish Yagnik, UPES
Dr. Neeraj Anad, UPES
Mrs. Madhuben Sharma, A.P., UPES
Mr. Somyadeep Baksi, A.P., UPES

Venue Arrangement Committee
Mr. Venkat Krishnakanth, A.P., UPES
Mr. B. P. Yadav, A.P. (SG), UPES
Dr. S. M. Tauseef, A.P. (SG) and Head Environment Research Institute (R&D), UPES
Mr. Iqbal, AP, UPES

Transport Committee
Mr. Anjul Khandoori, UPES
Dr. Rishi Diwan, A.P. (SG), UPES
Mr. Amit Mondal, UPES
Mr. Shival Dubey, A.P., UPES

Member of International Technical Panel
Prof. S. A. Abbasi, Pondicherry University, India
Prof. K. Kesava Rao, Indian Institute of Science (IISc) Bangalore, India
Prof. F. I. Khan, Memorial University, Newfound Land, Canada
Prof. R. Rusli, Universiti Teknologi Petronas, Canada
Prof. A. Tugnoli, Universitá Di Bologna, Italy
Dr. V. Garaniya, National Centre for Maritime Engineering and Hydrodynamics, University of Tasmania, Australia
Prof. R. Abbassi, University of Tasmania, Australia
Ms. A. Roychowdhury, Executive Director, Centre for Science and Environment (CSE), New Delhi, India
Dr. Savita, Director, Forest Research Institute (FRI), and Vice-Chancellor, FRI Deemed University, Dehradun, India
Dr. Anurag Garg, Indian Institute of Technology (IIT) Bombay, India
Prof. S. Mukherji, Indian Institute of Technology (IIT) Bombay, India
Dr. M. K. Chandel, Indian Institute of Technology (IIT) Bombay, India
Dr. S. Kushari, Indian Institute of Technology (IIT) Kanpur, India
Prof. D. S. Arya, Indian Institute of Technology (IIT) Roorkee, India
Prof. A. Gairola, Indian Institute of Technology (IIT) Roorkee, India
Prof. I. D. Mall, Indian Institute of Technology (IIT) Roorkee, India
Prof. S. N. Upadhyay, Former Director, Indian Institute of Technology (IIT) BHU, Varanasi, India
Prof. I. M. Mishra, Indian School of Mines (ISM), Dhanbad, India
Dr. B. Sengupta, Former Member Secretary, Central Pollution Control Board (CPCB), India
Prof. S. Shariatipour, Coventry University, UK
Prof. S. Ahmed, Memorial University, Newfound Land, Canada
Prof. H. Gaber, University of Ontario, Canada
Prof. Javed Ahmad, Dean Faculty of Science, Jamia Hamdard, India
Dr. Waddal Shihab M.A. Ghanem, Emirates National Oil Company Ltd. (ENOC) L.L.C, Dubai, United Arab Emirates (UAE)
Dr. P. Sojan Lal, ADCO, Abu Dhabi, United Arab Emirates (UAE)
Dr. C. Rezzaei, KOC, Kuwait

Mr. Shakeel H. Kadr, Executive Director, CCPS
Mr. Hari Kumar Polavarapu, Director, ENOC
Hafedh Al-Qassab, GM, The Bahrain Petroleum Company, Awali, Kingdom of Bahrain
Prof. D. Rashtchian, Sharif University of Technology, Iran
Dr. B. Abdolhamidzadeh, Sharif University of Technology, Iran
Prof. S. Ayatollahi, Sharif University of Technology, Iran
Dr. M. Tasaduq, UT Southwestern Medical Centre at Dallas, USA
Prof. A. S. Rangwala, Worcester Polytechnic Institute, USA
Ms. Erica Poff, Government Affairs and Outreach Manager, Board of Certified Safety Professionals, USA
Dr. Ramkrishna Akula, Emirates Gas LCC, Dubai, United Arab Emirates (UAE)
Dr. M. Yunus, Vice-Chancellor, Jawahar University, India
Dr. T. Yogeshwara, Vice President, JSW Steel, India
Dr. Shajid Jamal, Oil and Natural Gas Corporation of India (ONGC), New Delhi, India
Dr. Asha Rajvanshi, Professor and Head, Wildlife Institute of India, India
Dr. T. K. Joshi, Director, Centre for Occupational and Environmental Health, Maulana Azad Medical College, India
Mr. Fareed Bin Ali Al-Hinai, HSE and Sustainable Development Team Leader, PDO, Oman
Prof. Suresh Jain, Teri University, New Delhi, India
Dr. Elangawan, Deputy Director, Director General Factory and Services Labour Institute, Mumbai, India
Dr. Piyoosh Rautela, Executive Director, Disaster Mitigation and Management Centre, Dehradun, India
Dr. Suresh Kumar, Scientist, Centre Pulp and Paper Research Institute, Saharanpur, India
Dr. Seyed M. Shariatipour, Faculty Coventry University, UK
Prof. Sumeet S. Aphale, University of Aberdeen, UK
Prof. Shephard Ndlovu, University of Central Lancashire, UK

Contents

Editors and Contributors

About the Editors

Dr. N. A. Siddiqui is currently Professor at the University of Petroleum and Energy Studies (UPES), India (2005–present). He has a master's degree in Environmental Sciences from Lucknow University, India (1996), and a Ph.D. in Environmental Biology from Awadhesh Pratap Singh (APS) University, Rewa, India. Additionally, he has a number of professional certifications, viz NEBOSH diploma, NEBOSH–IGCNE, IOSH Managing Safely, Lead Auditor OHSAS 18001, Lead Auditor ISO 14001, and GRAD-IOSH. His research interests include occupational health, safety, and environment. He has co-authored more than 100 papers and 8 books. He has supervised 17 Ph.D. and 200 M.Tech. theses. He has organized a number of national and international seminars/conferences/MDPs. He has also worked on a number of major R&D projects.

Dr. S. M. Tauseef is an Assistant Professor (SG) and Head of the Environmental Research Institute at the University of Petroleum and Energy Studies (UPES), India (2015–present). He has an MS in Chemical Engineering (Process Design) from Sharif University of Technology, Iran (2008), and a Ph.D. in Environmental Engineering from Pondicherry University (PU) (2011). Prior to joining the UPES, he worked as a Pool Officer [Council of Scientific and Industrial Research (CSIR)] at PU (2012–2015). His research interests include process safety, especially forecasting and consequence assessment of accidents in chemical process industries. He has developed methodologies for the application of computational fluid dynamics (CFD) to dense gas dispersion, vapor cloud explosions, single and multiple pool fires. His other interest has been in solid waste management, especially the design of continuously operable vermireactor systems for assorted wastes and high-rate anaerobic digesters. So far, he has published 25 journal articles, co-authored 3 books, filed 3 patents, and presented papers in 21 conferences.

Prof. S. A. Abbasi is Professor Emeritus (UGC) and Founder Head of the Centre for Pollution Control and Environmental Engineering, Pondicherry University, India. His main calling has been environmental engineering since 1970—a field to which he has contributed 42 books, 9 patents, and over 400 research papers. In 1992, he started the, now globally recognized, process safety research program at Pondicherry University. His lifetime achievements have been recognized in the form of several national and international awards. He has been elected Fellow of the Institution of Engineers (1993), Indian Institute of Chemical Engineers (1999), and National Academy of Sciences (2003). From 1984 onward, he has been invited frequently to chair technical sessions or present plenary lectures in international conferences around the globe.

Dr. Ali S. Rangwala is Associate Professor in the Department of Fire Protection Engineering at Worcester Polytechnic Institute (WPI), Worcester, USA (2006–present). He has a B.S. in Electrical Engineering from the Government College of Engineering, Pune, India (2000), an M.S. in Fire Protection Engineering from the University of Maryland, College Park (2002), and a Ph.D. in Mechanical and Aerospace Engineering from the University of California, San Diego (2006). His research interests include deflagration of combustible dust clouds, ignition behavior of combustible dust layers, in situ burning of oil, spread of an oil slick in channels, velocity measuring techniques in fire-induced flows, and flame propagation and burning rate behavior of condensed fuel surfaces. In the last 10 years, he has published over 30 journal articles and has presented in over 30 conferences. He teaches three graduate courses: Explosion Protection, Industrial Fire Protection, and Combustion in the Department of Fire Protection Engineering, Worcester Polytechnic Institute, Worcester, USA.

Contributors

Tasneem Abbasi Centre for Pollution Control and Environmental Engineering, Pondicherry University, Kalapet, Pondicherry, India

Bahman Abdolhamidzadeh Department of Chemical and Petroleum Engineering, Center for Process Design, Safety and Loss Prevention (CPSL), Sharif University of Technology, Tehran, Iran

B. Abhishek Shastri Department of Health, Safety and Environment Engineering, University of Petroleum and Energy Studies, Bidholi, Dehradun, India

Aditya Agarwal University of Petroleum and Energy Studies, Bidholi, Dehradun, India

Sushaban Choudhary University of Petroleum and Energy Studies, Dehradun, India

Aman Dalmia Department of Aerospace Engineering, University of Petroleum and Energy Studies, Dehradun, Uttarakhand, India

Rishi Dewan University of Petroleum and Energy Studies, Dehradun, India

Shagufta Ejaz HSE Design, Technip India Ltd., Noida, India

Anita Gehlot University of Petroleum and Energy Studies, Dehradun, India

Tannaz Gharebaghian Faculty of Environment and Energy, Science and Research Branch, Islamic Azad University, Tehran, Iran

Beporam Iftekhar Hussain Department of Mechanical Engineering, Bapatla Engineering College, Bapatla, Guntur, Andhra Pradesh, India

Jeevani Jayasinghe Department of Electronics, Wayamba University of Sri Lanka, Kuliyapitiya, Sri Lanka

Joji John Department of Health, Safety and Environment Engineering, University of Petroleum and Energy Studies, Dehradun, India

Vipin Khattri Faculty of Computer Science & Engineering, Shri Ramswaroop Memorial University, Barabanki, India

Vamsi Krishna Kudapa Department of Petroleum Engineering, University of Petroleum and Energy Studies, Dehradun, India

Dharani Kumar University of Petroleum and Energy Studies, Bidholi via Premnagar, Dehradun, India

Rishabh Kumar Department of Aerospace Engineering, University of Petroleum and Energy Studies, Dehradun, Uttarakhand, India

Akshay Maan University of Petroleum and Energy Studies, Dehradun, India

Ashwin Malviya Vedanta Limited, New Delhi, India

Raj Gaurav Mishra Department of Electronics, Instrumentation and Control, CoES, University of Petroleum and Energy Studies, Dehradun, India

Euginia Diana Mukhim Centre for Pollution Control and Environmental Engineering, Pondicherry University, Kalapet, Pondicherry, India

Abhishek Nandan Department of Health Safety and Environment Engineering, University of Petroleum and Energy Studies, Dehradun, India

Behnaz Nosraty Faculty of Environment and Energy, Science and Research Branch, Islamic Azad University, Tehran, Iran

Appil Ora University of Petroleum and Energy Studies, Dehradun, India

Saurabh Pandey Department of Aerospace Engineering, University of Petroleum and Energy Studies, Dehradun, Uttarakhand, India

Praveen Pitta University of Petroleum and Energy Studies, Dehradun, India

Farshad Rahimi Department of Chemical and Petroleum Engineering, Center for Process Design, Safety and Loss Prevention (CPSL), Sharif University of Technology, Tehran, Iran

Davood Rashtchian Department of Chemical and Petroleum Engineering, Center for Process Design, Safety and Loss Prevention (CPSL), Sharif University of Technology, Tehran, Iran

Cyrus Rezaei HSE Group, Kuwait Oil Company, Ahmadi, Kuwait

Mir Safiulla Department of Mechanical Engineering, Ghousia College of Engineering, Ramanagaram, Karnataka, India

R. Sahadev Department of Health, Safety and Environment Engineering, University of Petroleum and Energy Studies, Bidholi, Dehradun, India

Suvek Salankar College of Engineering Studies, UPES, Dehradun, India

Dipti Sharath Department of Health Safety and Environment Engineering, University of Petroleum and Energy Studies, Dehradun, India

Mayank Sharma University of Petroleum and Energy Studies, Dehradun, India

Pushpa Sharma Department of Petroleum Engineering, University of Petroleum and Energy Studies, Dehradun, India

R. K. Sharma India Glycol, Kashipur, India

Vishal Sharma University of Petroleum and Energy Studies, Bidholi, Dehradun, India

R. Shristi Vidusha Weatherford Drilling Services, Dubai, UAE

Vishakha Shukla University of Petroleum and Energy Studies, Bidholi, Dehradun, India

Nihal A. Siddiqui Department of Health, Safety & Environment, University of Petroleum and Energy Studies, Dehradun, Uttarakhand, India

Deepak Kumar Singh Department of Computer Science, Integral University, Lucknow, India

Rajesh Singh University of Petroleum and Energy Studies, Dehradun, India

Vishal Kumar Singh Vedanta Limited, New Delhi, India

Y. Sivaji Raghav Department of Health, Safety and Environment Engineering, University of Petroleum and Energy Studies, Bidholi, Dehradun, India

Rohit Smkaria University of Petroleum and Energy Studies, Dehradun, India

S. M. Tauseef Environmental Research Institute, University of Petroleum and Energy Studies (UPES), Dehradun, Uttarakhand, India

Cherish Thomas Department of Health, Safety and Environment Engineering, University of Petroleum and Energy Studies, Dehradun, India

Sagarkumar Vaghasia Department of Health, Safety and Environment Engineering, University of Petroleum and Energy Studies, Bidholi, Dehradun, India

Ayush Vidyadharan Department of Health, Safety and Environment Engineering, University of Petroleum and Energy Studies, Dehradun, India

Bikarama Prasad Yadav Department of Health, Safety and Environment Engineering, University of Petroleum and Energy Studies, Bidholi, Dehradun, India

Jitendra Yadav University of Petroleum and Energy Studies, Dehradun, India

Practical Aspects of Application of Bayesian Networks to Cause and Effect Modeling in Process Safety

G. Unnikrishnan, Cyrus Rezaei and Nihal A. Siddiqui

Abstract Cause and effect scenarios in process safety are commonly modeled using fault tree analysis, event tree analysis, and/or bow tie methods. These can be readily mapped into Bayesian networks, and there have been several applications of the same. Bayesian network offers several advantages including easy visualization, updating as well as forward and backward calculations. However, there are several practical aspects that are to be kept in mind while modeling with Bayesian networks. This includes the increase in number of parent nodes and state entries in conditional probability tables, the use of equations, and difficulties in populating the same meaningfully with probability values. This paper will discuss the above factors in cause and effect modeling with Bayesian networks including the use of object-oriented Bayesian network and Noisy gates to handle the large number of parents and will be useful for researchers in the subject.

Keywords Cause and effect · Process safety · Bayesian network
Noisy OR

G. Unnikrishnan (✉)
Project Support Services Group, Kuwait Oil Company, PO Box 9758,
61008 Ahmadi, Kuwait
e-mail: ukrishnan@koc.com

C. Rezaei
HSE Group, Kuwait Oil Company, PO Box 9758, 61008 Ahmadi, Kuwait
e-mail: CRezaei@kockw.com

N. A. Siddiqui
Department of Health, Safety & Environment, University of Petroleum
and Energy Studies, Dehradun, Uttarakhand, India
e-mail: nihal@ddn.upes.ac.in

N. A. Siddiqui et al. (eds.), *Advances in Fire and Process Safety*,
Springer Transactions in Civil and Environmental Engineering,
https://doi.org/10.1007/978-981-10-7281-9_1

1 Introduction

Process safety accidents are commonly analyzed using cause and effect principles. Most common models of cause and effect use fault tree (FT), event tree (ET), or bow tie methods. Recently, there has been considerable interest in utilizing Bayesian methods to understand complex systems involving cause and effect. Mapping of the FT to Bayesian network (BN) [1], bow tie to BN [2], and ET to BN [3] has been described by researchers previously. To a certain extent, BN has been used to analyze accidents in process industries [4–9]. FT and ET use OR, AND, and other gates to model causes and effects or consequences. However, when these are mapped to BN, the resulting BN is sometimes quite large which is then difficult to understand and parameterize. Populating the conditional probability table (CPT) of the effect (child) nodes having large number of causes (parents) also poses a problem. On the other hand, industrial process systems are complex and require consideration for all important causes. Accidents in such systems involve immediate visible causes, many invisible secondary and primary root causes. How can such numerous causes that are probabilistic in nature be modeled? BN models offer several advantages in handling such situations. This paper will describe certain salient aspects of using BN to model cause and effect including the use of Noisy gate.

2 Basics of Fault Tree Analysis (FTA), Event Tree Analysis (ETA), and Bow Tie

2.1 *Fault Tree Analysis (FTA)*

Fault tree analysis (FTA) is a top-down deductive method for analyzing possible causes with the top event representing the hazard event or systems' failure. Event tree analysis (ETA) is a inductive method that can describe the logical progress a hazard event, for example, loss of containment to its eventual consequences (pool fire, vapor cloud explosion, boiling liquid expanding vapor explosion, toxic gas dispersion, etc.) considering the success or failure of a condition or a prevention (safety) barrier. Both these methods are well known and therefore not described further.

2.2 *Bow Tie Analysis*

Bow tie analysis is a combination of FTA and ETA. See Fig. 1. Left-hand side of the bow tie is the fault tree leading up to the top event (hazard), and the right side is the event tree showing the sequential progression of the hazard to its eventual consequences. Probabilities are computed based on the principles followed in FTA and ETA.

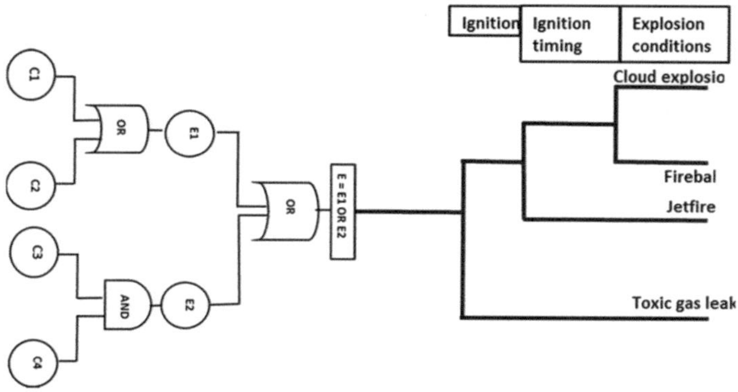

Fig. 1 Bow tie (combination of ET and FT)

2.3 *Bayesian Network (BN)*

BN consists of nodes and directed lines that can represent causes and effects relationships in an easily understandable way. Nodes represent system variables and arc the dependencies between nodes. The effect (child node *A*) is dependent (conditional) on the cause (parent node *B*). The probability of *A* happening once *B* has occurred $P(A|B)$ is calculated using the Bayes formula

$$P(A|B) = \frac{P(B|A) * P(A)}{P(B)} \tag{1}$$

where

$$P(B) = P(B|A) * P(A) + P(B|A') * P(A')$$

$P(A)$ Probability of *A* happening,
$P(B)$ Probability of *B* happening,
$P(A')$ Probability of *A* not happening.

$P(A|B)$ is called the posterior probability computed based on the likelihood function $P(B|A)$ and prior probability value $P(A)$. $P(B)$ is the normalizing factor calculated from sum of probability of occurrence and non-occurrence of *A*. Further details on BN are available in well-known references [10–12]. The conditional probability for the child (dependent) node is encoded in the conditional probability table (CPT) of the respective nodes. CPT defines the nature of the dependency of the parent nodes to the child nodes, while the parent nodes themselves contain the probability values of whether it will happen or not.

See Fig. 2a–d for cases of BN and corresponding simulation diagrams with one cause (parent) and effect (child) and 10 causes and one effect and corresponding using Netica software.

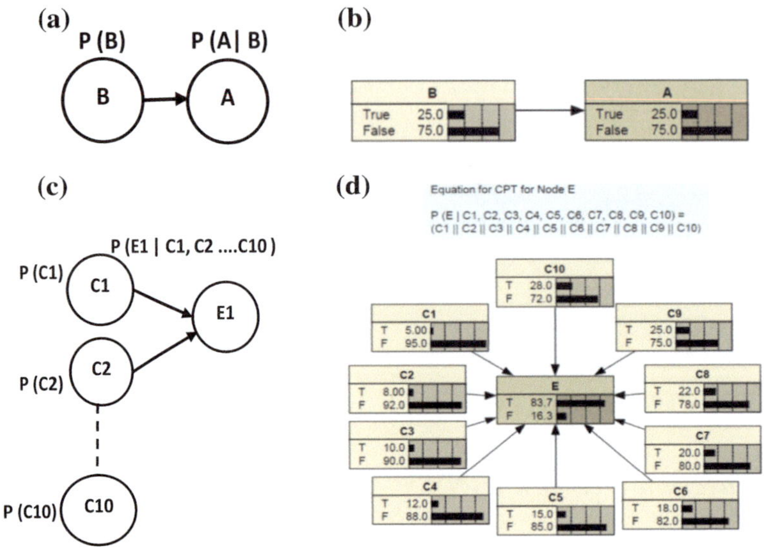

Fig. 2 Bayesian networks and corresponding simulation models

Table 1 A portion of conditional probability table for node *E* with OR gate in Fig. 2d

E	*C*1	*C*2	*C*3	*C*4	*C*5	*C*6	*C*7	*C*8	*C*9	*C*10
T	T	T	T	T	T	T	T	T	T	T
T	T	T	T	T	T	T	T	T	T	F
T	T	T	T	T	T	T	T	T	F	T
T	T	T	T	T	T	T	T	T	F	F
–	–	–	–	–	–	–	–	–	–	–

The CPT for the node *E* in Fig. 2d is quite large since it has 10 parents and has to take $2^{10} = 1024$ entries. Table 1 shows a portion of the conditional probability table for node *E* in Fig. 2d.

3 Mapping of FT and ET and Bow Tie to Bayesian Networks

References in [1–3] describe mapping of fault tree, bow tie, and event tree, respectively, to BN. These papers also note the advantages of BN in updating of the probabilities based on actual data as well as the forward (predictive) and backward (diagnostics) calculation capabilities. However, the examples used in the papers are generally amenable to conversion to BN easily. In actual industrial situations, modeling the system can become difficult due to the following:

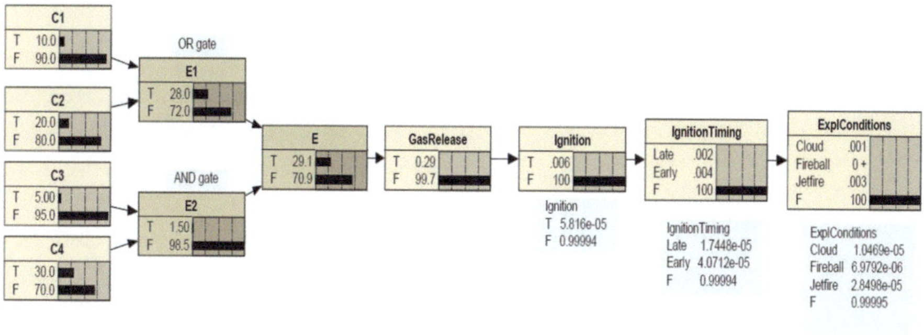

Fig. 3 Bayesian network for bow tie given in Fig. 1

a. High number of entries required in the CPT due to huge number of possible combinations: As noted above, if there are 10 causes to an effect (consequence) with binary states True (T) and False (F), the CPT has 2^{10} (1024) entries. Manual input is difficult and tedious. Equations have to be employed which require several considerations.
b. It is difficult to define the CPTs in a case stated above. BNs have to be decomposed and factored to include intermediate (predecessor) consequence nodes, which can make the BN quite large.
c. It is easier to use discrete probability values for parent nodes. When probability distributions are used, the same has to be converted to discrete levels and made amenable to further calculations.
d. Independence of causes which is a critical assumption sometimes may not be valid.
e. There are situations where effects can occur even when all causes are false (due to unknown reason) or when effects need not occur even if all the subsystems fail (the system continues to function for an unknown reason).

BN computations need software to fully map and simulate the model. Manual calculations are time consuming even for a moderate BN. See Fig. 3 depicting the BN model in Netica software for bow tie given in Fig. 1.

The following section describes practical ways of taking up the difficulties noted above.

4 Practical Aspects in Mapping FT and ET to BN

4.1 Factored BN

As noted in Sect. 3a and b above, it is possible to avoid combinatorial explosion by using factored BN. This will mean adding an intermediate consequence or predecessor node, which may result in a large BN. An example is shown in Fig. 4a, b.

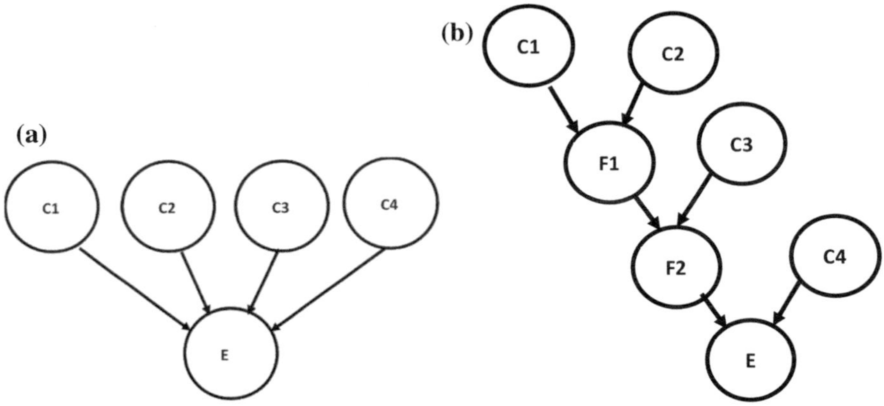

Fig. 4 **a** Bayesian network with four parents and **b** the same BN factored with intermediate nodes

The large number of entries in the CPT of node E due to four number of parents can be avoided by factoring the connections as shown in Fig. 4b, where each of the intermediate nodes $F1$, $F2$, and $F3$ has only two parents.

4.2 Use of Equations

Alternatively, equations can be used to populate the CPTs with True (T) or False (F) in line with the OR and AND gates. As an example for the BN showing 10 causes (parents) in Fig. 2b connected to the single cause (child) node through OR gate, the CPT can be filled by the equation shown in typical symbols where || represents the OR gate.

$$P(E|C1, C2\ldots C10) = (E, C1||C2\ldots||C10) \tag{2}$$

Similarly, when there is an AND gate connecting 10 causes to a single child effect, the CPT for the effect node will be

$$P(E|C1, C2\ldots C10) = (E, C1 \,\&\&\, C2\ldots \&\& \, C10) \tag{3}$$

However, it is important to note that such equations can only populate the CPT with deterministic T or F state and cannot add probabilistic values.

4.3 Use of Probability Distributions

Use of probability distributions for describing a parent prior state is more realistic and requires equations, a typical example of which is shown below based on Fig. 3, assuming that node *C*1 has normal probability distribution with mean 35 and standard deviation 11. Symbology is from Netica software.

$$P(E|C1) = \text{NormalDist}(C1, 35, 11) \quad (4)$$

The above equation will populate the CPT of the cause node *C*1 with values obtained from the normal distribution specified above. However, in order to use the values of CPT further, the values have to be placed in ranges called discretization levels. If node *C*1 has an overall range of 0–80, then suitable levels could be starting from 0 at 8, 16, 24, 32, 40, 48, 56, 64, 72, and 80, which are implemented in the node *C*1 levels. The defined levels are *L*1 = 0–8, *L*2 = 8–16, *L*3 = 16–24, *L*4 = 24–32, *L*5 = 32–40, *L*6 = 40–48, *L*7 = 48–56, *L*8 = 56–64, *L*9 = 64–72, and *L*10 = 72–80. See Fig. 5.

Without specifying such levels (syntax of which could differ between softwares), further use of the CPT values is not possible. The node *R* in Fig. 5 contains an

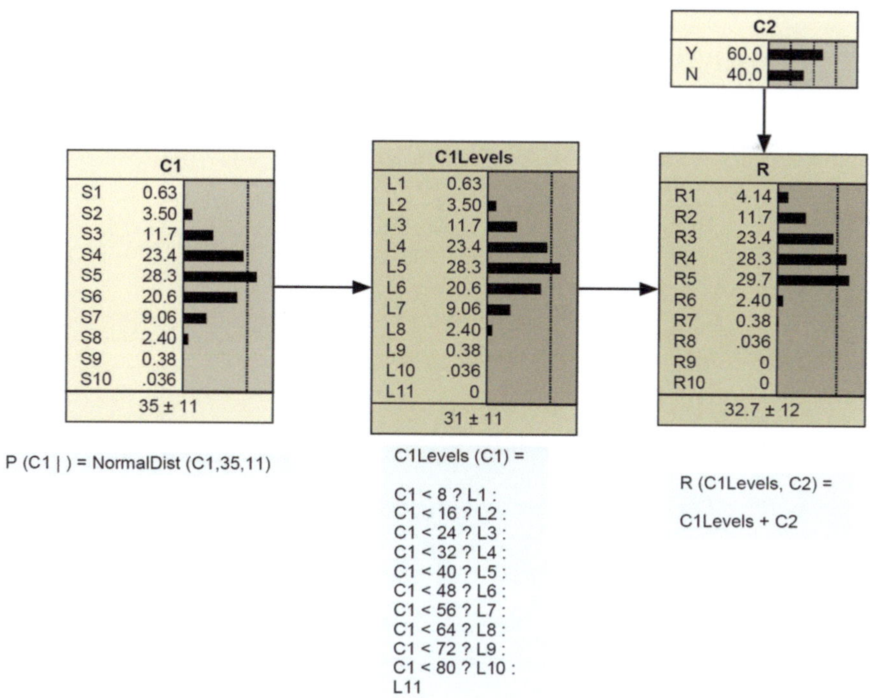

Fig. 5 Specifying normal distribution and discretization levels

equation to add the nodes $C1$Levels and $C2$. The values are added, and the probability of each addition in a particular range is displayed at the node. Other commonly used distributions for causes are binomial and beta distributions.

It is important to note that sufficient granularity has to be maintained in the discretization levels to yield reasonably accurate results.

4.4 Object-Oriented Bayesian Network (OOBN)

OOBN contains, in addition to normal nodes, 'instance nodes,' which is another complete sub-BN with interfaces that facilitates flow of information. Such models are useful when the requirements call for modeling repetitive standard pattern structures. The use of OOBN helps in constructing large and complex BN. The terminologies used in OOBN are same as those used in object-oriented programming languages, which include encapsulation, inheritance, and hierarchy. A simple OOBN for disease progression adapted from HUGIN [13] is shown below in Fig. 6a, b for illustration purposes.

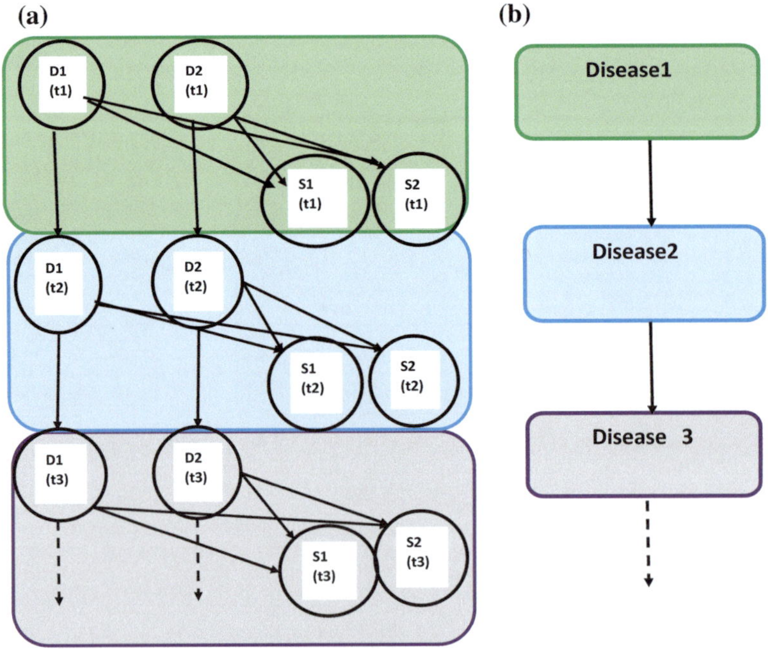

Fig. 6 Specifying normal distribution and discretization levels. **a** BN representing three time slices of disease progression and **b** condensed boundaries show the specification of OOBN for BN on the right

$D1$ and $D2$ are diseases that progress over time (t) such that $D1$ ($t2$) the next time slice depends on the current $D1$ ($t1$). $S1$ and $S2$ represent the symptoms of the disease. Suppose we want to know the probability of the disease and its symptoms after 10 time slices, then creating this repeatedly over 10 time slices is tedious work. With the key assumption that the transition probabilities $PD1$ $(t2)|D1$ $(t1)$ are identical in all time slices, construction of the time-sliced model becomes an easier work.

4.5 *Noisy-OR/AND Gates*

If there are numerous causes (parent nodes) as predecessors of an effect (child node), with unknown probability values, then modeling such a system realistically with BN becomes difficult. In such situation, Noisy-OR and/or Noisy-AND gates can be used [1]. Noisy-OR gates have been applied in BN developed for the HEPAR II project for diagnosis of liver disorders [14] with good results. It is quite easy to use Noisy-OR gate when the system is binary and follows Boolean logic as indicated below.

4.5.1 Noisy-OR Gate

Noisy-OR gate and its distribution can be used when there are several possible causes (parents) for an event, any of which can cause the event by itself, but only with a certain probability. Then, the parents have a characteristic called 'independence of causal influence,' which means that they have no influence from any other cause or parent. Noisy-OR gate reduces the data requirements and is quite useful when limited data is available. Additionally, a probability leak factor can be included in the distribution equation to take care of the fact that the event can occur spontaneously without any of the known causes being true. Typical equation for Noisy-OR gate is given below. Refer Netica help file (2010) [15] for details.

$$P(E|C1, C2, C3, \ldots C10) = \text{NoisyOrDist}(E, \text{leak factor}, C1, p1, C2, p2, C3, p3, \ldots C10, p10) \tag{5}$$

where

leak factor: fraction representing the probability that the effect (consequence) can occur by itself,

$C1$ to $C10$: represents causes, $p1$ to $p10$ representing the probability that the effects will occur due to the respective cause by itself, probability of all other causes being 0.

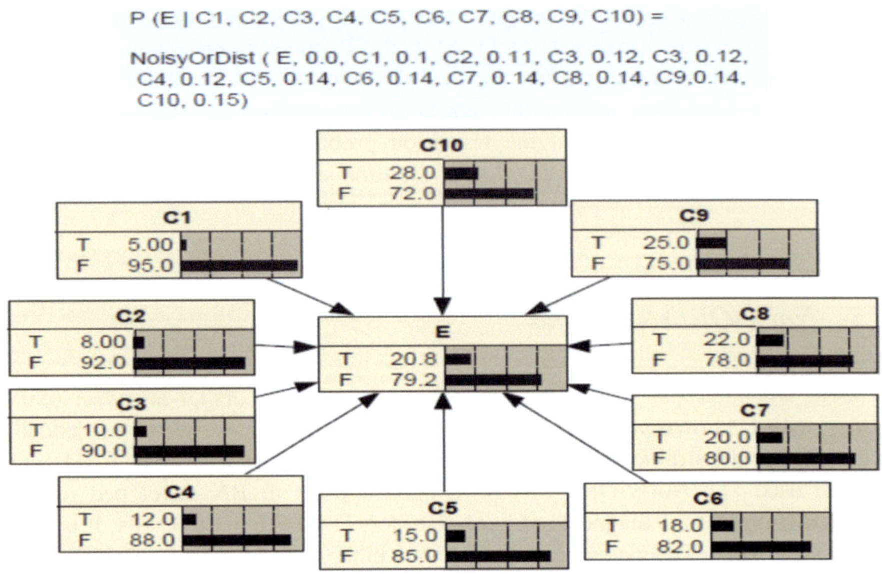

Fig. 7 BN model with 10 causes and one effect with Noisy-OR gate

Figure 7 shows the BN model for 10 causes and a single effect with Noisy-OR gate specification for the node E, and Eq. (6) defines the CPT table.

$$P(E|C1, C2, C3, \ldots C10) = \text{NoisyOrDist}(E, 0.0, C1, 0.1, C2, 0.11, C3, 0.12, C4, 0.12, C5, 0.14, C6, 0.14, C7, 0.14, C8, 0.14, C9, 0.14, C10, 0.15) \quad (6)$$

Table 2 shows a portion of the node E CPT having 1024 entries that has been automatically filled by use of the Noisy-OR distribution equation.

Table 2 A portion of the node E CPT having 1024 entries

E											
T	F	*C*1	*C*2	*C*3	*C*4	*C*5	*C*6	*C*7	*C*8	*C*9	*C*10
0.781731	0.218269	T	T	T	T	T	T	T	T	T	T
0.743213	0.256787	T	T	T	T	T	T	T	T	T	F
0.746199	0.253801	T	T	T	T	T	T	T	T	F	T
0.701411	0.298589	T	T	T	T	T	T	T	T	F	F
0.746199	0.253801	T	T	T	T	T	T	T	F	T	T
0.701411	0.298589	T	T	T	T	T	T	T	F	T	F
0.704883	0.295117	T	T	T	T	T	T	T	F	F	T
–	–	–	–	–	–	–	–	–	–	–	–

4.5.2 Noisy-AND Gate

Noisy-AND gate is the counterpart of Noisy-OR gate. It is used where there are several possible causes for an effect through an AND gate, each of which has to have a certain probability that will be necessary for the event to occur. It can also model a situation where the effect may not occur even when all causes are true through a similar leak factor.

One of the key advantages of Noisy-OR and Noisy-AND distribution is that the parameters can be fine-tuned based on actual data.

4.5.3 Noisy Max

Noisy Max is an extension of Noisy OR and can be used when the parents have more than two states. It represents a particular type of influence a child node can have with its parents, in which each parent contributes probabilistically to the child to a certain extent, with the effect being the maximum of all those amounts. Noisy Max has been used successfully in medical diagnostics [16, 17]. For details, readers are referred to Netica [14]. The method can be readily adopted to cause and effect in accident scenarios.

5 Conclusion

The paper describes the practical aspects that are to be considered when FTA, ETA, and bow tie are mapped to BN. The huge number of entries in the CPT is a problem when handling BN with large number of causes (parents). But it can be overcome through the use of factorization and equations which requires discretization and careful definition of the discretization levels. Situations with large number of causes with uncertainties can be modeled by the use of Noisy-OR, Noisy-AND, or Noisy-Max gates. Thus, BN offers more flexibility than any of the traditional methods.

References

1. Bobio, A., L. Portinale, M. Minichino, and E. Ciancamerla. 2001. Improving the analysis of dependable systems by mapping fault trees into Bayesian networks. *Reliability Engineering and System Safety* 71: 249–260.
2. Khakzad, N., F. Khan, and P. Amyotte. 2013. Dynamic safety analysis of process systems by mapping bow-tie into Bayesian networks. *Process Safety and Environmental Protection* 91: 46–53.
3. Unnikrishnan, G., H. Shrihari, N. Siddiqui. 2014. Application of Bayesian methods to event trees with case studies. *Reliability: Theory & Applications* 9: 32–45.

4. Kalantarnia, M., F. Khan, and K. Hawboldt. 2009. Dynamic risk assessment using fault tree and Bayesian theory. *Journal of Loss Prevention in the Process Industries* 22: 600–606.
5. Kalantarnia, M., F. Khan, and K. Hawboldt. 2010. Modelling of BP Texas city refinery accident using dynamic risk assessment approach. *Process Safety and Environmental Protection* 88: 191–199.
6. Wang, Y.F., M. Xie, K.S. Chin, and X.J. Fu. 2013. Accident analysis based on Bayesian network and evidential reasoning approach. *Journal of Loss Prevention in the Process Industries* 26: 10–21.
7. Tan, Q., G. Chen, L. Zhang, J. Fu, and Z. Li. 2014. Dynamic accident modeling for high-sulfur natural gas gathering station. *Process Safety and Environmental Protection* 92: 565–576.
8. Khakzad, N., F. Khan, and P. Amyotte. 2013. Risk based design of process systems using discrete-time Bayesian networks. *Reliability Engineering and System Safety* 109: 5–17.
9. Ale, B., C. van Gulijk, D. Hanea, P. Hudson, and P.H. Lin. 2014. Towards BBN based risk modelling of process plants. *Safety Science* 69: 48–56.
10. Pourret, O., P. Naim, and B. Marcot (eds.). 2008. *Bayesian networks: A practical guide to applications*. Chichester: Wiley.
11. Kjærulf, U.B., and M.L. Anders. 2005. *Probabilistic networks—an introduction to Bayesian networks and influence diagrams*. USA: Springer.
12. Neapolitan, R.E. 2003. *Learning Bayesian networks*. Prentice Hall.
13. Accessed at Internet. http://download.hugin.com/webdocs/manuals/Htmlhelp/oobn_pane.html.
14. Onisko, A., M.J. Druzdzel, and H. Wasyluk. 2001. Learning Bayesian network parameters from small data sets: application of Noisy-OR gates. *International Journal of Approximate Reasoning* 27: 165–182.
15. Norsys Software Corporation, Netica. 2015. www.norysys.com.
16. Kraaijeveld, P., and M. Druzdzel. GeNIeRate: An interactive generator of diagnostic Bayesian network models. Accessed at Internet. http://www.pitt.edu/ ~ druzdzel/psfiles/dx05.pdf.
17. Wiegerink, W., B. Kappen, and W. Burgers. Bayesian networks for expert systems, theory and practical applications. Accessed at Internet. www.snn.ru.nl/v2/serve.php?doc= WiegerinckKappenBurgersIcisFinal.

Application of Computational Analysis for Risk Assessment of Chlorine Gas from Tank in Chlorine Production Unit: A Case Study

A. K. Dash, M. K. Pradhan and R. Singh

Abstract Occurrence of accidental toxic chemical release may happen in industries because of accident and if inhaled will cause loss of human life, damage to environment. Chlorine is one of the toxic gases, if released accidentally than the dispersion distance of the gas must be known by the people working around it to take decisions for safe evacuation distance for the general public. In the current analysis, a chlorine tank has been considered as the source of leakage and a computational technique has been employed to evaluate the dispersion distance of the gas in the downwind direction with different concentration (0.5, 2, 20 ppm). The dispersion distances are affected by the meteorological parameters of a place such as wind velocity, humidity, and ambient temperature, etc. Finally, risk assessment has been done and the distance traveled by the gas in different concentration has been found and listed down the consequences for different level of concentration.

Keywords Risk assessment · Dispersion distance · Chlorine plant
Chlorine · Safety

1 Introduction

The scientific term of risk assessment which has been an important tool for establishing safety codes and crisis preparedness is known as a "methodology, for determination of risks and their social evaluation both by qualitatively and quantitatively" [1]. Hazard quantification for a process-based industry producing chemicals is a critical domain to safeguard ecosystem and human from the damages

A. K. Dash (✉) · R. Singh
BML Munjal University, Gurgaon, India
e-mail: dash.amiya@gmail.com

R. Singh
e-mail: ranbirsheoran@gmail.com

M. K. Pradhan
Directorate of Factory & Boilers, Govt. of Odisha, Bhubaneswar, India

N. A. Siddiqui et al. (eds.), *Advances in Fire and Process Safety*,
Springer Transactions in Civil and Environmental Engineering,
https://doi.org/10.1007/978-981-10-7281-9_2

being made to them. Listing down the fallout of the hazard is an integral part of risk assessment [2]. Fallout analysis is qualitative/quantitative result of a situation. Fallout analysis is also used for other purposes such as hazop, hazan study [3]. Chlorine is a harmful industrial element which is used in various processes of industries, stored and transported in bulk. As the chlorine gas is very toxic and a potential threat to the life of human being, this had been used as weapon during world war. This chemical chlorine is a necessary evil for today's life requirement required continuously mostly in urban areas. It is generally applied as water disinfectant to supply water for domestic purpose in urban and rural areas. It also functions as chemical agent in paper industry. Chlorine is usually stored in liquid form in cylinders. The capacity of each cylinder varies from 100, 500 to 1,000 kg. If chlorine in a cylinder was not handled and preserved properly, a valve or bottle in a cylinder was likely to be damaged or corroded, especially in the case of improper operation, a release accident would occur suddenly. From previous releases accidents, a release of, i.e. 1–10 kg/min release, is around 38% of the total number of chlorine release accidents [4]. The chlorine gas is a poisoning gas and when goes into the human bodies through respiratory systems may lead to loss of human life. From various standards, the occupational exposure threshold of the gas is 1 mg/m^3. Even a small amount of chlorine can quickly reach the threshold of acute poisoning and cause casualties [5]. So, in this paper, we applied computational fluids dynamics (CFD) method to study a typical small amount of chlorine release in a factory. Our purpose was to investigate the flow field under continuous and instantaneous leak on this scale which served for the emergency preparedness if accident occurred. Especially, heavier-than-air and obstruction effects by the industrial facilities in each scenario were investigated. We finally conducted accident causal analysis and proposed related risk reduction measures and accident prevention strategy. The residue of this paper is presented as per the following. We stated the theoretical foundation for heavy gas dispersion problem formally and discussed three categories models before, in Sect. 2. In Sect. 3, we introduced an accidental release of chlorine in a small zone, and with specific conditions, the gaseous form of chlorine will exert toxic effect on a dense population living and working nearby. In Sect. 4, we conducted an accident causal analysis using fishbone diagram and put forward risk reduction measures accordingly. In Sect. 5, we concluded the paper and pointed out future research directions. Hundred people were killed in a terrorist attack at Washington D.C by chlorine and nearly 2.5 million people were affected [6]. During an analysis in Groningen province, it was analyzed that if mass is exposed to the deadly gas for 45 min, 17,800 casualties and 5,000 deaths may be caused [7]. Chlorine railcar releases in industrial and urban areas and its consequence analysis also used a CFD method similar to this paper to investigate its dispersion process [8, 9]. Also, the effect of obstruction of buildings upon dispersion under flashing releases was studied in detail especially [1]. More research fruits on this focused on the area of street canyon [2, 3, 10–16].

From the review of the papers, it is observed that the Gaussian model is used for measuring the dispersion distance of the toxic gas. So in this research, we have used the computational model for evaluating the risk area for a chlorine tank for a leakage in the valve of the tank.

2 A Theoretical Basis

Since the chlorine gas is denser than air, the model for dispersion of the gas can follow heavy gas dispersion. The investigation for heavy gas dispersion has started since 1960s. From last few decades, a lot of researches are being conducted to develop a number of heavy gas dispersion models. As per the models, they are categorised: empirical, box-type, and computational fluid dynamics model [1]. From all, empirical model is found to be the simple one. By conducting various field experiments and using different simplifications and assumptions, concentrations and the important parameters are measured. A successful B&M technique was developed in 1983 [11]. It is primarily used for long-distance, flat, and open-space dispersion [12]. According to box-type model with application of heavier-than-air effect, gas clouds can form certain types of shape and distributions were observed in cloud parameters like concentration, the quality of flux, enthalpy by uniform or Gaussian distribution [14]. Insufficient chlorine amount was reported for small release or leakage (1–10 kg/min) to form a cloud. Chlorine release process is greatly impacted by external conditions and obstructions.

3 Computational Technique

Input data:

Emission of pollutant is a function of atmospheric stability and conditions in addition to the dependence on wind speed and direction. The available information should be analyzed and the information relating to stable conditions should be selected for investigating and assessing the maximum risk. In this study, the meteorological information related to city of Ganjam, Odisha, India was collected from Meteorological Department of India. This information was analyzed for different times of year and atmospheric conditions relating to high stability were considered. In order to calculate social risk level resulting from pollutant emission, it is necessary to assess population density and distribution in the risky place. Population distribution is a function of two factors of time and place and is different during day or nighttime as well as public places and buildings. Also, in risk assessment, it is necessary to consider the population groups which are at different risk levels (on-site and out-of-site) [10].

Chemical data:

Name of Chemical	Chlorine
Molecular Weight	70.91 g/mol

AEGL-1 (60 min)	0.5 ppm; AEGL-2 (60 min): 2 ppm; AEGL-3 (60 min): 20 ppm
IDLH	10 ppm
Ambient Boiling Point	−34.1 °C
Freezing Point	−101.0 °C

Meteorological information:

	Summer (day)	Rainy (day)	Winter (day)
Average speed of wind (m/s)	7.22	10.55	9.44
Average direction of wind (from)	N	NE	S
Humidity (%)	82	84	79
Max ambient air temp. (°C)	43.2	35.8	31.9

Tank information:

No. of Tanks	1
Capacity	100 MT
Tank Diameter	2.97 m
Tank Length	13.65 m
Tank Mounting	Horizontal

Local information:

Latitude	N19° 23.001′
Longitude	E085° 03.094′
Site Elevation	10 m Above Mean Sea Level

4 Results and Discussions

The results obtained from the analysis have been depicted in the current section. The dispersion model used for estimating the range of chlorine emission was ALOHA. ALOHA is an atmospheric dispersion model to assess emission of pollutant and predict rate of dispersion and has the following advantages: determining distance of chemical emission using physical properties and toxicity rate of the substance; ability to perform calculation at due times; drawing the affected zones in terms of concentration of chemical substance; ability to predict concentration of the chemical indoors and outdoors during emission of the pollutant; drawing concentration of chlorine gas at specified points after 1 h of pollutant emission in terms of indoor concentration, concentration rate at ground level, and concentration rate inside buildings. Figures 1,2, and 3 show results of ALOHA dispersion model in

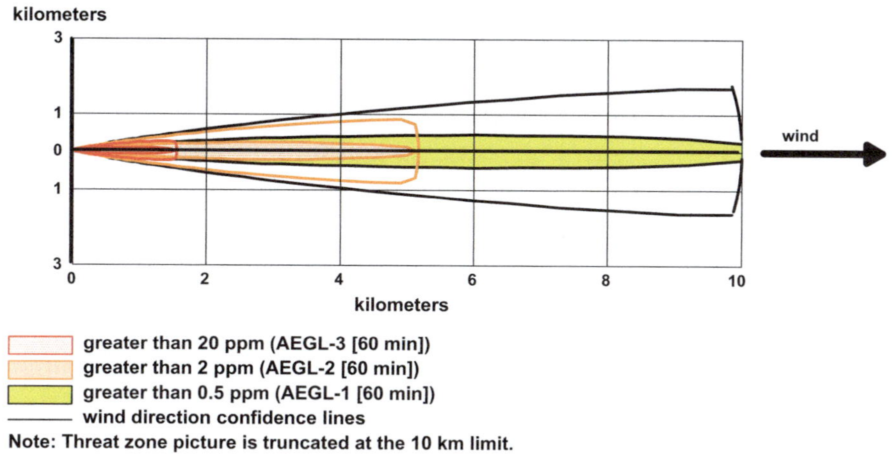

Fig. 1 Results of ALOHA dispersion model in summer season

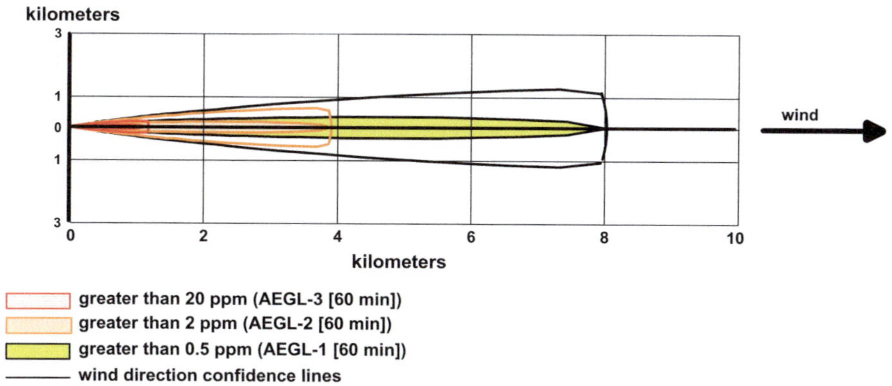

Fig. 2 Results of ALOHA dispersion model in rainy season

summer, rainy, and winter seasons, respectively. Tables 1, 2, and 3 present toxic significance level and indications of ALOHA dispersion model in summer, rainy, and winter seasons, respectively.

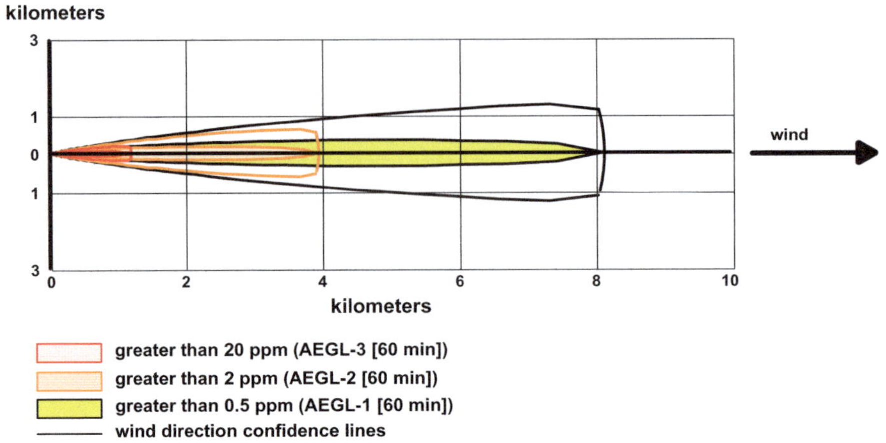

Fig. 3 Results of ALOHA dispersion model in winter season

Table 1 Toxic significance level and indications of ALOHA dispersion model in summer season

Toxic significance level	Value (ppm)	Experience distance (km)	Indication
Yellow	0.5	10	Population could experience notable discomfort
Orange	2	5.2	Non-reversible/serious of long-lasting effects/decreased ability to escape
Red	20	1.5	People could face death/life-threatening health conditions

Table 2 Toxic significance level and indications of ALOHA dispersion model in rainy season

Toxic significance level	Value (ppm)	Experience distance (km)	Indication
Yellow	0.5	8	Population could experience notable discomfort
Orange	2	3.9	Non-reversible/serious of long-lasting effects/decreased ability to escape
Red	20	1.2	People could face death/life-threatening health conditions

Table 3 Toxic significance level and indications of ALOHA dispersion model in winter season

Toxic significance level	Value (ppm)	Experience distance (km)	Indication
Yellow	0.5	8.1	Population could experience notable discomfort
Orange	2	3.9	Non-reversible/serious of long-lasting effects/decreased ability to escape
Red	20	1.2	People could face death/life-threatening health conditions

5 Conclusion

Emission risk of chlorine gas is inevitable as a highly applied substance in different process of plant operation. The dispersion distance is significantly affected by local data such as latitude, longitude and its height from mean sea level. The wind velocity and ambient air temperature also play an important role in finding the dispersion distance along the downwind direction. If the threatened zones by the chlorine emission in Jayashree Chemical Plant, Ganjam can be generally classified, the following can be given: concentration rates of chlorine cloud of above 20, 2 and 0.5 ppm are toxic with radius of 1.5, 5.2 and 10.0 km from chorine emission center, respectively, in summer season. It is suggested for Jayashree Chemical Plant to have some solutions on its agenda for replacing modern treatment methods through using nanosciences, using ozone instead of chorine or hypo or transferring the water treatment plant to a sparsely populated zone in future. Also, administration of Jayashree Chemical Plant is recommended to inform all the risk-exposed zones of chlorine emission at the emission time by alarms or any other suitable information methods. The densely populated centres which should be informed at the time of chlorine emission.

References

1. Khan, F.I., and S.A. Abbasi. 2001. Risk analysis of a typical chemical industry using ORA procedure. *Journal of Loss Prevention in the Process Industries* 14: 43–59.
2. Casal, J. 2007. *Evaluation of the effects and consequences of major accidents in industrial plants*, 1st ed. Amsterdam: Industrial Safety Series, Elsevier Science.
3. American Institute of Chemical Engineers, Center of Chemical Process Safety. 1999. *Guideline for consequence analysis of chemical releases*, 3rd ed. New York.
4. Drogaris, G. 1993. Learning from major accidents involving dangerous substances. *Safety Science* 16: 89–113.
5. Christou, M. 1999. Analysis and control of major accidents from the intermediate temporary storage of dangerous substances in marshalling yards and port areas. *Journal of Loss Prevention in the Process Industries* 12: 109–119.

6. Egidi, D., F.P. Foraboschi, G.E. Spadoni, and A. Amendola. 1995. The ARIPAR project: Analysis of the major accident risks connected with industrial and transportation activities in the Ravenna area. *Reliability Engineering and System Safety* 49: 75–89.
7. Hubert, P., and P. Pages. 1989. Risk management for hazardous materials transportation: A local study in Lyons. *Risk Analysis* 9 (4): 445–451.
8. Deaves, D.M., S. Gilham, B.H. Mitchell, P. Woodburn, and A.M. Shepherd. 2001. Modeling of catastrophic flashing releases. *Journal of Hazardous Materials* A88: 1–32.
9. Rao, P.G., and K.V. Raghavan. 1996. Hazard and risk potential of chemical handling at ports. *Journal of Loss Prevention in the Process Industries* 9 (3): 199–204.
10. Hanna, S, J. Chang. 2008. Gaps in toxic industrial chemical (TIC) model system. In *12th Conference on Harmonization within Atmospheric*. New York.
11. Yu, Q., Y. Zhang, X. Wang, W.C. Ma, and L.M. Chen. 2009. Safety distance assessment of industrial toxic releases based on frequency and consequence: A case study in Shanghai China. *Journal of Hazardous Materials* 168: 955–961.
12. Galal-Gorchev, H. 2007. Chlorine in water disinfection. *Pure & Applied Chemistry* 68 (9): 1731–1735.
13. Hanna, S., and S. Dharmavaram. 2007. Comparison of six widely-used dense gas dispersion models for three actual railcar accidents. In *ITM2007*. Portugal.
14. Pandya, N., and E. Marsden. 2008. Toxic release dispersion modeling with PHAST: Parametric sensitivity analysis. In *3rd International Conference on Safety & Environment in Process Industry*. Italy.
15. Moussa, M.I., and T. Eid. 2007. Risk management for chlorine producing factory in Egypt. *Australian Journal of Basic & Applied Science* 1 (3): 239–248.
16. Guarnaccia, J., and T. Hoppe. 2007. Off-site toxic consequence assessment: A simplified modelling procedure and case study. *Journal of Hazardous Materials* 159 (1): 177–184.

Modeling of Gas Flow Within the Shale Fracture Network

Vamsi Krishna Kudapa, D. K. Gupta and Pushpa Sharma

Abstract At present, shale gas has become the greatest source around worldwide in energy supply. For gas production from shale at economic rates, new technologies like horizontal wellbore drilling and incused fracturing are implemented in all shale gas reservoirs. Rate of gas production from shale reservoirs basically depends upon the reservoir properties and the hydraulic fracture properties. This paper concentrates on the modeling and simulation of gas flow within the fracture network. The reservoir model is represented in the form of cube, which contains the pore spaces where the adsorbed gas and free gas are stored. Free gas is present in matrix pore spaces, and in the fractures, adsorbed gas is present on the surface of the kerogen. Once the pressure difference is created, then the gas will flow from the matrix pores to the fractures and to the horizontal wellbore. The horizontal wellbore is connected to a set reservoir cube representation. In this work, an updated dual porosity model has been considered for modeling the gas flow in the reservoir. The basic assumption involved in this model is, one pore space represents the matrix, and the other pore space represents the hydraulic fractures. For describing the gas flow behavior within the matrix, a nonlinear PDE has been developed which is then compiled using JAVA to calculate the pressure variation across the reservoir. In the matrix, the gas flow is considered in all three dimensions, i.e., 3D flow. Accordingly, a three-dimensional reservoir model is developed. Several flow mechanisms like slip flow, non-Darcy flow, and Darcy flow are considered along with the parameters Langmuir pressure and Langmuir volume in this model. The flow of gas in the matrix is considered as single-phase flow. The production data are estimated for a period of three years, and the obtained results are validated using CMG-IMEX reservoir simulator.

V. K. Kudapa (✉) · D. K. Gupta · P. Sharma
Department of Petroleum Engineering, University of Petroleum and Energy Studies, Dehradun, India
e-mail: vkkudapa@ddn.upes.ac.in

D. K. Gupta
e-mail: dkgupta@ddn.upes.ac.in

P. Sharma
e-mail: pushpa.sharma@ddn.upes.ac.in

N. A. Siddiqui et al. (eds.), *Advances in Fire and Process Safety*, Springer Transactions in Civil and Environmental Engineering,
https://doi.org/10.1007/978-981-10-7281-9_3

Keywords Shale · Matrix · Desorption · Non-Darcy flow · Slip flow
Single-phase flow

1 Introduction: General History

Pressure variation differentiation and production rate estimation by modeling the gas flow in unconventional gas reservoirs are being done for the past few years. The transport of gas in the shale reservoirs exhibits complex behavior, as it contains different sizes of pores like micropores, nanopores, and macropores [1–10]. Several scholars have worked on the gas transport mechanism between the matrix and the hydraulic fractures. Most of the researchers assumed that Darcy's flow is followed in the matrix pores, but in actual the flow deviates from Darcy's law due to very low permeability. The free gas flow and the adsorbed gas flow will take place at the same time once the pressure in the fracture gets reduced [11, 12]. As the permeability of the shale reservoir varies with respect to location, a uniform permeability named apparent gas permeability has been introduced to represent the gas flow in the nanopores [2, 13–15].

The rate of gas production from shale gas reservoirs will mainly depend upon the fractures interconnectivity. Most of the assumptions state that the fractures are sourced by the matrix system only. But with the present literature, it is unknown that the gas flow in the fracture network is sourced by the matrix system [16–19]. Several authors had revealed that the gas flow in the matrix is Darcy law due to the pressure difference between the matrix and the fractures. Ozkan et al. [20, 21] stated that the gas flow in the nanopores is negligible. A detailed research has to be done for studying the gas flow behavior in the shale matrix.

As per Javadpour et al. [10], the gas flow through nanopores in shale reservoirs follows Knudsen diffusion and slip flow. In micropores, the flow will be Darcy's flow, desorption from the surface of the kerogen. Here, our concentration is on desorption, Darcy's law, and Non-Darcy's flow process. The behavior of gas desorption in coal bed methane is linked with desorption in shale gas reservoir.

In this paper, we presented an updated dual porosity model which includes the free gas and the adsorbed gas. The gas is from the pore spaces in the matrix, and the adsorbed gas is from the surface of the kerogen. The entire reservoir has been divided into 9 * 9 * 9 cubical (Matrix) blocks. Between these cubical blocks, we have the fractures. The compressibility of the reservoir is also considered in addition to the effect of rock compression due to reduction of pore pressure.

2 Mass Balance Flow Equation for the Gas Flow in the Shale Matrix

In this work, a new updated dual-mechanism model had been introduced for developing a simulation model for shale gas reservoir. Similar to dual porosity model, shale reservoir contains matrix and natural fractures along with hydraulic fractures. As the natural fractures are not uniform throughout the shale reservoir, the assumption of gas flow from matrix to natural fracture will not be applicable in all the cases. Based on intensity of natural fracture, the effective matrix permeability will be enhanced. In order to overcome this situation, in present work, matrix pores and natural fractures as single porous zone and the hydraulic fractures as the second porous zone have been assumed. Based on aforesaid assumption, a reservoir model has been developed. The following are the assumptions that are considered while developing this model.

(1) The flows of gas from the matrix to hydraulic fracture and then from hydraulic fracture to horizontal wellbore.
(2) Only single-phase flow (only gas flow) in the matrix.
(3) Two-phase flow (Gas + Water) in the hydraulic fracture is assumed.
(4) No gas is flown directly from the matrix to the horizontal wellbore.
(5) The only source of gas for wellbore is the hydraulic fracture.

The pictorial description of the model is shown in Fig. 1.

The pictorial description of matrix block with gas flow in it is shown in Fig. 2.

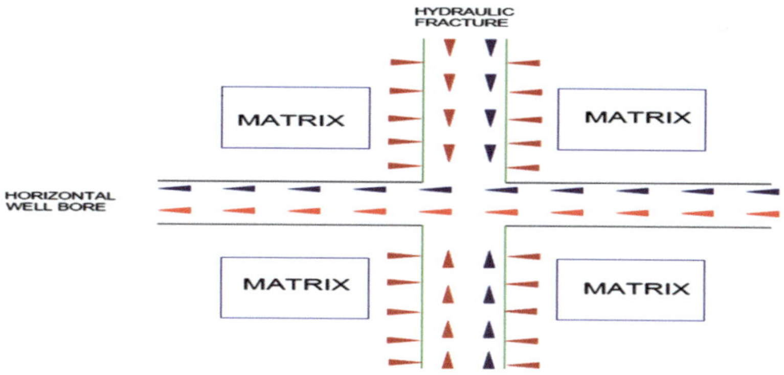

Fig. 1 Reservoir model representing the flow of gas from matrix to hydraulic fractures and from hydraulic fracture to wellbore

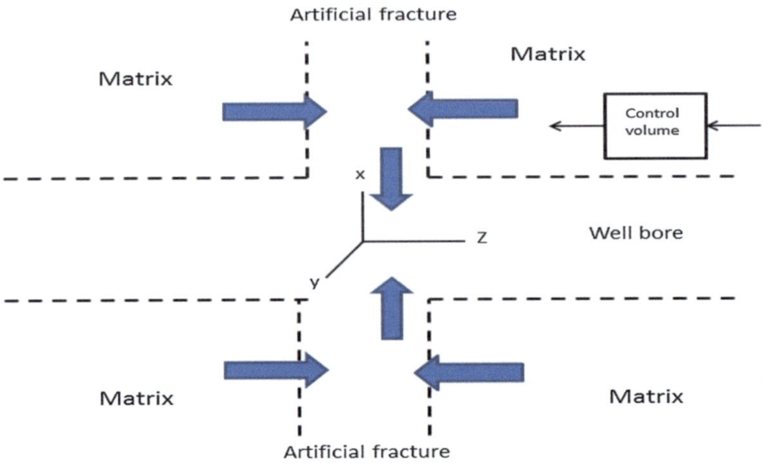

Fig. 2 Pictorial representation of gas flow from matrix block to fracture

2.1 Assumptions

(1) Flow of water in the matrix is negligible.
(2) Considering the process as isothermal process, i.e., constant temperature.
(3) Gas flow through matrix has been considered to be of Darcy type with incorporation of Klinkenberg effect for slippage or non-viscous flow or molecular flow through tiny pores.
(4) Since natural fractures are believed to be discrete, they do not significantly contribute to gas flow on regional scale by themselves. Their effect can be incorporated in the matrix effective permeability or porosity.

2.1.1 Mass Balance Equation

In the present work, the pores present inside the matrix filled with free gas and the pores present on the surface of the matrix filled with adsorbed gas have been considered for developing mass balance equation.

The control volume of the matrix is: $-\Delta x * \Delta y * \Delta z$ (negative sign indicates the gas flow is in negative *x*-, *y*-, and *z*-direction).

Mass balance equation is given as,

$$\begin{aligned}&\text{Mass of free gas in} - \text{Mass of free gas out} + \text{Mass of gas desorbed/generated}\\&\quad = \text{Mass rate of change of gas (free and adsorbed) in the control volume.}\end{aligned}$$

$$\begin{aligned}
&- \left(\upsilon_{g_{mX}} \rho_{g_m|_{X+\Delta X}} \Delta Y \Delta Z\right) - \left(-\upsilon_{g_{mX}} \rho_{g_m|_X} \Delta Y \Delta Z\right) + \left(-\upsilon_{g_{mX}} \rho_{g_m|_{Y+\Delta Y}} \Delta X \Delta Z\right) \\
&- \left(-\upsilon_{g_{mY}} \rho_{g_m|_Y} \Delta X \Delta Z\right) + \left(-\upsilon_{g_{mZ}} \rho_{g_m|_{Z+\Delta Z}} \Delta X \Delta Y\right) - \left(-\upsilon_{g_{mZ}} \rho_{g_m|_Z} \Delta X \Delta Y\right) \\
&+ \frac{\Delta\left((\Delta X \Delta Y \Delta Z.(1-\emptyset_m).\rho_m.\rho_{g_s}.V_d\right)}{\Delta t} \\
&= \frac{\Delta\left((\Delta X \Delta Y \Delta Z.(1-\emptyset_m).\rho_m.\rho_{g_s}.V_a + \Delta X \Delta Y \Delta Z.\emptyset_m.S_{g_m}\rho_{g_m}\right)}{\Delta t},
\end{aligned} \tag{1}$$

where $V_a = V_L - V_d$

V_a Remaining adsorbed gas volume at standard conditions.
V_d Desorbed gas volume.
V_L Langmuir Volume, i.e., maximum amount of gas adsorbed per unit mass of rock in volume.

Now, dividing Eq. 1 by $(\Delta X\, \Delta Y\, \Delta Z)$, we got

$$\begin{aligned}
&\frac{\left(-\upsilon_{g_{mX}} \rho_g|_{X+\Delta X} + \upsilon_{g_{mX}} \rho_g|_X\right)}{\Delta X} + \frac{\left(-\upsilon_{g_{mY}} \rho_g|_{Y+\Delta Y} + \upsilon_{g_{mY}} \rho_g|_Y\right)}{\Delta Y} \\
&+ \frac{\left(-\upsilon_{g_{mZ}} \rho_g|_{Z+\Delta Z} + \upsilon_{g_{mZ}} \rho_g|_Z\right)}{\Delta Z} + \frac{\Delta\left((1-\emptyset_m).\rho_m.\rho_{g_s}.V_d\right)}{\Delta t} \\
&= \frac{\Delta\left((1-\emptyset_m).\rho_m.\rho_{g_s}.V_a + \emptyset_m.S_{g_m}\rho_{g_m}\right)}{\Delta t}.
\end{aligned} \tag{2}$$

Now taking limit for ΔX, ΔY, ΔZ, and $\Delta t \to 0$, the equation can be written as

$$\begin{aligned}
&\frac{\partial\left(\upsilon_{g_{mX}} \rho_{g_m}\right)}{\partial X} + \frac{\partial\left(\upsilon_{g_{mY}} \rho_{g_m}\right)}{\partial Y} + \frac{\partial\left(\upsilon_{g_{mZ}} \rho_{g_m}\right)}{\partial Z} \\
&= \frac{\partial\left[((1-\emptyset_m).\rho_m.\rho_{g_s}.V_a + \emptyset_m.S_{g_m}\rho_{g_m}\right]}{\partial t} - \frac{\partial\left[((1-\emptyset_m).\rho_m.\rho_{g_s}.V_d\right]}{\partial t}.
\end{aligned} \tag{3}$$

Now considering Darcy's Law,

$$q_{gX} = \frac{-kA}{\mu} \cdot \frac{\mathrm{d}P}{\mathrm{d}X}.$$

For X-direction,

$$\frac{\text{Gas Volume}}{\text{time}} = \frac{(-k.\Delta y.\Delta z)}{\mu} \frac{\partial(P_m)}{\partial X}.$$

- $$\frac{(\Delta x.\Delta y.\Delta z)}{t} = \frac{(-k.\Delta y.\Delta z)}{\mu}\frac{\partial\,(P_m)}{\partial X}.$$

- $$\frac{\Delta x}{t} = \frac{(-k)}{\mu}\frac{\partial(P_m)}{\partial X}. \tag{4}$$

But velocity $\left(\upsilon_{g_{mX}}\right) = \frac{\text{Displacement}}{\text{time}}$,

- $$\text{time}\,(t) = \frac{\text{Displacement}\,(\Delta X)}{\upsilon_{g_{mX}}}.$$

Now substituting time (t) in Eq. 4, I got

- $$\upsilon_{g_{mX}} = \frac{-k_m k_{rg}}{\mu_g}\frac{\partial(P_m)}{\partial X} \quad \text{for } x\text{-direction}.$$

- $$\upsilon_{g_{mY}} = \frac{-k_m k_{rg}}{\mu_g}\frac{\partial(P_m)}{\partial Y} \quad \text{for } y\text{-direction}.$$

- $$\upsilon_{g_{mZ}} = \frac{-k_m k_{rg}}{\mu_g}\frac{\partial(P_m)}{\partial Z} \quad \text{for } z\text{-direction}.$$

where

k_m	Effective matrix permeability.
k_m	$k_\infty\left(1 + \frac{b}{P_m}\right)$; including Klinkenberg effect.
k_∞	Equivalent liquid permeability of matrix.
S_g	Gas saturation in rock pore = constant (S_g = 1).
ρ_{g_s}	Standard gas density = constant.
ρ_m	Rock density = constant.
V_d	standard volume of desorbed gas per unit rock mass

Now substituting all the above terms in Eq. 3, we get

- $$\frac{\partial\left(\frac{-k_m k_{rg}}{\mu_g}\frac{\partial(P_m)}{\partial X}\rho_{g_m}\right)}{\partial X}+\frac{\partial\left(\frac{-k_m k_{rg}}{\mu_g}\frac{\partial(P_m)}{\partial y}\rho_{g_m}\right)}{\partial Y}+\frac{\partial\left(\frac{-k_m k_{rg}}{\mu_g}\frac{\partial(P_m)}{\partial Z}\rho_{g_m}\right)}{\partial Z} = \frac{\partial\left[(1-\emptyset_m).\rho_m.\rho_{g_s}.V_a+\emptyset_m.S_{g_m}\rho_{g_m}\right]}{\partial t}-\frac{\partial\left[(1-\emptyset_m).\rho_m.\rho_{g_s}.V_d\right]}{\partial t}.$$

- $$\left[\frac{\partial\left(\frac{-k_m k_{rg}}{\mu_g}\frac{\partial(P_m)}{\partial X}\rho_{g_m}\right)}{\partial X}+\frac{\partial\left(\frac{-k_m k_{rg}}{\mu_g}\frac{\partial(P_m)}{\partial Y}\rho_{g_m}\right)}{\partial Y}+\frac{\partial\left(\frac{-k_m k_{rg}}{\mu_g}\frac{\partial(P_m)}{\partial Z}\rho_{g_m}\right)}{\partial Z}\right] = S_{g_m}\frac{\partial(\emptyset_m\rho_{g_m})}{\partial t}+\rho_m\rho_{g_s}\frac{\partial((1-\emptyset_m)(V_L-V_d))}{\partial t}-\rho_m\rho_{g_s}\frac{\partial((1-\emptyset_m)(V_d))}{\partial t}. \tag{5}$$

From energy of state,

$$\text{Formation volume factor }(B_g)=\frac{\rho_{g_{sc}}}{\alpha_c\rho_{g_m}},$$

where

α_c Volume conversion factor = 5.6145 $\frac{\text{BTU}}{ft^3}$.
$\rho_{g_{sc}}$ Density of gas at standard conditions.
ρ_{g_m} Density of gas in the matrix.

- $$\rho_{g_m}=\frac{\rho_{g_{sc}}}{\alpha_{cB_g}}$$

Now, substituting ρ_{g_m} in Eq. 5, we get

- $$\left[\frac{\partial\left(\frac{k_m k_{rg}}{\mu_g}\frac{\partial(P_m)}{\partial x}\frac{\rho_{g_{sc}}}{\alpha_{cB_g}}\right)}{\partial x}+\frac{\partial\left(\frac{k_m k_{rg}}{\mu_g}\frac{\partial(P_m)}{\partial y}\frac{\rho_{g_{sc}}}{\alpha_{cB_g}}\right)}{\partial y}+\frac{\partial\left(\frac{k_m k_{rg}}{\mu_g}\frac{\partial(P_m)}{\partial z}\frac{\rho_{g_{sc}}}{\alpha_{cB_g}}\right)}{\partial z}\right] = -S_{g_m}\frac{\partial\left(\emptyset_m\frac{\rho_{g_{sc}}}{\alpha_{cB_g}}\right)}{\partial t}-\rho_m\rho_{g_s}\frac{\partial((1-\emptyset_m)(V_L-V_d))}{\partial t}+\rho_m\rho_{g_s}\frac{\partial((1-\emptyset_m)(V_d))}{\partial t}. \tag{6}$$

Now multiplying Eq. 6 with bulk volume, i.e., $\Delta x.\Delta y.\Delta z$

- $$\left[\frac{\partial\left(\frac{k_m\beta_c k_{rg}A_x}{\mu_g}\frac{\partial(P_m)}{\partial x}\frac{\rho_{gsc}}{\alpha_{cB_g}}\right)}{\partial x}\Delta x+\frac{\partial\left(\frac{k_m\beta_c k_{rg}A_y}{\mu_g}\frac{\partial(P_m)}{\partial y}\frac{\rho_{gsc}}{\alpha_{cB_g}}\right)}{\partial y}\Delta y+\frac{\partial\left(\frac{k_m\beta_c k_{rg}A_z}{\mu_g}\frac{\partial(P_m)}{\partial z}\frac{\rho_{gsc}}{\alpha_{cB_g}}\right)}{\partial z}\Delta z\right]$$
$$=-V_b S_{g_m}\frac{\partial\left(\emptyset_m\frac{\rho_{gsc}}{\alpha_{cB_g}}\right)}{\partial t}-V_b\rho_m\rho_{g_s}\frac{\partial((1-\emptyset_m)(V_L-V_d))}{\partial t}+V_b\rho_m\rho_{g_s}\frac{\partial((1-\emptyset_m)(V_d))}{\partial t}. \quad (7)$$

where β_c = Transmissibility conversion factor = $1.127\,\frac{scf}{D-psi}$.
Now consider the R.H.S in Eq. 7,

- $$\begin{aligned}\frac{\partial((1-\emptyset_m)(V_L-V_d))}{\partial t}&=\frac{\partial((1-\emptyset_m)(V_a))}{\partial t}\\&=V_a\frac{\partial(1-\emptyset_m)}{\partial t}+(1-\emptyset_m)\frac{\partial(V_a)}{\partial t}\\&=\left[V_a\frac{\partial(1-\emptyset_m)}{\partial p_m}+(1-\emptyset_m)\frac{\partial(V_a)}{\partial p_m}\right]\frac{\partial p_m}{\partial t}.\end{aligned} \quad (8)$$

- Consider, $\frac{\partial(V_a)}{\partial p_m}$

$$\frac{\partial(V_a)}{\partial p_m}=\frac{\partial\left(\frac{V_L P_m}{P_L+P_m}\right)}{\partial p_m}.$$

- $$V_L\frac{\partial\left(\frac{P_m}{P_L+P_m}\right)}{\partial p_m}=V_L\frac{\partial\left(P_m.(P_L+P_m)^{-1}\right)}{\partial p_m},$$

- $$V_L\left[(P_L+P_m)^{-1}+P_m\frac{\partial(P_L+P_m)^{-1}}{\partial P_m}\right],$$

- $$V_L\left[(P_L+P_m)^{-1}+P_m\frac{-1}{\left((P_L+P_m)^2\right)}\right],$$

- $$V_L\left[\frac{1}{(P_L+P_m)}-\frac{P_m}{(P_L+P_m)^2}\right],$$

- $$\frac{P_LV_L}{(P_L+P_m)^2},$$

- $$\frac{\partial(V_a)}{\partial p_m}=\frac{P_LV_L}{(P_L+P_m)^2}.$$

From literature, $\emptyset_m=\emptyset_o e^{c_m(P_m-P_o)}$.

- $$\begin{aligned}\frac{\partial\emptyset_m}{\partial P_m}&=\emptyset_o e^{c_m(P_m-P_o)}c_m\\&=\emptyset_m.c_m.\end{aligned}$$

- $$\frac{\partial(1-\emptyset_m)}{\partial P_m}=-\frac{\partial\emptyset_m}{\partial P_m}=-\emptyset_m.c_m.$$

Now substituting $\frac{\partial(V_a)}{\partial p_m}$ and $\frac{\partial(1-\emptyset_m)}{\partial P_m}$ in Eq. 8, we get

- $$\frac{\partial((1-\emptyset_m)(V_L-V_d))}{\partial t}=\left[V_a(-\emptyset_m.c_m)+(1-\emptyset_m)\frac{P_LV_L}{(P_L+P_m)^2}\right]\frac{\partial p_m}{\partial t}. \quad (9)$$

Considering the term

$$\frac{\partial((1-\emptyset_m)(V_d))}{\partial t}$$

$$
\begin{aligned}
\frac{\partial((1-\emptyset_m)(V_d))}{\partial t} &= \frac{\partial((1-\emptyset_m)(V_d))}{\partial p_m}\frac{\partial p_m}{\partial t} \\
&= \left\{V_d\frac{\partial(1-\emptyset_m)}{\partial p_m} + (1-\emptyset_m)\frac{\partial(V_d)}{\partial p_m}\right\}\frac{\partial p_m}{\partial t} \\
V_d &= V_L - V_a \\
&= V_L - \frac{V_L P_m}{P_L + P_m} \\
&= \frac{V_L P_L}{P_L + P_m} \\
\frac{\partial(V_d)}{\partial p_m} &= \frac{\partial\left(\frac{V_L P_L}{P_L + P_m}\right)}{\partial p_m} \\
&= \frac{\partial\left((P_L V_L)(P_L + P_m)^{-1}\right)}{\partial p_m} \\
&= -\frac{P_L V_L}{(P_L + P_m)^2}.
\end{aligned}
\tag{10}
$$

From Eq. 10,

$$
\begin{aligned}
\frac{\partial((1-\emptyset_m)(V_d))}{\partial p_m}\frac{\partial p_m}{\partial t} &= \left\{V_d(-\emptyset_m c_m) + (1-\emptyset_m)\left(-\frac{P_L V_L}{(P_L + P_m)^2}\right)\right\}\frac{\partial p_m}{\partial t} \\
&= \left\{-\frac{P_L V_L}{(P_L + P_m)^2} + \emptyset_m\frac{P_L V_L}{(P_L + P_m)^2} - V_d\emptyset_m c_m\right\}\frac{\partial p_m}{\partial t}.
\end{aligned}
$$

Substituting $\frac{\partial((1-\emptyset_m)(V_L - V_d))}{\partial t}$ and $\frac{\partial((1-\emptyset_m)(V_d))}{\partial t}$ in Eq. 7, We get

$$
\begin{aligned}
&\left[\frac{\partial\left(\frac{k_m\beta_c k_{rg}A_x}{\mu_g}\frac{\partial(P_m)}{\partial x}\frac{\rho_{gsc}}{\alpha_c B_g}\right)}{\partial x}\Delta x + \frac{\partial\left(\frac{k_m\beta_c k_{rgA_y}}{\mu_g}\frac{\partial(P_m)}{\partial y}\frac{\rho_{gsc}}{\alpha_c B_g}\right)}{\partial y}\Delta y + \frac{\partial\left(\frac{k_m\beta_c k_{rgA_z}}{\mu_g}\frac{\partial(P_m)}{\partial z}\frac{\rho_{gsc}}{\alpha_c B_g}\right)}{\partial z}\Delta z\right] \\
&= \left[\frac{-V_b S_{g_m}\rho_{g_{sc}}\emptyset_m c_m}{\alpha_c B_g} - V_b\rho_m\rho_{g_{sc}}\emptyset_m c_m(V_d - V_a) + \frac{2P_L V_L V_b(1-\emptyset_m)}{(P_L + P)^2}\right]\frac{\partial p_m}{\partial t}.
\end{aligned}
\tag{11}
$$

Equation 11 represents the gas flow within the matrix.

The above equation is a nonlinear PDE equation which is to be solved for determining the variation of pressure with time.

The following relations are used for calculating K_m, μ_g, $\rho_{g_{sc}}$, B_g, C_m, V_a, and V_d (which will vary with respect to pressure)

$$\text{Klinkenberg Effect}\,(K_m) = K_{\text{darcy}}\left(1 + \frac{b_k}{P_m}\right)$$

where Klinkenberg coefficient $(b_k) = 12.639\,K_{\text{darcy}}^{-0.33}$.

Viscosity (μ_g) is calculated by using the following correlation

$$\ln\left(T_{pr}\frac{\mu_g}{\mu_1}\right) = a_0 + a_1P_{pr} + a_2P_{pr}^2 + a_3P_{pr}^3 + T_{pr}\left(a_4 + a_5P_{pr} + a_6P_{pr}^2 + a_7P_{pr}^3\right)$$
$$+ T_{pr}^2\left(a_8 + a_9P_{pr} + a_{10}P_{pr}^2 + a_{11}P_{pr}^3\right) + T_{pr}^3\left(a_{12} + a_{13}P_{pr} + a_{14}P_{pr}^2 + a_{15}P_{pr}^3\right).$$

Density of gas at standard conditions $(\rho_{g_{sc}})$

- $$\rho_{g_{sc}} = \frac{P_{sc}M_a}{Z_{sc}RT_{sc}}.$$

Gas formation volume factor (B_g):$B_g = 0.02827\frac{ZT}{P_m}$.

Now, for solving Eq. 11 for the entire reservoir, the reservoir is divided into several blocks, i.e., matrix blocks. In this model, we have divided the entire reservoir into 9 * 9 * 9 3-D reservoir as represented in Fig. 3.

2.1.2 Discretization Method

The developed equation representing the flow of gas in matrix blocks is nonlinear partial differential equation (PDE). For discretization of this nonlinear PDE, finite difference method has been used (Fig. 4).

By applying finite difference method, Eq. 11 can be written as

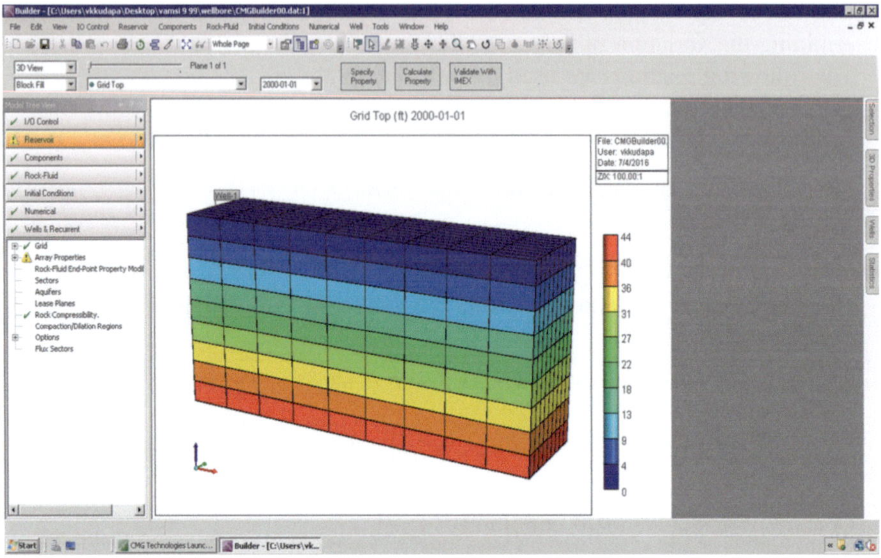

Fig. 3 Pictorial representation of the 3-D reservoir

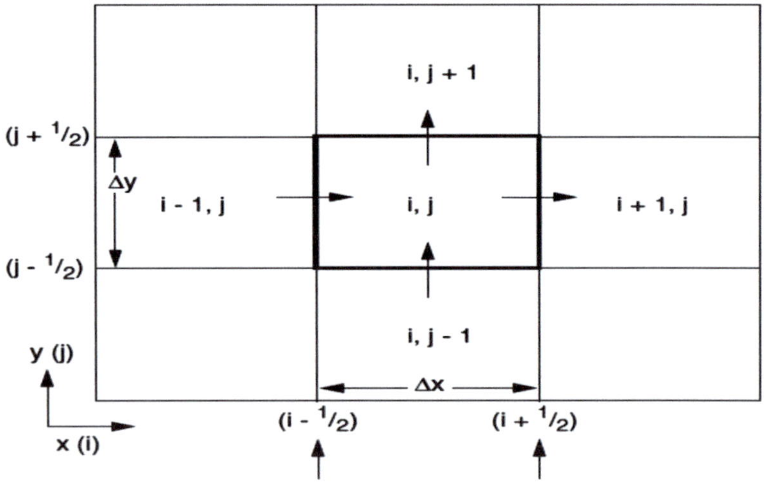

Fig. 4 Discretization and notation indication for a 3-D pressure equation

$$\bullet \quad \left[\left(\frac{k_m\beta_c k_{rg}A_x}{\mu_g}\frac{\rho_{g_{sc}}}{\alpha_{cB_g}}\frac{1}{\Delta x}\right)_{I+\frac{1}{2},J,K}\left(P^{n+1}_{I+1,J,K}-P^{n+1}_{I,J,K}\right)-\left(\frac{k_m\beta_c k_{rg}A_x}{\mu_g}\frac{\rho_{g_{sc}}}{\alpha_{cB_g}}\frac{1}{\Delta x}\right)_{I-\frac{1}{2},J,K}\left(P^{n+1}_{I,J,K}-P^{n+1}_{I-1,J,K}\right)\right]$$
$$+\left[\left(\frac{k_m\beta_c k_{rg}A_y}{\mu_g}\frac{\rho_{g_{sc}}}{\alpha_{cB_g}}\frac{1}{\Delta y}\right)_{I,J+\frac{1}{2},K}\left(P^{n+1}_{I,J+1,K}-P^{n+1}_{I,J,K}\right)-\left(\frac{k_m\beta_c k_{rg}A_y}{\mu_g}\frac{\rho_{g_{sc}}}{\alpha_{cB_g}}\frac{1}{\Delta y}\right)_{I,J-\frac{1}{2},K}\left(P^{n+1}_{I,J,K}-P^{n+1}_{I,J-1,K}\right)\right]$$
$$+\left[\left(\frac{k_m\beta_c k_{rg}A_z}{\mu_g}\frac{\rho_{g_{sc}}}{\alpha_{cB_g}}\frac{1}{\Delta z}\right)_{I,J,K+\frac{1}{2}}\left(P^{n+1}_{I,J,K+1}-P^{n+1}_{I,J,K}\right)-\left(\frac{k_m\beta_c k_{rg}A_z}{\mu_g}\frac{\rho_{g_{sc}}}{\alpha_{cB_g}}\frac{1}{\Delta z}\right)_{I,J,K-\frac{1}{2}}\left(P^{n+1}_{I,J,K}-P^{n+1}_{I,J,K-1}\right)\right]$$
$$=\left[\frac{-V_bS_{g_m}\rho_{g_{sc}}\emptyset_m c_m}{\alpha_c B_g}-V_b\rho_m\rho_{g_{sc}}\emptyset_m c_m(V_d-V_a)+\frac{2P_LV_LV_b(1-\emptyset_m)}{(P_L+P)^2}\right]\frac{\left(P^{n+1}_{I,J,K}-P^{n}_{I,J,K}\right)}{\Delta t}. \tag{12}$$

As we have chosen the spatial discretization terms at new time interval, i.e., $(n + 1)$, the applied finite difference method can be considered has implicit finite difference method.

Considering transmissibility $\left(T_{gx}\right)=\frac{\beta_c k_{rg}}{\mu_g}\frac{\rho_{gsc}}{\alpha_{cB_g}}$ in all direction

Equation 12 becomes

$$\bullet \quad \left[\left(\frac{K_mA_{xT_{gx}}}{\Delta x}\right)_{I+\frac{1}{2},J,K}\left(P^{n+1}_{I+1,J,K}-P^{n+1}_{I,J,K}\right)-\left(\frac{K_mA_{xT_{gx}}}{\Delta x}\right)_{I-\frac{1}{2},J,K}\left(P^{n+1}_{I,J,K}-P^{n+1}_{I-1,J,K}\right)\right]$$
$$+\left[\left(\frac{K_mA_{yT_{gy}}}{\Delta y}\right)_{I,J+\frac{1}{2},K}\left(P^{n+1}_{I,J+1,K}-P^{n+1}_{I,J,K}\right)-\left(\frac{K_mA_{yT_{gy}}}{\Delta y}\right)_{I,J-\frac{1}{2},K}\left(P^{n+1}_{I,J,K}-P^{n+1}_{I,J-1,K}\right)\right]$$
$$+\left[\left(\frac{K_mA_{zT_{gz}}}{\Delta z}\right)_{I,J,K+\frac{1}{2}}\left(P^{n+1}_{I,J,K+1}-P^{n+1}_{I,J,K}\right)-\left(\frac{K_mA_{zT_{gz}}}{\Delta z}\right)_{I,J,K-\frac{1}{2}}\left(P^{n+1}_{I,J,K}-P^{n+1}_{I,J,K-1}\right)\right]$$
$$=\left[\frac{-V_bS_{g_m}\rho_{g_{sc}}\emptyset_m c_m}{\alpha_c B_g}-V_b\rho_m\rho_{g_{sc}}\emptyset_m c_m(V_d-V_a)+\frac{2P_LV_LV_b(1-\emptyset_m)}{(P_L+P)^2}\right]\frac{\left(P^{n+1}_{I,J,K}-P^{n}_{I,J,K}\right)}{\Delta t}. \tag{13}$$

Writing Eq. 13 as

$$\bullet \quad \lambda_{g_{I+\frac{1}{2},J,K}}\left(P^{n+1}_{I+1,J,K}-P^{n+1}_{I,J,K}\right)-\lambda_{g_{I-\frac{1}{2},J,K}}+\lambda_{g_{I,J+\frac{1}{2},K}}\left(P^{n+1}_{I,J+1,K}-P^{n+1}_{I,J,K}\right)-\lambda_{g_{I,J-\frac{1}{2},K}}\left(P^{n+1}_{I,J,K}-P^{n+1}_{I,J-1,K}\right)$$
$$+\lambda_{g_{I,J,K+\frac{1}{2}}}\left(P^{n+1}_{I,J,K+1}-P^{n+1}_{I,J,K}\right)-\lambda_{g_{i,j,k-\frac{1}{2}}}\left(P^{n+1}_{I,J,K}-P^{n+1}_{I,J,K-1}\right)=X_{I,J,K}\left(P^{n+1}_{I,J,K}-P^{n}_{I,J,K}\right) \tag{14}$$

where

$$\lambda_{g_{I+\frac{1}{2},J,K}} = \left(\frac{K_m A_{xT_{gx}}}{\Delta x}\right)_{I+\frac{1}{2},J,K},$$

$$\lambda_{g_{I-\frac{1}{2},J,K}} = \left(\frac{K_m A_{xT_{gx}}}{\Delta x}\right)_{I-\frac{1}{2},J,K},$$

$$\lambda_{g_{I,J+\frac{1}{2},K}} = \left(\frac{K_m A_{yT_{gy}}}{\Delta y}\right)_{I,J+\frac{1}{2},K},$$

$$\lambda_{g_{I,J-\frac{1}{2},K}} = \left(\frac{K_m A_{yT_{gy}}}{\Delta y}\right)_{I,J-\frac{1}{2},K},$$

$$\lambda_{g_{I,J,K+\frac{1}{2}}} = \left(\frac{K_m A_{zT_{gz}}}{\Delta z}\right)_{I,J,K+\frac{1}{2}},$$

$$\lambda_{g_{I,J,K-\frac{1}{2}}} = \left(\frac{K_m A_{zT_{gz}}}{\Delta z}\right)_{I,J,K-\frac{1}{2}},$$

$$X_{I,J,K} = \left[\frac{-V_b S_{gm} \rho_{gsc} \emptyset_m c_m}{\alpha_c B_g} - V_b \rho_m \rho_{g_{sc}} \emptyset_m c_m (V_d - V_a) + \frac{2P_L V_L V_b (1-\emptyset_m)}{(P_L+P)^2}\right].$$

Equation 14 can be written as

$$\begin{aligned} &B_{I,J,K} P^{n+1}_{I,J,K-1} + S_{I,J,K} P^{n+1}_{I,J-1,K} + W_{I,J,K} P^{n+1}_{I-1,J,K} + C_{I,J,K} P^{n+1}_{I,J,K} \\ &\quad + E_{I,J,K} P^{n+1}_{I+1,J,K} + N_{I,J,K} P^{n+1}_{I,J+1,K} + A_{I,J,K} P^{n+1}_{I,J,K+1} = Q_{I,J,K}, \end{aligned} \tag{15}$$

where

$$B_{I,J,K} = \lambda_{g_{I,J,K-\frac{1}{2}}},$$

$$S_{I,J,K} = \lambda_{g_{I,J-\frac{1}{2},K}},$$

$$W_{I,J,K} = \lambda_{g_{I-\frac{1}{2},J,K}},$$

$$C_{I,J,K} = -\left[\lambda_{g_{I+\frac{1}{2},J,K}} + \lambda_{g_{I-\frac{1}{2},J,K}} + \lambda_{g_{I,J+\frac{1}{2},K}} + \lambda_{g_{I,J-\frac{1}{2},K}} + \lambda_{g_{I,J,K+\frac{1}{2}}} + \lambda_{g_{I,J,K-\frac{1}{2}}}. + X_{I,J,K}\right].$$

$$E_{I,J,K} = \lambda_{g_{I+\frac{1}{2},J,K}},$$

$$N_{I,J,K} = \lambda_{g_{I,J+\frac{1}{2},K}},$$

$$A_{I,J,K} = \lambda_{g_{I,J,K+\frac{1}{2}}},$$

$$Q_{I,J,K} = \left(-X_{I,J,K}\right)P^{n}_{I,J,K}.$$

Equation 15 is applied for the entire reservoir in the $N_1 * N_2 * N_3$ 3-D reservoir. The obtained equations are compiled using JAVA.

In this case, the wellbore is placed horizontally in the fifth layer, and the gas will flow from the adjacent matrixes to the wellbore. The pictorial representation of model is shown in Fig. 5.

The detailed input properties of the shale reservoir are listed in Table 1.

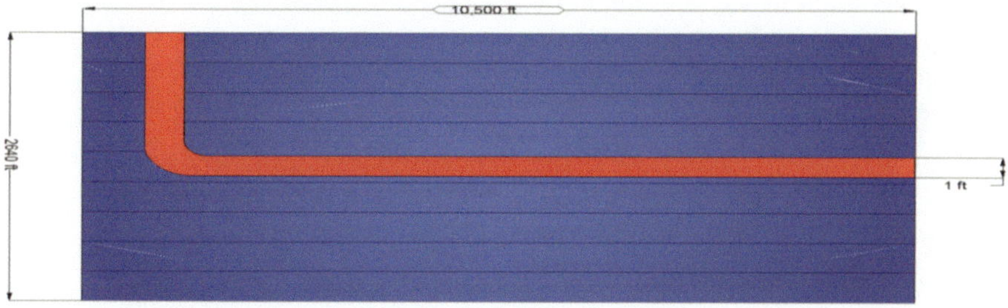

Fig. 5 Pictorial representation of wellbore in the fifth layer from top

Table 1 Parameters used for calculating the gas flow in the matrix

Parameter	Value	Units
Matrix dimensions	1166 * 293 * 5.55	ft
Reservoir porosity	0.07	
Reservoir permeability	0.0002	md
Reservoir temperature	240	F^0
Reservoir thickness	50	ft
Horizontal wellbore length	9000	ft
Wellbore diameter	1	ft
Wellbore pressure	100	psi
Reservoir pressure	3800	psi
Fracture spacing	1	ft
Gas specific gravity	0.68	
Gas composition:	$CH_4 = 0.85$ $CO_2 = 0.08$ $N_2 = 0.04$ $H_2S = 0.03$	

3 Pressure Variation in Matrix Blocks

A nonlinear PDE, which represents the gas low behavior in the matrix, is developed and then compiled using JAVA to get the pressure variation in all the blocks. Figure 6 represents the shale reservoir with no hydraulic fractures.

The model has been divided into 9 * 9 * 9 blocks. Each block will be having different pressure drops during the gas production, but the rate of pressure variation in all matrix blocks is almost same. The variation of pressure in all the matrix blocks is shown in Fig. 7.

Figure 7 represents that the pressure variation in all the blocks is almost similar. As a result, for calculating the flow rate through the horizontal wellbore, the pressure variation in the surrounding blocks can be considered.

In oil and gas industry, Borisov proposed formula 16 for representing the gas flow rate in the horizontal wellbore.

$$q_g = \frac{\left[-2\Pi\beta_c \mathrm{d}k\left(p_{i,j,k} - p_{wf}\right)\right]}{\mu B\left[\ln\left(\frac{\sqrt{A}}{r_w}\right) + \ln\left(C_H\right) + S - \frac{3}{4}\right]}. \tag{16}$$

where, $k = \sqrt{(k_x k_z)}$.

$d = \Delta y.$

$A = c.h = \Delta x.\Delta z.$

$$\ln c_H = 6.28\frac{c}{h}\sqrt{\frac{k_z}{k_x}}\left[\frac{1}{3} - \frac{X_o}{C} + \left(\frac{X_o}{C}\right)^2\right] - \ln\left(\sin\frac{\pi Z_0}{h}\right) - 0.5\left[\ln\left(\frac{c}{h}\sqrt{\frac{K_z}{K_x}}\right)\right] - 1.088.$$

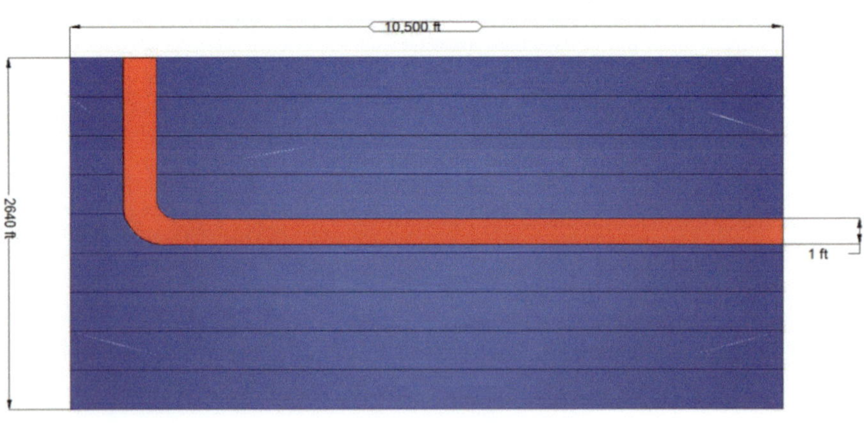

Fig. 6 Shale reservoir with horizontal wellbore and no hydraulic fractures

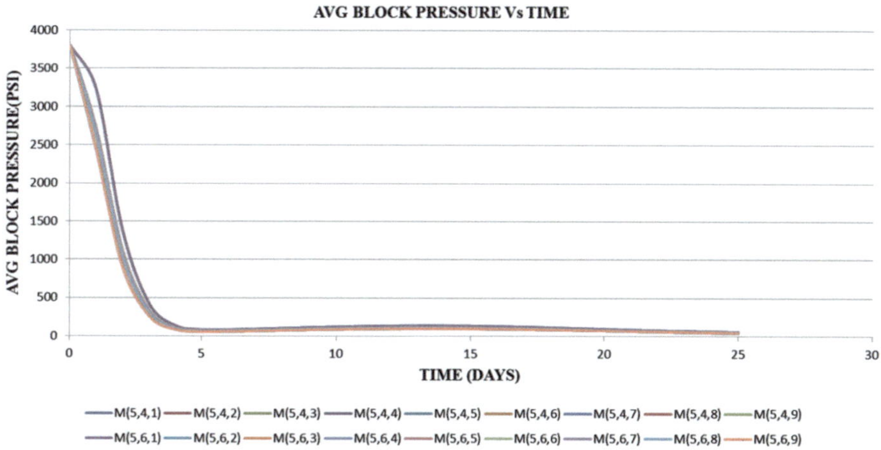

Fig. 7 Individual matrix block pressure variation with time

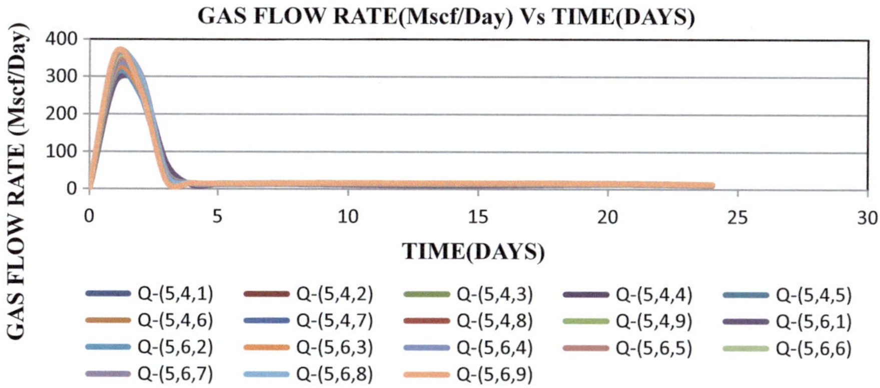

Fig. 8 Gas flow rate variation with respect to time in different matrix blocks

$p_{i,j,k}$ = pressure in the surrounding blocks of horizontal wellbore, psi.

p_{wf} = well flowing pressure in the horizontal wellbore, psi.

r_w = radius of the wellbore, ft.
S = Skin factor.

Figure 8 indicates the flow rate of gas from the matrix block to the horizontal wellbore.

4 Simulation of Shale Gas Reservoir Model

A case study with no hydraulic fracture was performed with the purpose of validating the derived analytical model with the help of a numerical reservoir simulator of CMG-IMEX (CMG-IMEX, 2008). The basic reservoir and the horizontal wellbore parameters are shown in Table 2. The reservoir is expected to be filled with only gas under the condition of residual water saturation.

The model in three-dimensional is 10,500 ft * 2640 ft. Along *X*-direction, the model is divided into nine layers, and height of each layer is of 5.55 ft (Fig. 9). The initial layer starts at 6800 ft from the bottom. A horizontal wellbore has been considered in fifth layer (Fig. 10). Natural fractures presence is assumed in this reservoir. The horizontal wellbore is of 10,000 ft in length.

Figures 11 and 12 give the variation of pressure values between 1 year and 3 years around layer 5 through which horizontal wellbore is drilled. From the above two simulation graphs, it is stated that at the end of both first-year and third-year simulation, the pressure drop variations inside the reservoir do not touch the boundaries.

Table 2 List of reservoir parameters

Parameters	Values	Unit
Reservoir dimensions	10,500 * 2640 * 50	ft
Pressure	3800	psi
Reservoir depth	6800	ft
Temperature	300	°F
Total compressibility	3e−06	psi^{-1}
Langmuir pressure	535	psi
Langmuir volume	197	SCF/ton
Reservoir permeability (md)	0.00020	md
Matrix porosity (%)	0.070	
Natural fracture permeability	0.03	md
Horizontal wellbore length	10,000	ft
Wellbore pressure	200	psi

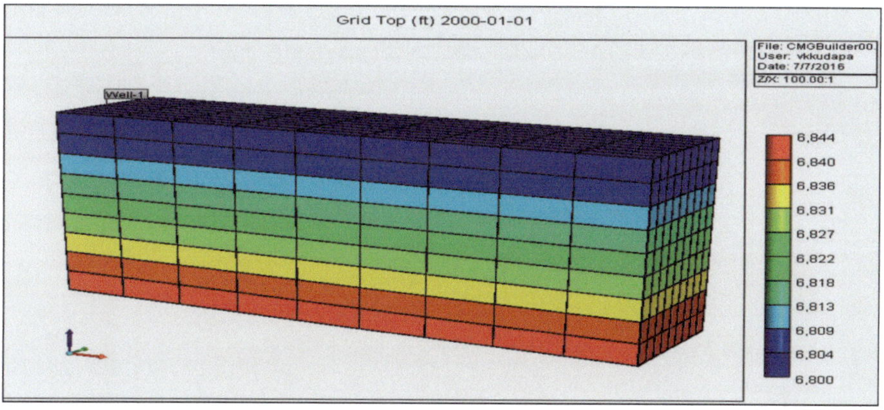

Fig. 9 3-D representation of the shale gas reservoir model

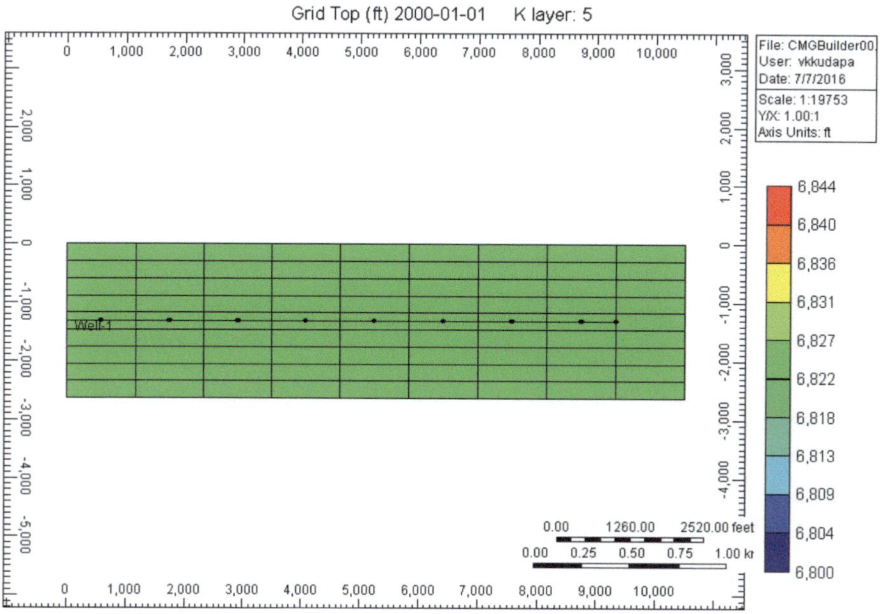

Fig. 10 Reservoir with a horizontal wellbore in fifth layer from top

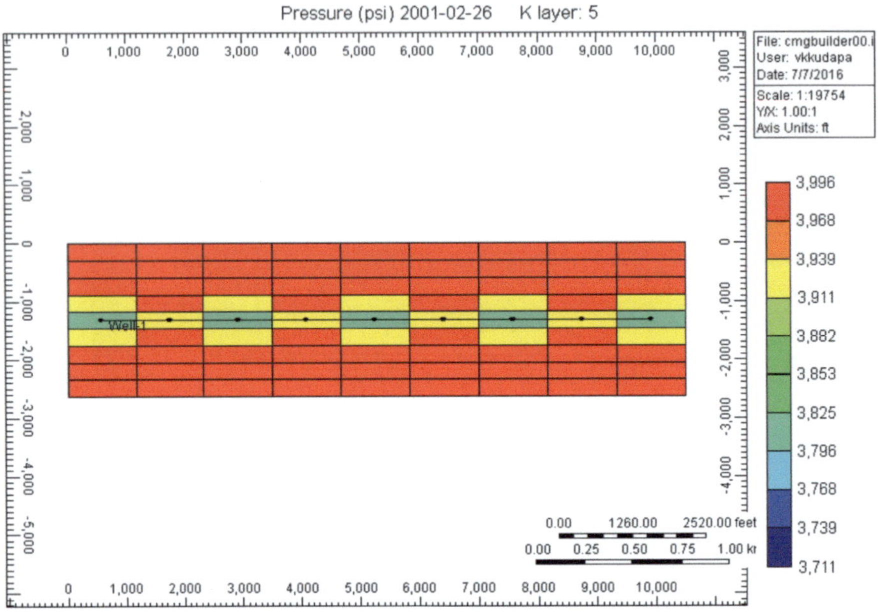

Fig. 11 Pressure variation across fifth layer after first-year production

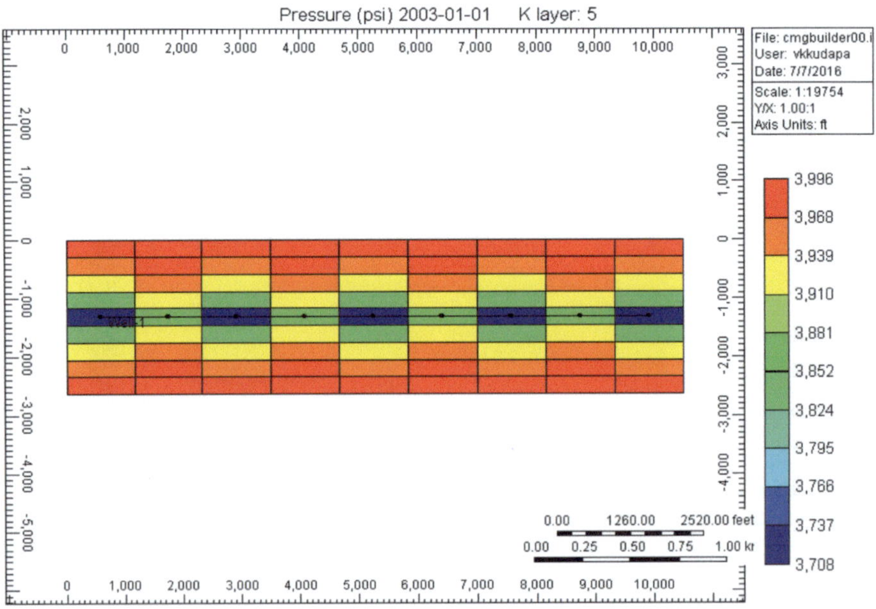

Fig. 12 Distribution of pressure across layer 5 after third-year production

5 Results

5.1 Shale Reservoir Model

Figure 13 gives the detailed information about the gas flow rate from the matrix to the wellbore. At the start of the production, the free gas which is present in the pores of the matrix will start flowing into the wellbore because of which the gas flow rates will increase drastically initially. Now, once the free gas content is reduced in the matrix, the adsorbed gas starts desorbing and now this adsorbed gas starts flowing into the wellbore, the rate of desorption of gas from the surface of the matrix mainly depends upon the pressure variation in the matrix pores. As the rate of desorption is low, the gas flow rates will start decreasing and at some time it becomes constant. As in this case we have assumed the flow is only from the matrix to the wellbore, the extent of constant rate production will be high, which can be seen in Fig. 13.

5.2 CMG-IMEX Simulator Model

Figure 14 gives the rate of gas production from the shale gas reservoir model developed using CMG-IMEX Simulator.

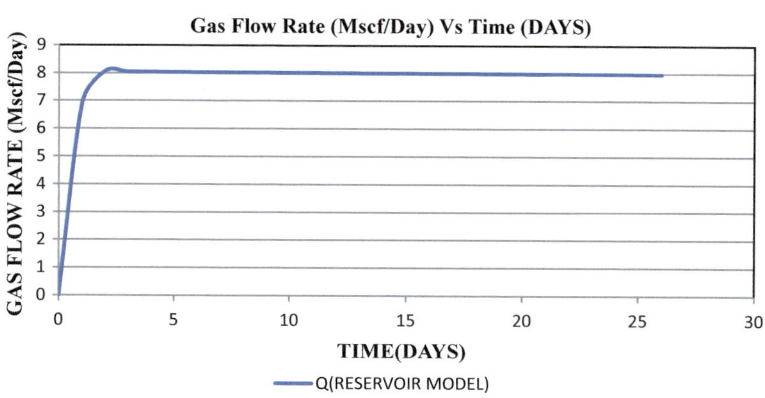

Fig. 13 Graph representing the time varying gas flow rate—reservoir model

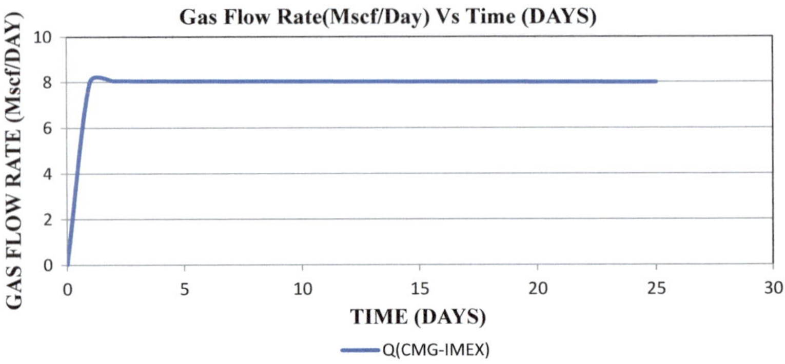

Fig. 14 Graph representing the time varying gas flow rate (Mscf/day)-CMG-IMEX

6 Conclusion and Future Work

In this paper, an updated dual porosity model is introduced which represents a hydraulically fractured shale gas reservoir. This study concludes that

(1) The effect of matrix fractures for gas production from shale reservoirs is negligible. As, is seen in the results that the gas rates into the wellbore is almost the same.
(2) Shale gas is produced for longer periods as major portion of gas is in adsorbed state.
(3) The role of hydraulic fractures on production of gas from shale reservoirs is studied.
(4) Also, we can study the effect of various parameters like fracture spacing, width, matrix permeability, fracture permeability, and porosity.

References

1. Bumb, A.C., and C.R. McKee. 1988. March 1, Gas-well testing in the presence of desorption for coalbed methane and Devonian shale. SPE-15227-PA.
2. Carlson, E.S., and J.C. Mercer. 1991. Devonian shale gas production: Mechanisms and simple models. JPT (April 1991), 476–482. SPE-19311-PA.
3. Cipolla, C.L., E.P. Lolon, J.C. Erdle, and B. Rubin. 2010. Reservoir modeling in shale-gas reservoirs. SPEREE, August, 638–653. SPE-125530-PA.
4. Daniels, F., and R.A. Alberty. 1957. *Physical Chemistry*, 524. New York: John Wiley & Sons Inc.
5. Brown M, E. Ozkan, R. Raghavan, and H. Kazemi. 2009. Practical solutions for pressure transient responses of fractured horizontal wells in unconventional reservoirs. SPE 125043 presented at the SPE Annual Technical Conference and Exhibition, 4–7 October, New Orleams, LA.

6. Chipperfield, S.T., J.R. Wong, D.S. Warner, C.L. Cipolla, M.J. Mayerhofer, E.P. Lolon, and N.R. Warpinski. 2007. Shear dilation diagnostics: A new approach for evaluating tight gas stimulation treatments. In *SPE 106289 presented at the SPE Hydraulic Fracturing Technology Conference*, 29–31 January, College Station, TX.
7. De-swaan_O, A. 1976. Analytical solution for determining naturally fractured reservoir properties by well testing. *Society of Petroleum Engineers Journal (June), Trans AIME* 261: 117–122.
8. Ertekin, T., G.R. King, and F.C. Schwerer. 1986. Dynamic gas slippage: A unique mechanism approach to the flow of gas in tight formations. *SPE Formation Evaluation* 1 (2): 43–52.
9. Gutierrez, M., L.E. Oino, and R. Nygard. 2000. Stress-dependent permeability of a de-mineralised fracture in shale. *Marine and Petroleum Geology* 17: 895–907.
10. Javadpour, F., D. Fisher, and M. Unsworth. 2007. Nano scale gas flow in shale gas sediments. *Journal of Canadian Petroleum Technology* 46 (10): 55–61.
11. Javadpour, F. 2009. Nano pores and apparent permeability of gas flow in mud rocks (shales and siltstone). *Journal of Canadian Petroleum Technology* 48 (8): 16–21.
12. Ozkan, E., M. Brown, R. Raghavan, and H. Kazemi. 2009. Comparison of fractured horizontal well performance in conventional and unconventional reservoirs. SPE 121290, SPE Western Regional Meeting, 24–26 March 2009, San Jose, California.
13. Kazemi, H. 1969. Pressure transient analysis of naturally fractured reservoirs with uniform fracture distributions. *Society of Petroleum Engineers Journal Trans., AIME* 246 (10): 451–461.
14. Klinkenberg, L.J. 1941. The permeability of porous media to liquids and gases. *Drill and Production Practice*. API. 200–213.
15. Medeiros Jr., F., E. Ozkan, and H. Kazemi. 2008. Productivity and drainage area of fractured horizontal wells in tight gas reservoirs. *SPE Reservoir Evaluation & Engineering* 11 (5): 902–911.
16. Raghavan, R. and L.Y. Chin. 2004. Productivity changes in reservoirs with stress dependent permeability. SPE Reservoir Evaluation and Engineering (Aug.) 308–315.
17. Seera, K., A.C. Reynolds, and R. Raghavan. 1983. New pressure transient analysis methods for naturally fractured reservoirs. *Journal of Petroleum Technology*. 35 (10): 2271–2283.
18. Walsh J.B. 1981. Effect of pore pressure and confining pressure on fracture permeability. *International Journal of Rock Mechanics Mining Science and Geomechanics Abstract*, 429–435.
19. Warren, J.E., and P.J. Root. 1963. The behavior of naturally fractured reservoirs Trans., AIME. *Society of Petroleum Engineers Journal* 228 (9): 245–255.
20. E. Ozkan, SPE, Colorado School of Mines. 2010. Modeling of fluid transfer from shale matrix to fracture network. In *SPE Annual Technical Conference and Exhibition*.
21. Wang, Liang, Armando Torres, Xu Li Xiang, Akhona Naido Fei, and Wu Wensi. 2015. A technical review on shale gas production and unconventional reservoir modeling. *Natural Resources* 6: 141–151.

Quantitative Assessment of Risk Caused by Domino Accidents in Chemical Process Industries

S. Sanjay Pramanathan, S. M. Tauseef, Dharani Kumar and P. N. K. Mohanty

Abstract The use of a quantitative assessment to study the domino accidents can help in deriving a more perceptible and more steadfast result than in that of a qualitative assessment. The data required for the study is derived from various risk assessment studies previously taken up in chemical process industries. The methodology followed in this paper would help in the determination of the maximum-credible accident scenarios (MCAS) from a list of several credible accident scenarios obtained for a definite scope or different escalation scenarios to a secondary accident scenario from a primary accident. The most credible accident scenario is determined based on some potential factors—financial loss, fatalities, ecosystem damage that consider site-specific information for population density, asset density of the site, population distribution, damage area, importance factor, etc. The damage radii and other possible consequences are determined by modeling with the help of a comprehensive process hazard analysis software tool. Considering a maximum-credible accident scenario as a primary event, the various escalations to its possible secondary and tertiary scenarios are evaluated for their frequencies and severity of their consequences.

Keywords MCAS · Domino effect · Quantitative assessment · Failure frequency Consequences analysis

S. M. Tauseef (✉) · D. Kumar
University of Petroleum and Energy Studies, Bidholi via Premnagar, Dehradun 248007, India
e-mail: sm.tauseef@gmail.com

P. N. K. Mohanty
CPCL, Chennai Petroleum Corporation Limited, Chennai, India

S. Sanjay Pramanathan
Weatherford Drilling International, Dubai, UAE

N. A. Siddiqui et al. (eds.), *Advances in Fire and Process Safety*, Springer Transactions in Civil and Environmental Engineering,
https://doi.org/10.1007/978-981-10-7281-9_4

1 Introduction

Domino Effect is a cascade of events in which the consequences of a previous accident are increased by following one(s), spatially as well as temporally, leading to a major accident [1–5]. The possible consequences of multiple accidents, either at the secondary or tertiary level, are usually neglected due to the complexity and the uncertainty involved in the study. The regular or usual risk assessments we carry out like the HZOP, QRA, etc., only consider the stand-alone failure frequency of each equipment individually deriving various possible ways in which it may fail and how frequent it may happen, and by considering just the severity and frequency of the incident, the risk involved in the operation of the equipment is estimated. Even though the study's result gives a probable risk involved, during a real-life accident scenario the possibility of an accident to be a stand-alone accident is extremely rare. Every accident has its own consequences that may trigger another accident either directly or indirectly, and this primary accident's damage is increased by following one(s), as well spatially as temporally, leading to a major accident. To study the possible domino events that may occur and to be aware of the actual risk involved when considering these domino scenarios, a domino specific Quantitative Assessment of Risk is to be carried out. For a Quantitative Assessment of Risk to be carried out, various possible accident scenarios are to be listed and categorized according to their credibility based on various factors like financial losses, fatalities, ecosystem damage, etc. The methodology used for this is maximum-credible accident scenarios (MCAS) [6]. For a domino impact to occur, the outcomes from the primary accident, called the primary event, should be of an adequate magnitude in order to cause damage to other equipment for the accident to escalate. The different types of consequences that may trigger the secondary events are:

1. Overpressure
2. Heat radiation or flames impingement
3. Missiles
4. A combination of the above.

From the above consequences that are actually possible, only the scenarios of overpressure are considered.

2 Problem Definition

The onus of managing the risk levels at the minimum has become mandatory on the part of the industries especially chemical industries in recent times. This is not only due to statutory requirements but also because of public awareness. It is, indeed, imperative that risk has to be assessed quantitatively to employ safer operating conditions, prepare effective emergency plan, and bring down the residual risk of the plant to the minimum. The broad title of risk analysis includes quantitative and qualitative risk assessment. The advantage of these techniques in these could be employed combined or independent. Quantitative risk assessment (QRA) refers to three techniques, viz., maximum-credible accident and consequence analysis (MCAC), Reliability Analysis, and Risk Estimation.

Quantitative risk analysis (QRA)—this method is comprised of four steps: hazard identification, frequency estimation, consequence analysis, and measure of risk.

3 Maximum-Credible Accident Scenarios (MCAS)

This is a study by the risk analyst based on the past accidents of similar nature, engineering judgment, and sound knowledge in employing the tool of choice (software, etc.). This technique involves [6]:

- Identification of Hazardous chemicals, unit process, and unit operations
- Creating and screening of accident scenarios
- Short-listing the events
- Estimation of source strength in case of release of a chemical
- Using engineering judgment to assess the damage potential in terms of
  - Thermal radiation
  - Explosion
  - Toxic dispersion

A credible accident is defined as: an accident that is within the realm of possibility (i.e., probability higher than 1×10^{-6}/year) and has a propensity to cause significant damage (at least one fatality). The methodology is to identify the most credible accident scenario.

3.1 Methodology for MCAS

Initial step to carry out a MCAS analysis is to develop all the possible accident scenarios at the unit taken as the scope for the project. Once listed, the credible scenarios are categorized and taken for the analysis once all the scenarios are studied and the similar scenarios are grouped.

In the second step, calculation of the damage radii for every scenario is carried out using the software PHAST 6.7.

The next step is to estimate the probability and severity of consequences of each accident scenario; this is done by the study of literature for the frequency of occurrence of similar events in the past under similar conditions [7–9].

The chemical accident scenarios can be broadly classified under three main categories, namely fire and explosion, toxic or corrosive release and dispersion, and a combination of these categories.

Once the damage radii and stand-alone failure probabilities are estimated, the procedure to estimate their effect on various potential factors that influence the severity of each accident is carried out. Potential factors considered are financial loss (A), fatalities (B), ecosystem damage (C)—which consider site-specific information for population density, asset density of the site, population distribution, damage area, importance factor, etc. The analysis of MCAS considers only the fire and explosion factors of the accident scenarios as the scope for this case study. The methodology adopted for this analysis is from the literature by the Author Faisal I. Khan, Memorial Univ. of Newfoundland on Use Maximum-Credible Accident Scenarios for Realistic and Reliable Risk Assessment [6].

4 Case Study: Oil Movement and Storage Section

The Process Plant, for the process of storage and refining of Crude Oil into its various by-products, with a vast plant area surrounded by a densely populated municipality residential area. The main products of the company are LPG, Motor Spirit, Superior Kerosene, Aviation Turbine Fuel, High Speed Diesel, Naphtha, Bitumen, Lube Base Stocks, Paraffin Wax, Fuel Oil, Hexane, and Petrochemical feedstocks. Hence, large-capacity storage of the raw materials and products is necessary within the company premises. The movement and storage of these materials in the tank farms are regulated and monitored by the OM & S department of the company.

The study was initially performed on different top stand-alone accident scenarios, and data collected on each of those accident scenarios were used to model each of the scenarios separately to estimate the risk and probable consequences of each. Once the damage radius of a definite threshold value of 14.7 kW/m^2 was estimated for each of the explosion models, most credible accident scenario of them was identified. The scenario with credibility within the maximum-credibility zone ($L = 0.5–1.0$) was considered as the primary event for the domino accident analysis.

Primary Event 1

A compressed gas storage Sphere of LPG (20 D-709) Liquefied Petroleum Gas of 80% propane of 40 °C and 11 bar pressure. The Sphere is connected with one input line, one outflow line, a pressure-relief valve, and other conventional safety devices.

Table 1 Resulting overpressures and their corresponding probable damage

Overpressures (kPa)	Distance (m)	Damage
20	261	100% damage of atmospheric tanks/rupture of heavy equipment
13	333	Minor damage to atmospheric tank
2	1326	Minor damage to cooling tower

The vessel is located in the compressed gas storage area along with various other high-pressurized vessels of large capacity. The total source strength of the vessel is 1,250,000 kg, which is one of the largest capacity compressed gas storage vessel present in the plant. The population density of the plant area and its surrounding where the effects of the explosion can be felt (approx. 2000 m radius) is 9.41×10^{-4} persons/m^2, and the asset density around the unit is 3.184 \$/m^2. There is a local water body present close to the plant site which is used by the population of the local municipality, and the company site is also present close to the cost line too on which the livelihood of a large population depends.

Resulting Overpressures
See Table 1.

Scenario 1
Fireball due to catastrophe failure of Sphere 20 D-721 is a Propane storage Sphere 161 m from Sphere 20 D-709

Mass	1.6×10^5kg
Pressure	12 bar
Temperature	40 °C

Scenario 2
Fireball due to catastrophe failure of Sphere 20 D-710 is a LPG storage Sphere 26 m from Sphere 20 D-709

Mass	1.2×10^5kg
Pressure	11 bar
Temperature	40 °C

Scenario 3
Fireball due to catastrophe failure of Sphere 20 D-711 is a LPG storage Sphere 38.7 m from Sphere 20 D-709

Mass	1.2×10^5kg
Pressure	11 bar
Temperature	40 °C

Scenario 4
Fireball due to catastrophe failure of Sphere 20 D-706 is a LPG storage Sphere 63 m from Sphere 20 D-709

Mass	1.2×10^5kg
Pressure	11 bar
Temperature	40 °C

Data Setting for BLEVE in Sphere 20 D-709
See Table 2 and Fig. 1.

Credibility Factors for the Scenarios in the Study
See Tables 3 and 4.

The four Scenarios mentioned above are secondary scenarios that may occur in the event of the primary event BLEVE blast of Sphere 709 followed by a fireball during the event. The secondary events may subsequently add to the increase in the damage if they may trigger a tertiary event. The blast radius with the overpressure of 0.2 bar can potentially lead to the deformation of the atmospheric tank—data

Table 2 Data setting for BLEVE in Sphere 20 D-709

Parameters	Value
Chemical involved	LPG
Quantity of chemical involved	1,250,000 kg
Phase of chemical	Liquid
Unit operation	Storage
Operating temperature, T	40 °C
Operating pressure	11 bar
Site population density (within of 2000 m radius)	9.401083×10^4
Asset density (within 500 m radius)	3.1849 $/m^2
Population distribution factor	0.3
Weather probability factor	1.5 F
Importance factor	1

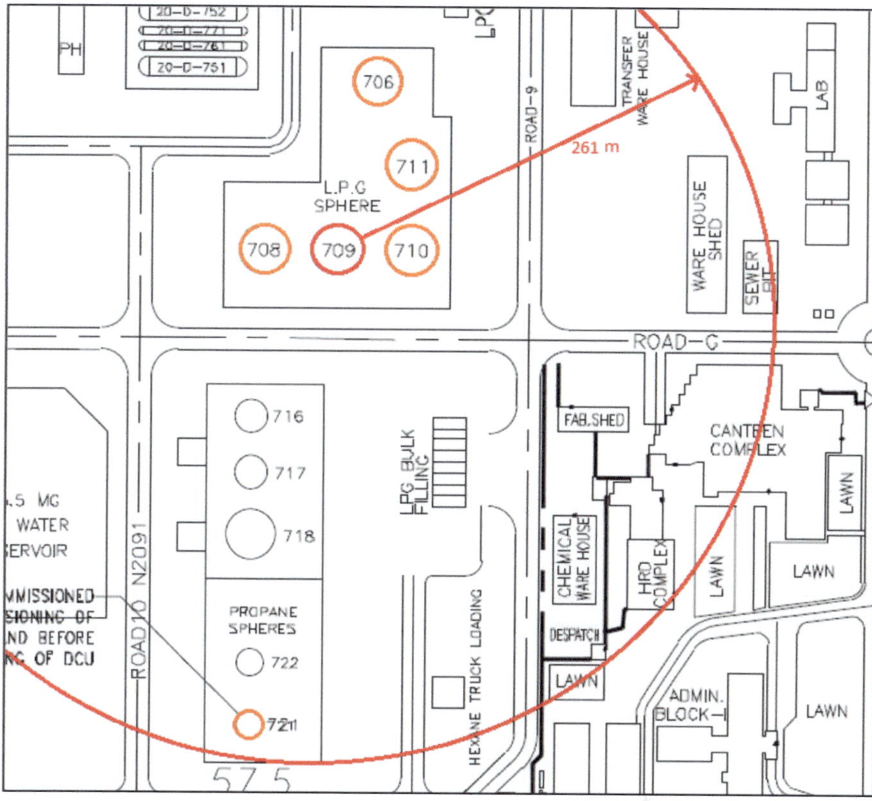

Fig. 1 Primary and potential secondary scenarios considered in the study

reported from the literature for damage to process equipment due to pressure loads [7].

Tertiary Event

An atmospheric storage tank (20 D-107) of Crude Oil is stored at 45 °C and 1 bar pressure. The tank is connected with one input and output line, a pressure-relief valve, and other conventional safety devices. The vessel is located in the tank farm area under the OM & S departments' constant monitoring along with various other tanks with large inventory storage of material like High Speed Diesel, Naphtha, Aviation fuel, kerosene, etc. The total source strength of the vessel is 50176×10^3 kg. The population density of the plant area and its surrounding where the effects of the explosion can be felt (approx. 2000 m radius) is 3.41×10^{-3} persons/m^2, and the Asset density (Radius 500 m) around the unit is 088 \$/m^2. There is a local water body present close to the plant site which is used by the population of the local municipality, and the company site is also present close to the cost line too on which the livelihood of a large population depends.

Table 3 Credibility factors for the scenarios in the study

Accident scenario	Damage radius	Frequency of occurrence (/year)	Fire and explosion				Credibility L
			A	B	C	L_1	
Scenario 1 Tank 721	425.2 m	1.0×10^{-6}	1.44×10^{-5}	0.062	5.6×10^{-4}	0.06	Uncertainty zone
Scenario 2 Tank 710	1326 m	7.0×10^{-5}	0.123	1	0.386	1	Maximum-credibility zone
Scenario 3 Tank 711	1326 m	7.0×10^{-5}	0.1	1	0.38	1	Maximum-credibility zone
Scenario 4 Tank 706	1326 m	7.0×10^{-5}	0.12	1	0.38	1	Maximum-credibility zone

Table 4 Secondary and tertiary events due to domino effect

Parameters	Secondary scenarios and their likely impacts				Tertiary event
	Scenario 1: 721	Scenario 2: 710	Scenario 3: 711	Scenario 4: 706	Scenario 1.1: tank 20-D-107
Explosion					
Total energy released, kJ	34,349,290	2.62E+05	2.62E+05	2.62E+05	NA
Peak overpressure developed kPa	21	20.6	20.6	20.6	
Fire					
Radius of fireball, m	153.82	300.03	300.03	300.03	
Duration of fireball, s	18.69	31.09	31.09	31.09	
Energy released by fireball, kJ	34,349,290	262,077,202	262,077,202	262,077,202	
Radius of pool fire, m					71 m—14.7 kW/m^2
Burning area m^2	567,988.4	2,414,348	2,414,348	2,414,348	1,5828
Checking for the possibility of escalation of accident (Domino)					
Location of unit from primary event, m	161	26	38.7	63	360 m to primary event 99 m to secondary scenario-721 propane sphere
Domino effect due to overpressure					
Peak overpressure, bar	0.2	0.2	0.2	0.2	0.2
Distance at peak overpressure level, m	132	261	261	261	Within the 137 m radius of scenario 1
Probability of domino effect due to overpressure	1	1	1	1	1

The atmospheric storage tank (20 D-107) of Crude Oil which is at a distance of 360 m from Sphere 709 undergoes primary accident-explosion. The damage radius, with the peak overpressure of 20 kPa, for the primary event is 261 m. The secondary scenario 1, BLEVE of 721, is calculated to have a damage radius of 132 m. The two accidents, primary and secondary, have a total damage radius of 393 m with peak overpressure of 0.2 bar or 20 kPa. This will result in the tertiary event of deformation of the atmospheric tank 20 D-107. This may lead to a pool fire with a catastrophic release of the tanks' total containments Fig. 2.

The secondary scenario 1, the explosion of Propane storage Sphere 721 will bring in the Administration block of the company into the highly hazardous zone (Overpressure—0.2 bar). The population in this area is always high during the daytime, and these buildings are outside the designated plant area; hence, they are not constructed to sustain such explosions and are not equipped with necessary safeguards [10]. This scenario will increase the risk many times from the normally assumed value, because of the possible escalations of thermal radiations and overpressures in real-life scenarios. When the primary and secondary events take place simultaneously, the chances for the damage are very high.

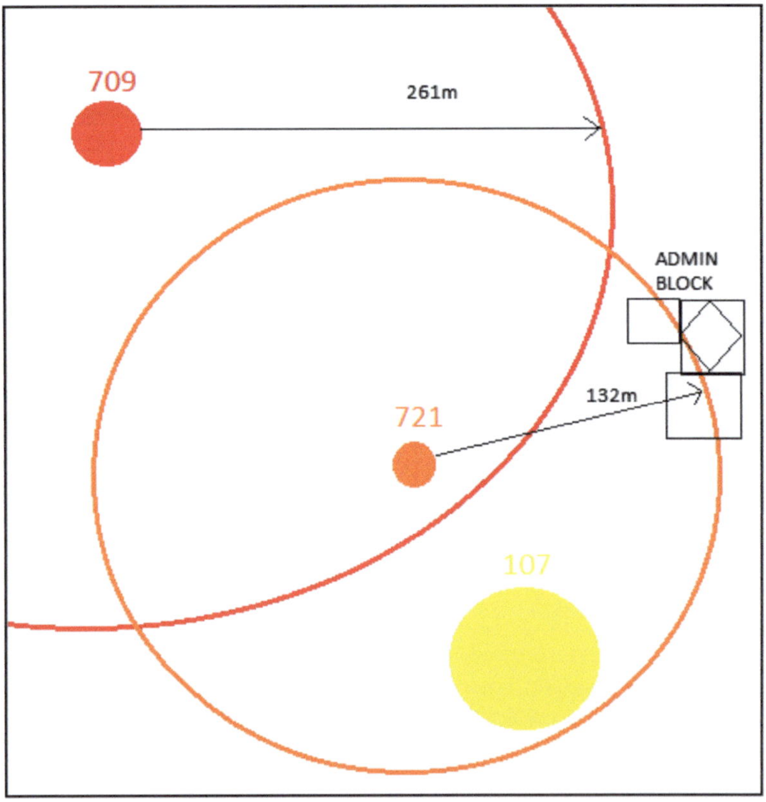

Fig. 2 Escalation of secondary scenario 1 of tank 721 to possible tertiary scenarios

5 Conclusion

The quantitative assessment of risk caused by domino accidents in chemical process industries gives a better understanding on how the usually neglected consideration of the possible domino accidents in risk assessment studies may be of vital importance. domino accidents study gives more of real-life accident sequences scenarios unlike the usual stand-alone accident scenarios, which are less likely to take place. The case study specifies two types of tertiary event escalations, where a Critical Unit or entity is included in the hazardous zone and a scenario where the damage is increased spatially as temporally, leading to a major accident.

References

1. Christian Delvosalle. 1996. Domino effects phenomena: Definition, overview and classification. Paper presented at the first European Symposium on Domino effects, Leuven, Sept 1996.
2. Mukhima, Euginia Diana, Tasneem Abbasia, S.M. Tauseef, and S.A. Abbasi. 2017. Domino effect in chemical process industries triggered by overpressure—Formulation of equipment-specific probits 106: 263–273.
3. Salzano, Ernesto, Giacomo Antonioni, Gabriele Landucci, and Valerio Cozzani. 2013. Domino effects related to explosions in the framework of land use planning, *Chemical Engineering Transactions*, vol. 31.
4. Antonioni, Giacomo, Gigliola Spadoni, and Valerio Cozzani. 2009. Application of domino effect quantitative risk assessment to an extended industrial area. *Journal of Loss Prevention in the Process Industries* 22: 614–624.
5. Genserik Reniers and Valerio Cozzani (eds.). 2013. *Domino effects in the process industries: Modelling, prevention and managing*, Elsevier.
6. Faisal, I. 2011. *Khan, use maximum-credible accident scenarios for realistic and reliable risk assessment*. Cepmagazine: Memorial University of Newfoundland.
7. Cozzani, Valerio, and Ernesto Salzano. 2004. The quantitative assessment of domino effect caused by overpressure part II. Case studies. *Journal of Hazardous Materials* A107: 81–94.
8. Cozzani, Valerio, Gianfilippo Gubinelli, Giacomo Antonioni, Gigliola Spadoni, and Severino Zanelli. 2005. The assessment of risk caused by domino effect in quantitative area risk analysis. *Journal of Hazardous Materials A* 127: 14–30.
9. Landucci, Gabriele, Gianfilippo Gubinelli, Giacomo Antonioni, and Valerio Cozzani. 2009. The assessment of the damage probability of storage tanks in domino events triggered by fire. *Accident Analysis and Prevention* 41: 1206–1215.
10. Gavin Towler, R.K. Sinnott. Chemical engineering design: Principles, practice and economics of plant and process design.

An Assessment of the Appropriateness of the Prescribed "Safe Distances" for Siting Hazardous Process Units to Prevent Domino Effect

Euginia Diana Mukhim, Tasneem Abbasi, S. M. Tauseef and S. A. Abbasi

Abstract One of the most dangerous consequences of an equipment failure in process industries resulting in a fire and/or explosion is the triggering off of the "domino effect" or "chain of accidents." An accident in a unit may trigger an accident in another unit which, in turn, may lead to yet other accidents. To prevent such an occurance, it is imperative that hazardous units are kept safe distances away from each other. Of the types of accidents that can occur in chemical process industries, the boiling liquid expanding vapor explosion (BLEVE) is perhaps the most destructive of the forms of accidental explosions that can occur during the manufacture, storage, or transportation of chemicals in a pressure-liquefied state. The resulting blast wave, missiles, and fire or toxic release can cause great damage to life and property. There are several standards and codes which prescribe minimum safe distances to be maintained between equipment to prevent the domino effect from occurring. Of these the code prescribed by the National Fire Protection Agency (NFPA), USA, is among the ones most widely adopted for locating hazardous units. This paper evaluates the effectives of the safe distances prescribed by the NFPA 58 code in preventing a BLEVE in one vessel from causing other vessels and nearby structures to fail. The study reveals that the distances presented by NFPA are not sufficient to prevent the domino effect when the primary accident is a BLEVE.

Keywords Safe distance · Minimum separation distance · BLEVE
Domino effect · NFPA · Chain of accidents

E. D. Mukhim · T. Abbasi (✉) · S. A. Abbasi
Centre for Pollution Control and Environmental Engineering, Pondicherry University, Kalapet 605014, Pondicherry, India
e-mail: t.abbasi.cpe@pondiuni.edu.in

S. M. Tauseef
Department of Health, Safety and Environment, University of Petroleum and Energy Studies (UPES), Dehradun 248007, Uttarakhand, India

N. A. Siddiqui et al. (eds.), *Advances in Fire and Process Safety*,
Springer Transactions in Civil and Environmental Engineering,
https://doi.org/10.1007/978-981-10-7281-9_5

1 Introduction

Accidents in chemical process industries are seldom one-incident events [1–5]. More often than not, they begin with the occurrence of a primary incident, which then sets off other incidents, leading to a knock-on effect of a chain of accidents occurring, much like an array of dominos toppling over when one of them is made to fall [6–10]. Thus, what perhaps begins as a minor innocuous looking event can spiral very quickly into a major catastrophe [11–14]. The accident is propagated from one unit to other units by what is termed in the literature as "escalation vectors." The escalation vectors are the manifestations of the accident in the form of one or more of the following: blast waves, heat radiation, and missiles/projectiles arising from bursting or shattering vessels and/or structures, and even toxic release which can incapacitate service personnel leading to adverse situations which have the potential to cause further accidents. An incident/accident in a unit can result in the generation of one or more of these escalation vectors. The range of effectiveness of the escalation vector—and thus the extent of damage they can cause—depends on the magnitude of the escalation vector [15–16]. Further, the magnitude of the escalation vector, and thus its effectiveness, decreases as the distance between the originating point of the escalation vector—the primary unit which is failing—and the other unit(s) in its vicinity increases. Closer the other unit(s) are to the failing unit, the greater the chances that the escalation vectors shall be of a high enough magnitude to cause a significant degree of damage to the unit(s) [8, 17–20].

Thus, one of the strategies adopted to prevent the consequences of an incident/accident in one unit from spreading to other units is to space the units apart at prescribed minimum separation distances, also referred to as "safe distances." The expectation is that at these distances the escalation vector shall be a spent force by the time it arrives at and interacts with the nearby unit, thus reducing or eliminating the possibility of any damage/failure to it. How far away should a hazardous unit be kept from other hazardous units or structures so that in the event of failure of one unit the other units/structures are not affected and damaged?

Ideally, all hazardous units should be put as far apart from one another and from adjacent structures in order that the probability of failure of the nearby unit/structure is zero in the event of one of them failing. However, given the large number of units in a chemical process industry, it is not always economical nor process-wise feasible to have this arrangement, given the interconnections between the units in terms of piping and related appurtenances, in addition presures of restrictions to land area requirements. Therefore, clearly safety measures require that the distances between units, and between units and adjacent structures be maximized, but economic pressure compels that such distances should be minimized. How to work out the best deal out of this situation has been the topic of research for many years, and continues to be so.

In this study, the adequacy of safe distances to prevent one of the most destructive of explosions that can occur in process industries—the boiling liquid expanding vapor explosion (BLEVE)—is assessed. Any vessel carrying a pressure-liquefied gas (PLG) poses a potential BLEVE threat. A vessel/unit undergoing a BLEVE gives rise to blast waves, missiles/projectiles, and, if the material is flammable, a pool fire and/or a fire ball. If the material is non-flammable but toxic, it can give rise to a toxic release. The escalation vectors arising from a BLEVE have the potential to cause damage to several units simultaneously, leading to a cascading of accidents [3, 16]. Thus, the prevention and mitigation of the consequences of a BLEVE by the use of safe distances require closer examination.

There are several agencies that prescribe standards for minimum distances to be maintained between equipment/units and between units and structures. To name a few codes are those by National Fire Protection Association (NFPA), USA, Occupational Safety and Health Administration (OSHA), USA, the American Petroleum Institute (API), the Health and Safety Executive (HSE) of the UK, etc. In India, the Oil Industries Safety Directorate (OISD) lays down minimum safety requirements of LPG installations.

In this paper, we have carried out studies to check the effectiveness or not of the safe distances prescribed by the National Fire Protection Agency (NFPA), USA, to prevent an accident from a container carrying a PLG from causing damage to other PLG carrying container and/or a nearby structure. The NFPA 58 Liquefied Petroleum Gas Code [18] sets the minimum separation distances to be maintained between vessels containing LPGs and between a LPG containing vessel and nearby structures. The NFPA 58 code has been chosen for this study because it is very well known and perhaps the most widely adopted one.

2 Methodology

The NFPA 58 code [21] prescribes minimum distances to be maintained between LPG storage vessels and between LPG storage vessels and structures, for vessel volumes ranging from less than 0.5 m^3 water capacity to greater than 3785 m^3 water capacity. A minimum distance of separation of 1 m is prescribed between vessels of water capacity 1–7.6 m^3, and 1.5 m for vessels of water capacity between 7.6 and 114 m^3. For larger vessels, the standard specifies the distance to be "1/4th the sum of diameters of adjacent containers" (Table 1). The minimum distance to be maintained from buildings/group of buildings ranges from 3 m for vessels of water capacity 0.5–1.9 m^3 to 122 m for vessels of water capacity greater than 3785 m^3 water capacity (Table 1).

The effectiveness of the prescribed minimum separation distances was assessed for both cylindrical (horizontal) and spherical pressurized containers of volumes ranging from 0.5 to 5000 m^3 water capacities. The distance of separation between each of these vessel volumes (0.5, 1, 5, 10, 50, 100, 250, 500, 1000, 2000, 3000, 4000, 5000 m^3) and other vessels of volume 0.5, 10, 500, 1000, and 5000 m^3 are taken for the study. The effect of the failure of the larger vessel on the small vessel at the prescribed separation distance has been assessed by calculating the over-pressure that would be experienced by the smaller vessel as a result of the failure of the larger vessel. As the separation distance between the containers is prescribed as

Table 1 Minimum distance between LPG containers and buildings/structures prescribed by NFPA 58 [21]

Water capacity per container (m^3)	Minimum distances	
	Between above-ground containers and structures (m)	Between containers (m)
<0.5	0	0
0.5–1.0	3	0
1.0 + −1.9	3	1
1.9 + −7.6	7.6	1
7.6 + −114	15	1.5
114 + −265	23	¼ of sum of diameters of adjacent containers
265 + −341	30	
341 + −454	38	
454 + −757	61	
757 + −3785	91	
>3785	122	

"¼ of sum of diameters of adjacent containers" for vessels of volume greater than 265 m^3, for cylindrical vessels the assessments have been done at length to diameter (L/D) ratios of 2, 3, and 4. The vessels were taken to be 80% full with propane at the instant of failure, and the pressure at burst was taken to be 2000 kPa. The high level of fill and high burst pressures were chosen in order to facilitate testing of the safe distances at near worst-case scenarios. The prescribed distance to be maintained between container and building and between containers is obtained from Table 3.4.2.2 of the NFPA 58 code (Table 1). At this separation distance, the overpressure that would be experienced by the adjacent container/building is estimated using the methodology developed by Mukhim et al. [22]. This methodology requires the separation distance, mass of the vessel contents at the time of burst, and reduced pressure at time of burst as inputs. The overpressure is a function of these inputs and can be calculated form the correlations or read from the graphs given in the paper. Once the overpressure was determined, the probability of failure of the adjoining structure was determined using the probits reported in Mukhim et al. [23]. For determining the probability of failure of the adjacent vessel due to the overpressure resulting from the failing vessel, the probits reported in Mukhim et al. [24] were used. Table 2 gives the threshold values of overpressure that would result in

Table 2 Threshold overpressures for damage to pressure vessels and buildings/structures [23, 24]

Probability of damage (%)	Pressure vessels (kPa)		Building/structures (kPa)
	Spherical vessels	Horizontal cylindrical vessels	
10	45.5	37.6	1.5
50	59	43.5	5.6
99	110.5	56.7	60.7

10, 50, and 99% probability of failure of pressurized vessels and buildings/structures as per the method reported by Mukhim et al. [23, 24].

An accurate estimation of the efficacy of the prescribed minimum safe distances depends entirely on the accurate estimation of overpressure resulting from vessel failure, as the estimation of the probability of damage it can cause to other vessel(s) located at the prescribed safe distance follows from it. In this study, though the method used to estimate overpressure is based on empirical data [22], it can still contain inaccuracies wherein overpressures could be over or under estimated for vessel volumes outside of which the model was developed on. Thus, the study was also extended to several medium- to-large scale BLEVE tests in which data of overpressures measured by transducers located at various distances from the vessel are available. Data from the experiments carried out by Birk et al. [25, 26], Balke et al. [27], Johnson et al. [28], and Giesbrecht et al. [29] have been taken for the study. The probability of failure that any building/structure or nearby vessel would have were it located at the same distance from the vessel that the transducer is, is calculated using the same method as described above. The probability of failure at the separation distance prescribed by NFPA was also calculated for the test vessels for comparison.

3 Results and Discussion

The overpressure that the smaller vessel—cylindrical or spherical—shall experience when the larger vessel suffers a BLEVE is given in Table 3. The probability of failure of the vessel due to the failure of the adjacent vessel positioned at the minimum separation distance is given in Fig. 1 for spherical vessels and in Fig. 2 for cylindrical vessels of various length/diameter (*L*/*D*) ratios.

The magnitude of overpressure that the smaller vessel shall experience due to the failure of the larger vessel depends on the prescribed NFPA separation distance calculated based on the vessel of the larger volume (Table 2). When comparing the overpressure experienced by a vessel of volume 0.5 and 1 m^3 when placed at NFPA separation distances from vessels of volume ranging from 0.5 to 5000 m^3, it is seen that the overpressure experienced by the 0.5 and 1 m^3 vessels increases with increase in volume of the other vessel. A similar trend is seen when comparing vessels of diameter 0.5 and 10 m^3 versus vessels of volume greater than or equal to 250 m^3. For vessels of volume 500, 1000, and 5000 m^3, compared with all other vessels of volumes lower than the respective volume, the overpressure experienced by the smaller vessel decreases as the volume of the smaller vessel increases. This is because for vessels of volumes greater than 114 m^3, the NFPA prescribes the safe distance to be "1/4th the sum of diameters of adjacent containers." Hence, as the diameters of the two vessels being compared increases, the separation distance increases resulting in a lower overpressure being experienced by the smaller vessel.

For vessels of volumes 0.5–100 m^3, when placed at NFPA separation distances from vessels of volume 0.5 m^3, an alternating dip and rise is seen in the

Table 3 Overpressure experienced by the smaller vessel when the larger vessel undergoes a BLEVE

Volume of the failing vessel (m^3)	Overpressure generated (kPa) at the smaller vessels when the larger one fails																			
	0.5 m^3				10 m^3				500 m^3				1000 m^3				5000 m^3			
	Spherical	Cylindrical			Spherical	Cylindrical			Spherical	Cylindrical			Spherical	Cylindrical			Spherical	Cylindrical		
		L/ D = 2	L/ D = 3	L/ D = 4		L/ D = 2	L/ D = 3	L/ D = 4		L/ D = 2	L/ D = 3	L/ D = 4		L/ D = 2	L/ D = 3	L/ D = 4		L/ D= 2	L/ D = 3	L/ D = 4
0.5	107.6	152.5	152.5	152.5	137.5	137.5	137.5	137.5	287.9	420.8	484.1	534.7	293.6	429.1	493.7	545.2	303.2	443.1	509.8	563.1
1	94.4	94.4	94.4	94.4	137.5	137.5	137.5	137.5	281.0	410.7	472.5	521.9	287.9	420.8	484.1	534.7	299.6	437.9	503.8	556.4
5	164.7	164.7	164.7	164.7	137.5	137.5	137.5	137.5	259.6	379.4	436.5	482.1	269.8	394.4	453.7	501.1	287.9	420.8	484.1	534.7
10	137.5	137.5	137.5	137.5	137.5	137.5	137.5	137.5	247.8	362.1	416.6	460.1	259.6	379.4	436.5	482.1	281.0	410.7	472.5	521.9
50	239.7	239.7	239.7	239.7	239.7	239.7	239.7	239.7	214.0	312.8	359.9	397.5	229.6	335.6	386.0	426.4	259.6	379.4	436.5	482.1
100	304.5	304.5	304.5	304.5	304.5	304.5	304.5	304.5	197.2	288.2	331.5	366.2	214.0	312.8	359.9	397.5	247.8	362.1	416.6	460.1
250	281.0	410.7	472.5	521.9	234.3	342.4	393.9	435.1	173.4	253.5	291.6	322.1	191.5	279.9	322.0	355.7	229.6	335.6	386.0	426.4
500	287.9	420.8	484.1	534.7	247.8	362.1	416.6	460.1	154.9	226.4	260.5	287.7	173.4	253.5	291.6	322.1	214.0	312.8	359.9	397.5
1000	293.6	429.1	493.7	545.2	259.6	379.4	436.5	482.1	173.4	253.5	291.6	322.1	154.9	226.4	260.5	287.7	197.2	288.2	331.5	366.2
2000	298.3	436.0	501.5	554.0	269.8	394.4	453.7	501.1	191.5	279.9	322.0	355.7	173.4	253.5	291.6	322.1	179.3	262.1	301.5	333.0
3000	300.6	439.4	505.5	558.3	275.1	402.1	462.5	510.9	201.7	294.8	339.2	374.6	184.1	269.1	309.5	341.9	168.6	246.4	283.5	313.1
4000	302.1	441.6	508.0	561.1	278.5	407.1	468.3	517.2	208.7	305.1	351.0	387.7	191.5	279.9	322.0	355.7	160.9	235.2	270.5	298.8
5000	303.2	443.1	509.8	563.1	281.0	410.7	472.5	521.9	214.0	312.8	359.9	397.5	197.2	288.2	331.5	366.2	154.9	226.4	260.5	287.7

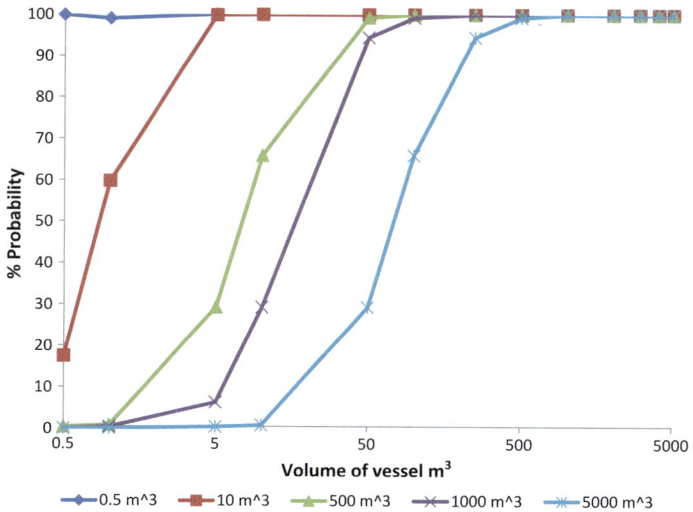

Fig. 1 Probability of damage to adjacent spherical vessels due to BLEVE as a function of the size of the failing vessel

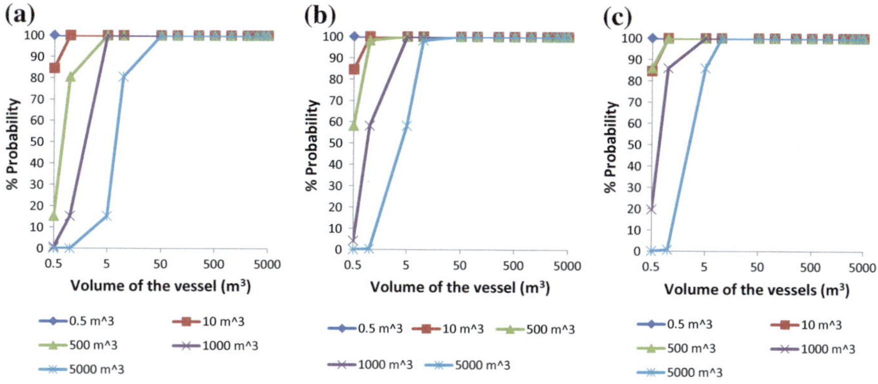

Fig. 2 Probability of failure of adjacent horizontal cylindrical vessels due to BLEVE as a function of the size of the failing vessel **a** *L*/*D* = 2. **b** *L*/*D* = 3. **c** *L*/*D* = 4

overpressure experienced by the latter as the volume of the other vessel increases. This is because of the manner in which the NFPA separation distances have been prescribed for vessels of volume less than 114 m^3. A single value of separation distance has been given for a range of volumes –1 m distance is prescribed between vessels of water capacity 1–7.6 m^3, and 1.5 m for vessels of water capacity between 7.6 and 114 m^3. Thus, the 0.5 m^3 vessel is to be kept at a 1 m distance from both the 1 and 5 m^3 vessel, and at a distance of 1.5 m from the vessels of volumes 10, 50, and 100 m^3. Thus, the 0.5 m^3 vessel experiences a smaller overpressure from a

10 m^3 vessel compared to from a 5 m^3 vessel. Similarly, when placed at the NFPA separation distance from a 250 m^3 vessel, it will experiences nearly the same overpressure when placed at the NFPA separation distance from a 5000 m^3 vessel!

In case of the 10 m^3 spherical and cylindrical vessels, when placed at the NFPA separation distance from vessels of volume 0.5–10 m^3, the overpressure experienced by the smaller vessel does not change even as the volume of the smaller vessel increases. This is because the standards specify a distance of 1.5 m for a vessel of size 10 m^3 when placed near vessels of volumes smaller than 10 m^3. Since they are all at the same fixed distance from the 10 m^3 vessel, they all experience the same overpressure irrespective of their volume.

For cylindrical vessels of up to 100 m^3 volume, the L/D ratio has no effect on the magnitude of overpressure experienced by the smaller vessel. This is because the vessel diameter is taken into account only for vessels of volumes greater than 114 m^3. For cylindrical pressure vessels of volume greater than 114 m^3, it is seen that an increase in the *L*/*D* ratio for the same vessel volume results in a higher overpressure experienced by the smaller vessel (Table 2). This is because as the *L*/*D* ratio increases, the diameter of the vessel reduces, leading to a smaller separation distance as per the "1/4th the sum of diameters of adjacent containers" specified by NFPA.

In terms of probability of damage to the smaller vessel that the overpressures represent, it is seen when comparing the threshold values given in Table 2, for all the vessels volumes—cylindrical or spherical—the probability of failure of the vessel is 100% at the NFPA separation distance for the smaller vessel. The probability of failure of the larger vessel due to failure of the smaller vessel, as shown in Fig. 1, indicates that as the vessel volume increases, the probability of failure of the larger vessel decreases. However, the damage probabilities even for a 5000 m^3 vessel volume crosses 50% when kept at the NFPA separation distance of a vessel as small as 100 m^3. For cylindrical vessels too (Fig. 2), the damage probability of the vessel decreases with increase in vessel volume of the larger vessel. The effect of the *L*/*D* ratio is seen clearly—larger *L*/*D* ratios for the same vessel volume translate to higher damage probabilities (Table 3).

In Table 4, assessments are carried out on data from BLEVE tests in which the location of the transducer from the vessel and the overpressure measured at the transducer is given. The prescribed NFPA separation distances for the vessel volumes used in the experiment along with the overpressure estimated to be experienced at that distance is given for comparison purposes. It is seen that at the distance the transducer is located, the overpressure recorded is low enough not to cause any damage whatsoever in case of the test vessels of volumes 0.4, 2, 5.6, and 100 m^3. However, were adjacent vessels located at distances prescribed by the NFPA, which for these particular test vessel volumes are much smaller than the transducer distance, the damage probability would have been 100%.

For assessing the effectiveness of the prescribed NFPA safe distances for preventing damage to buildings/structures, the overpressure experienced by a building/structure, and the associated probability of damage to it due to failure of vessels of volume ranging from 0.5 to 5000 m^3 is given in Table 5. Because there is only one

Table 4 Evaluating effectiveness of NFPA 58 separation distances between containers for overpressures measured from experiments on BLEVEs

Source	Volume of the vessel (m^3)	Location of the nearest transducer (m)	Separation distance prescribed by NFPA 58 (m)	Overpressure measured at the transducer (kPa)	Probability of failure at transducer distance (%)	Predicted overpressure at NFPA distance (kPa)	Probability of failure at NFPA distance (%)
Birk [11]	0.4	10	<1	2.36	0	71.86	100
	0.4	10	<1	2.69	0	71.86	100
	0.4	10	<1	2.73	0	71.86	100
	0.4	10	<1	3.68	0	71.86	100
	0.4	10	<1	4.95	0	71.86	100
	0.4	10	<1	6.32	0	71.86	100
	0.4	10	<1	7.12	0	71.86	100
	0.4	10	<1	8.20	0	71.86	100
	0.4	10	<1	8.44	0	71.86	100
	0.4	10	<1	10.37	0	71.86	100
Giesbrecht et al. [15]	1	5.33	1	69.60	100	431.17	100
	1	6.73	1	39.98	23	431.17	100
	1	8.67	1	35.18	3.12	431.17	100
	1	9.78	1	29.98	0.65	431.17	100
Birk et al. [12]	2	10	1	6.65	0	135.57	100
	2	10	1	3.97	0	135.57	100
	2	10	1	5.29	0	135.57	100
	2	10	1	5.02	0	135.57	100
	2	10	1	4.13	0	135.57	100
	2	10	1	13.11	0	135.57	100
	2	10	1	4.56	0	135.57	100

(continued)

Table 4 (continued)

Source	Volume of the vessel (m^3)	Location of the nearest transducer (m)	Separation distance prescribed by NFPA 58 (m)	Overpressure measured at the transducer (kPa)	Probability of failure at transducer distance (%)	Predicted overpressure at NFPA distance (kPa)	Probability of failure at NFPA distance (%)
Birk et al. [12]	2	10	1	4.15	0	135.57	100
	2	10	1	5.44	0	135.57	100
Johnson et al. [14]	5.6	25	1	6.2	0	237.98	100
	5.6	25	1	6.3	0	237.98	100
	5.6	25	1	5	0	237.98	100
	5.6	25	1	1	0	237.98	100
	5.6	25	1	2.3	0	237.98	100
	5.6	25	1	7	0	237.98	100
	10.8	25	1.5	8.2	0	144.95	100
Balke et al. [13]	45.36	100	1.5	2.5	0	173.26	100

Table 5 Overpressure experienced by buildings/structures located at the prescribed safe distances, and the probability of structural damage when the vessel BLEVEs

Volume of the vessel (m^3)	NFPA prescribed safe distance (m)	Overpressure experienced by the adjacent structure (kPa)	Probability of damage (%)
0.5	3	23.803	92.057
1	3	30.244	95.073
5	7.6	20.123	89.400
10	15	12.637	78.625
50	15	22.037	90.967
100	15	28.000	94.256
250	23	24.673	92.543
500	61	11.407	75.600
1000	91	9.575	70.100
2000	91	12.166	77.567
3000	91	13.995	81.350
4000	122	11.407	75.600
5000	122	12.321	78.000

separation distance given for a particular range of volume in the NFPA 58 code for distance to be maintained between vessel and structure, it is seen that the probability of damage does not show a steady increase or decrease as the vessel volume increases, rather it fluctuates as the vessel volume increases. However, for all the volumes covered in Table 5, the probability of damage is above 70%, going up to 95% for a vessel of volume 1 m^3! A similar damage probability to structures of 95% is seen for a vessel of volume 100 m^3. However, a 5000 m^3 vessel presents a 78% probability of damage at the prescribed safe distance. Thus, it is seen that a great deal of inconsistencies comes into the results due to the manner in which the

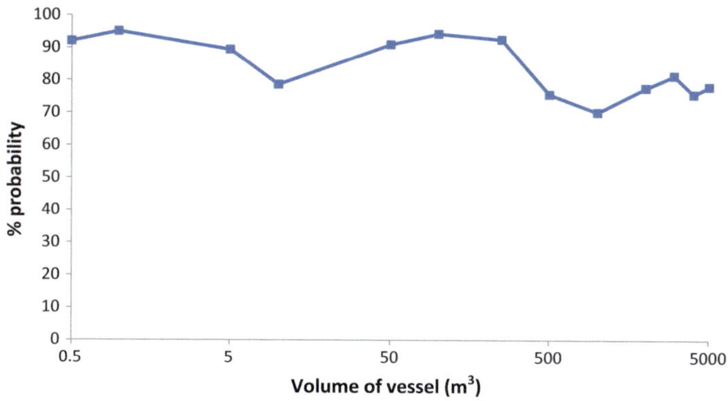

Fig. 3 Damage probabilities to structure due to BLEVE as a function of the size of the failing vessel

Table 6 Evaluating effectiveness of NFPA 58 separation distances between containers and adjacent structures/buildings for overpressures measured from experiments on BLEVE

Source	Volume of the vessel (m^3)	Location of the nearest transducer (m)	Separation distance prescribed by NFPA 58 (m)	Overpressure measured at transducer distance (kPa)	Probability of failure at transducer distance (%)	Predicted overpressure at NFPA prescribed distance (kPa)	Probability of failure at NFPA prescribed distance (%)
Birk [11]	0.4	10	<3	2.36	>19.80	71.86	>99.37
	0.4	10	<3	2.69	>23.72	71.86	>99.37
	0.4	10	<3	2.73	>24.18	71.86	>99.37
	0.4	10	<3	3.68	>33.93	71.86	>99.37
	0.4	10	<3	4.95	>45.27	71.86	>99.37
	0.4	10	<3	6.32	>54.53	71.86	>99.37
	0.4	10	<3	7.12	>59.15	71.86	>99.37
	0.4	10	<3	8.20	>64.40	71.86	>99.37
	0.4	10	<3	8.44	>65.45	71.86	>99.37
	0.4	10	<3	10.37	>72.70	71.86	>99.37
Giesbrecht et al. [15]	1	5.33	3	69.60	>99.31	117.25	>99.84
	1	6.73	3	39.98	>97.24	117.25	>99.84
	1	8.67	3	35.18	>96.35	117.25	>99.84
	1	9.78	3	29.98	>94.99	117.25	>99.84
Birk et al. [12]	2	10	7.6	6.65	>56.57	15.01	>83.33
	2	10	7.6	3.97	>36.74	15.01	>83.33
	2	10	7.6	5.29	>47.77	15.01	>83.33
	2	10	7.6	5.02	>45.73	15.01	>83.33
	2	10	7.6	4.13	>38.37	15.01	>83.33
	2	10	7.6	13.11	>79.68	15.01	>83.33

(continued)

Table 6 (continued)

Source	Volume of the vessel (m^3)	Location of the nearest transducer (m)	Separation distance prescribed by NFPA 58 (m)	Overpressure measured at transducer distance (kPa)	Probability of failure at transducer distance (%)	Predicted overpressure at NFPA prescribed distance (kPa)	Probability of failure at NFPA prescribed distance (%)
Birk et al. [12]	2	10	7.6	4.563	>41.93	15.01	>83.33
	2	10	7.6	4.15	>38.53	15.01	>83.33
	2	10	7.6	5.44	>49.00	15.01	>83.33
Johnson et al. [14]	5.6	25	7.6	6.2	>53.90	21.97	>90.92
	5.6	25	7.6	6.3	>54.47	21.97	>90.92
	5.6	25	7.6	5	>45.60	21.97	>90.92
	5.6	25	7.6	1	>4.58	21.97	>90.92
	5.6	25	7.6	2.3	>19.20	24.08	>92.21
	5.6	25	7.6	7	>58.57	21.97	>90.92
	10.8	25	15	8.2	>64.40	9.87	>71.10
Balke et al. [13]	45.36	100	15	2.5	>21.50	19.20	>88.48

safe distances are structured. Figure 3 depicts the probability of failure of the buildings/structures due to the failing vessels at various vessel volumes.

Similar assessments were carried out on data from BLEVE tests in which the location of the transducer from the vessel and the overpressure measured at the transducer is given (Table 6). It is seen that the overpressures that can have 20–99% damage probability are experienced even at distances several times higher than the prescribed minimum safe distances. The estimated damage probability at the prescribed distance is uniformly above 80% for all the cases, going up to 99% in some cases Birk et al. [22] and Giesbrecht et al. [26]. It may be noted that the transducer location distance is much higher than the separation distance prescribed by NFPA 58. Even at these transducer location distances the neighboring structures probability of damage/failure is high; thus, it is obvious that at the smaller NFPA 58 distances the damage probability would be several times higher. At the separation distance given by NFPA, the vessels will all fail catastrophically.

4 Conclusion

This study assesses the efficacy of the minimum separation distance, the "safe distance" prescribed by the National Fire Protection Agency (NFPA) code 58 in preventing damage to adjacent pressure vessels and buildings/structures due to a BLEVE in a nearby vessel. The assessments have been made for both cylindrical and spherical vessels filled with propane. It is seen that the distances are not adequate enough to prevent damage to adjacent vessels, unless the volume of the larger vessel is several times higher than the volume of the smaller vessel. The prescribed distances are also inadequate to prevent damage to adjacent structures. A possible reason for the inadequacy of the NFPA separation distances could be that the standards are based on preventing escalation of fire, rather than explosions.

5 Acknowledgement

SAA thanks the Council of Scientific and Industrial Research (CSIR), New Delhi, for the Emeritus Scientist grant (21(1034)/16/EMR-II).

References

1. Khan F.I., and S.A. Abbasi. 1998. Techniques and methodologies for risk analysis in chemical process industries. *Discovery Publishing House New Delhi* ix + 364 Pages.
2. Abbasi, T., E.V. Ramasamy, F.I. Khan, and S.A. Abbasi. 2013. Regional EIA and risk assessment in a fast developing country. Nova Science, New York. x+433 pages. ISBN 978-1-61942-234-6.
3. Khan, F.I., and S.A. Abbasi. 1997. Accident hazard index: A multi-attribute method for process industry hazard rating. *Process Safety and Environmental Protection* 75 (4): 217–224.
4. Abbasi, T., and S.A. Abbasi. 2008. The boiling liquid expanding vapour explosion (BLEVE) is fifty… and lives on! *Journal of Loss Prevention in the Process Industries* 21: 485–487.

5. Tauseef, S.M., and S.A. Abbasi. 2011. CFD-based simulation of dense gas dispersion in presence of obstacles. *Journal of Loss Prevention in the Process Industries* 24: 371–376.
6. Khan, F.I., and S.A. Abbasi. 1998. Models for domino effect analysis in process industries. *Process Safety Progress (AIChE)* 17: 107–123.
7. Khan, F.I., and S.A. Abbasi. 1998. DOMIEFECT: (DoMIno eFFECT) a user-friendly software for domino effect-analysis. *Environmental Modelling & Software* 13: 163–177.
8. Abdolhamidzadeh, B., T. Abbasi, D. Rashtchian, and S.A. Abbasi. 2010. A new method for assessing domino effect in chemical process industry. *Journal of Hazardous Materials* 182: 416–426.
9. Abdolhamidzadeh, B., D. Rashtchian, T. Abbasi, and S.A. Abbasi. 2011. Domino effect in process-industry accidents—an inventory of past events and identification of some patterns. *Journal of Loss Prevention in Process Industries (Elsevier)* 24 (5): 575–593.
10. Khan, F.I., and S.A. Abbasi. 1999. MAXCRED—a new software package for rapid risk assessment in chemical process industries. *Environmental Modeling & Software* 14 (1): 11–25.
11. Khan, F.I., and S.A. Abbasi. 2002. A criteria for developing credible accident scenarios for risk assessment. *Journal of Loss Prevention in the Process Industries* 15 (6): 467–475.
12. Khan, F.I., and S.A. Abbasi. 2001. Estimation of probabilities and likely consequences of a chain of accidents (domino effect) in Manali Industrial Complex. *Journal of Cleaner Production* 9 (6): 493–508.
13. Khan, F.I., and S.A. Abbasi. 2001. An assessment of the likelihood of occurrence, and the damage potential of domino effect in a typical cluster of industries. *Journal of Loss Prevention in the Process Industries* 14: 283–306.
14. Khan, F.I., and S.A. Abbasi. 1999. The worst chemical industry accident of 1990s: What happened and what might have been—a quantitative study. *Process Safety Progress* 18: 135–145.
15. Abbasi, T., and S.A. Abbasi. 2007. Accidental risk of superheated liquids and a framework for predicting the superheat limit. *Journal of Loss Prevention in the Process Industries* 20: 165–181.
16. Abbasi, T., H. Pasman, and S.A. Abbasi. 2010. A scheme for the classification of explosions in the chemical process industry. *Journal of Hazardous Materials* 174: 270–280.
17. Abdolhamidzadeh, B., T. Abbasi, D. Rashtchian, and S.A. Abbasi. 2010. Corrigendum to A new method for assessing domino effect in chemical process industry. *Journal of Hazardous Materials* 184: 1–3.
18. Khan, F.I., and S.A. Abbasi. 1997. OptHAZOP—an effective and optimal methodology for conducting HAZOP, study. *Journal of Loss Prevention in the Process Industries* 10: 191–204.
19. Khan, F.I., and S.A. Abbasi. 1997. TOPHAZOP: A knowledge-based software. *Journal of Loss Prevention in the Process Industries* 10: 333–343.
20. Vasanth S., S.M. Tauseef, T. Abbasi, and S.A. Abbasi. 2013. Assessment of four turbulence models in simulation of large-scale pool fires in the presence of wind using computational fluid dynamics. *Journal of Less Prevention in the Process Industries* 26: 1071–1084.
21. NFPA-58. 2001. *Standard for the storage and handling of liquefied petroleum gases*. 49 CFR 173.315. Date of action: 76 FR 3345, Jan. 19, 2011. An international codes and standards organization.
22. Mukhim, E.D., T. Abbasi, S.M. Tauseef, and S.A. Abbasi. 2017a. Estimation of overpressure from boiling liquid expanding vapour explosion (BLEVE). *Journal of Loss Prevention in Process Industries, communication.*
23. Mukhim, E.D., T. Abbasi, S.M. Tauseef and S.A. Abbasi. 2015. An assessment of the effectiveness of the NFPA 58 code in prescribing safe distances between LPG storage vessels and adjacent building/structures. *International Journal of Environmental Science and Engineering Research* 6 (2): 5–14.
24. Mukhim, E.D., T. Abbasi, S.M. Tauseef, and S.A. Abbasi. 2017b. Domino effect in chemical process industries triggered by overpressure-formulation of equipment-specific probits. *Process Safety and Environmental Protection* 106: 263–273.

25. Birk, A.M., 1997. *Fire tests of propane tanks to study BLEVEs and other thermal ruptures: detailed analysis of medium scale test results* (No. TP 12498E).
26. Birk, A.M., J.D.J. VanderSteen and C.R. Davison. 2003. *PRV field trials-the effects of fire conditions and PRV blowdown on propane tank survivability in a fire* (No. TP 14045E).
27. Balke, C., W. Heller, R. Konersmann, and J. Ludwig. 1999. *Study of the failure limits of a railway tank car filled with liquefied petroleum gas subjected to an open pool fire test, BAM project 3215*. Federal Institute for Materials Research and Testing (BAM).
28. Johnson, D.M., M.J. Pritchard, and M.J. Wickens. 1991. *Large scale experimental study of BLEVE: Contract report on CEC co-funded research project, Contract report 15367, Project M8411*. British Gas: Research & Technology division.
29. Giesbrecht, H., K. Hess, W. Leuckel, and B. Maurer. 1981. Analysis of explosion hazards on spontaneous release of inflammable gases into the atmosphere. *Germ Chem Eng* 4: 305–314.

Evaluating the Functionality of Industrial Emergency Operations Center (EOC) by Weighted Scoring Technique

Bahman Abdolhamidzadeh, Farshad Rahimi, Tannaz Gharebaghian, Behnaz Nosraty and Davood Rashtchian

Abstract In performing an efficient emergency response plan, one of the key elements is having a properly designed and well-equipped Emergency Operations Center (EOC). Lack of EOC functionality may even lead to the failure of the whole crisis management plan (CMP). Despite the significance of industrial EOC, there is not any guideline or well-established recommended practice in design and equipping these centers. In this paper, a comprehensive checklist giving recommendations on design and equipping EOCs is presented. Necessities in different fields such as configuration and layout, communication facilities, stationery and office supplies, welfare and eventually procedures are listed. Also, an evaluation method based on weighted scoring technique is offered to examine the functionality of EOC and compare the different design options.

Keywords Emergency Operations Center · Weighted scoring · Evaluation Functionality · Industrial crisis

1 Introduction

A crisis can be defined as any natural, man-made, accidental, or intentional undesired event that can severely affect the people, property, and/or the environment and as a result affects the business and reputation of the organization. The undesired effects depending on the type and size of the crisis can vary. These might be fatalities, serious and disabling injuries, significant property damage, and

B. Abdolhamidzadeh (✉) · F. Rahimi · D. Rashtchian
Department of Chemical and Petroleum Engineering, Center for Process Design, Safety and Loss Prevention (CPSL), Sharif University of Technology, Tehran, Iran
e-mail: Bahman_1788@yahoo.com

T. Gharebaghian · B. Nosraty
Faculty of Environment and Energy, Science and Research Branch, Islamic Azad University, Tehran, Iran

N. A. Siddiqui et al. (eds.), *Advances in Fire and Process Safety*, Springer Transactions in Civil and Environmental Engineering,
https://doi.org/10.1007/978-981-10-7281-9_6

destruction of the buildings and/or environmental contaminations which can severely undermine the organization's reputation and image.

Several evidences show that the trend of industrial crises around the world is increasing. In recent years, accidents such as Buncefield, 2005 [1], Gulf of Mexico, 2010 [7], or Fukushima, 2011 [6] can be classified as industrial crises which even ended up industrial disasters. That is why the importance of crisis management and emergency response plans is highlighted more than any time for the industry.

In order to make the organization ready for an effective response in case of any emergency, many companies develop a crisis management plan (CMP) in advance. CMP should be a dynamic, clearly defined, and documented plan of actions to be used at the time of a crisis. Typically, a plan will cover all the key personnel, resources, services, and actions required to implement and manage the crisis management plan [4].

An effective CMP is normally composed of many elements. These elements should be integrated in a comprehensive framework in order to guarantee the success of those plans. Some of these elements are hardware type, while there are many organizational and/or soft elements.

For effective management of any crisis, one of the vital elements is an Emergency Operations Center (EOC). EOCs are centers for communication, coordinating stuffs and supplies, facility and operational management where must have the capability of 24 h operation [5].

Past accident analysis (PAA) shows that deficiencies in EOCs may lead to the failure of the whole CMP. Even in a severe crisis like Three Miles Island accident in 1979 [8], impaired judgment in analyzing the data available in control council led to operator intervention that actually worsened the emergency case. Meshkati [9] has addressed the importance of human factors in interaction with automated systems of industrial control rooms that can be generalized to EOCs also.

The importance of EOCs is understood by industry more or less. Nowadays, in many industrial plants there is a center for dealing with potential emergency situations. During several site visits to Iranian oil and gas processing facilities by the authors, it was revealed that the majority of these facilities have dedicated a room or a specific space in the control rooms as EOC. But it was also found out that there is not any guideline or common recommendation in design and equipping these centers. The result is that the operability and functionality of those centers are highly under question in case of a real emergency situation.

Actually design, construction, and equipping an industrial EOC are quite a challenge. There is not any specific guideline or recommended practice in this order, available in the literature. There are some military standards such as [10, 14] or urban standards such as [11, 13, 15] regarding emergency response centers, but these are hardly expandable for industrial applications. As industrial EOCs are normally located in more compact areas and in many plants, are dual-use (e.g., an EOC on a manned oil platform).

To overcome the above-mentioned absence, in this paper authors have tried to prepare a comprehensive checklist based on standards for centers with similar usage, auditing several centers and observing the best practices and also learning from past accidents to cover all the necessary elements that shall exist in an EOC. To achieve this goal, authors used both references [3, 12] and experts to generate a checklist covering all of the features of an EOC. As the weights of these elements are not equal in the operability and functionality of these centers, an evaluation method based on weighted scoring techniques has also been presented. Application of this checklist and using the scoring method enables the analyst not only to identify the deficiencies of an EOC but also to rank different options against each other.

2 Emergency Operations Center (EOC)

Emergency Operations Center (EOC) is a central command and control facility in which several critical activities such as information gathering, communication, planning and organizing the activities of on-scene teams are performed. Although there are several contributing elements that can make an emergency response plan successful, EOC and its functionality is one of the most important ones.

There are some key characteristics that make an EOC a perfect one such as accessibility, survivability, flexibility, safety, and functionality. Although an EOC shall be located in a safe place, preferably far from sources of hazard inside an industrial plant, it should be also accessible. The sooner an EOC is mobilized, higher the chance of controlling the emergency situation would be. Survivability is defined for an EOC as its ability to remain operational for a long time despite the external events such as utility shutdown or process accidents. Flexibility of an EOC is its capability to allow the users to deal with different types of potential emergency scenarios inside a facility. Nowadays, EOCs are considered not only as command and control rooms and communication hubs, but also safe and pleasant working environment to accommodate personnel for longtime periods [15].

To meet the above-mentioned characteristics, EOCs should have certain physical and organizational conditions. There are many commonalities to any major office buildings but there are added complexities that make EOCs unique. Even the organizational structure of a company can influence the style and layout of the EOC [5, 14]. These requirements can be generally divided into the below fields.

2.1 Configuration and Layout

Configuration and layout of an EOC can directly affect the safety of this center itself. Also, proper layout will lead to better accessibility and saving the golden time in emergency cases. Even methods and materials used in construction of EOC can guarantee or jeopardize the safety of people and facilities inside the center. Internal space and configuration of tools also are other important factors necessary for perfect response to emergencies.

There are generally six basic configurations common for modern operation centers which are multipurpose, cluster, horseshoe, stadium, collaborative, and iris [5]. For industrial EOCs where dedicated space for response activities is not normally available, multipurpose layout seems to be a good choice. Although it is a flexible configuration, there are some disadvantages such as setup time requirement. Other layouts for EOCs and their advantages and disadvantages are fully discussed elsewhere [5, 14].

2.2 *Communication Facilities*

One of the very major activities that shall be performed properly inside an EOC is communication. This includes information gathering from accidents' site, transmitting commands and information to on-scene teams, and also communicating with external organizations and public. To achieve these goals normally advanced communication tools shall be provided. Also, parallel systems for diversification are highly recommended. As it was mentioned before, one of the basic characteristics of an EOC shall be survivability. Survivability in communication is defined as the ability of sending and receiving information even in abnormal conditions. So, backup power supplies for all communication tools shall be provided to fulfill this condition.

The evolution of EOCs like the field of emergency management is still ongoing and driven by technological improvements [5]. So, it is quiet common to find state-of-the-art technological tools such as GPS, satellite phones, and brand-new telecommunication tools in newly equipped centers.

2.3 *Stationery and Office Supplies*

Many of the office tools and stationery necessary in an EOC are similar to every major office. But there are some delicate points that make EOC special. Having at least two big whiteboards and markers to record the response steps and also situation monitoring is necessary. Having a big digital clock is highly recommended. All the common stationery that is used in a normal office is also useful here.

2.4 *Welfare*

In addition to being a safe and sophisticated center, EOC is a place in which people want to interact and perform sometimes for very longtime periods. So, it should be a comfortable place too. The lightening condition, cooling and heating systems, and even adequacy of the number of toilets will definitely show its effect on the performance of people engaged in emergency management inside EOC. Providing ergonomic furniture and also ergonomically designed control and monitoring councils are also key factors affecting the effectiveness of response.

2.5 Procedures and Audits

Based on several audits and during drills, authors found out that initial start-up of EOC in case of activation takes higher than expected times. Also, as some of the tools inside EOC are slightly different from regular office tools, this makes confusion while using those. So, it seems necessary to define very clear procedures from initial start-up and activation of EOC until deactivation phase. It is not only enough to define those procedures. Training and practicing is the key point to success. Based on the audits, those procedures can be modified, changed, or updated.

As it was mentioned before, many of the industrial EOCs are dual-use. In this case, procedures for quick transformation from normal to emergency configuration shall be defined and practiced.

There are much more details in every field mentioned above, and in every part there are a lot of sub-items. So, the authors have prepared a comprehensive checklist (Appendix) covering all the necessary items under each filed.

3 Weighted Scoring Technique

Application of weighted scoring technique is a common tool in multicriterion analysis and decision making [2]. In this method, different weights are assigned to each criterion which is important in the analysis. These weighting factors are used to define the level of importance of each criterion. Assigning the weighting factors is normally done by experts subjectively. It is recommended to keep the weight small. Next step is evaluating how well the existing factor meets the criterion. This leads to assigning a score for each item. A weighted score is calculated by multiplying the item score by the item weight. The final score is calculated by summing the weighted item scores up [2].

This process assigns numeric values to judgments. These judgments should not be merely subjective, but should reflect expert's views. These judgments should be supported by objective information and standards. That is why training of the auditor applying this method is necessary prior to the analysis. Due to its characteristics, this method works well in a case like evaluation of an EOC functionality and operability.

4 Weighted Scoring of EOC

In order to apply the weighted scoring method to analyze how perfect an EOC is, the first step is listing the necessary elements (both hardware and software) in such centers. As it was described before, there is not any ready-to-use guideline or recommendation mentioning specific criteria for industrial EOCs. So authors based on a comprehensive research based on standards for centers with similar usage (in different fields such as army or urban), auditing several centers and observing the best practices and also learning from past accidents constructed a checklist containing the necessary elements that shall exist in an EOC. Table 1 shows a part of this checklist covering items regarding configuration and layout of such center.

The complete checklist which has 124 items is available as Appendix of this paper. This checklist is a perfect tool to identify the potential shortcomings and deficiencies of an EOC when it is used just as an identification tool. But the items in this list have different weights in the overall perfection of the EOC. So in order to analyze how perfect and functional an EOC is, a ranking and evaluation method is required. Weighted scoring method is used for the first time in this paper to analyze an EOC.

Based on expert's opinion, four weight levels are defined as it is shown in Table 2 Then, a weight out of these four levels is assigned to every single element in the checklist.

Weight seven is assigned to items that are vital in an EOC. For instance in a gas refinery, the EOC should be safe against a probable vapor cloud explosion (VCE). So, being safe against explosion is necessary and gains weight seven. In the same EOC having ergonomic chairs is not necessary to guarantee the success of emergency response but it is favorable in this regard. All the 124 items are discussed one by one by a team of experts, and proper weight factor is assigned to those. In Table 2, there are two digits differences between weights in order to better demonstrating the importance of each level in the total scoring system.

Table 1 Sample items to be checked regarding configuration and layout of an EOC

Code	Item
D01	Safety of EOC against consequence of accidents
D01.1	Safety against process hazards (i.e., fire, explosion)
D01.2	Safety against toxic gas infiltration
D01.3	Safety against natural and ecological hazards (i.e., flood, earthquake, …)
D01.3.1	Resistance of EOC against the most violent flood in the past 500 years for newly constructed buildings and past 100 years for existing buildings (UFC 141-01, 3.5.1.3)
D01.3.2	Resistance of EOC against the most violent earthquake in the past 100 years (ASTM E2668, Sect. 6.6.1)

Table 2 Different weight levels and their meanings

Weight	Meaning
7	Necessary
5	Effective
3	Favorable
1	Optional

Table 3 Different score levels in analysis

Score	Meaning
6	Fully meets the criterion
4	Substantially meets the criterion
2	Partially meets the criterion
0	Does not meet the criterion

Next step is defining scores to show how well the exiting item meets the required criterion. Table 3 shows the different score levels defined by experts in this study.

Zero appears in the scores so if the auditor identifies that an element is totally missing, assign this score to that. Again the difference between each two score levels is two for better reflecting the differences. Now by adding additional columns to Table 1 to consider the weights and scores, the table becomes like Table 4.

For calculation of the score of each element in the checklist, its score determined by the auditor shall be multiplied by its weigh. Then, all the element scores are summed up as in Eq. (1) to give the final score.

$$I = \sum_{j=1}^{m} \sum_{i=1}^{n} S_{i,j}.w_{i,j} \quad (1)$$

In the above Eq. (1), the final score is calculated for the EOC under audit.

Having the final checklist in hand, a trained auditor can visit an EOC and, by checking and scoring each item, calculate a final score for the center under study. The final score reflects how perfect would be this EOC in time of emergency management. Table 5 shows a guideline for interpreting the final score.

It should be mentioned that threshold limits given in Table 5 can be used as a tool for evaluation of an EOC with the observations of the trained auditor. For instance, if the summation of the scores for an EOC falls between 471 and 110 while one of the “necessary” elements on an EOC is missing, then it is not tolerable anymore.

Table 4 Sample checklist with weights and score

Code	Item	Weight factor	Status				Weighted score
			Fully meets the criterion	Substantially meets the criterion	Partially meets the criterion	Does not meet the criterion	
D01	Safety of EOC against consequence of accidents						
D01.1	Safety against process hazards (i.e., fire, explosion)	7					
D01.2	Safety against toxic gas infiltration	7					
D01.3	Safety against natural and ecological hazards (i.e., flood, earthquake, …)	7					
D01.3.1	Resistance of EOC against the most violent flood in the past 500 years for newly constructed buildings and past 100 years for existing buildings (UFC 141-01, 3.5.1.3)	7					
D01.3.2	Resistance of EOC against the most violent earthquake in the past 100 years (ASTM E2668, Sect. 6.6.1)	7					

Table 5 Score thresholds for EOC evaluation

Final score (I)	Status
0–470	Unacceptable (requires immediate improvement)
471–1100	Tolerable* (needs improvement)
1101–2700	Acceptable

*Just if all the "Necessary" items based on the checklist exist

5 Conclusion

The extensive literature review revealed that despite the severe importance of industrial EOCs functionality in the success of any emergency plan, there is no well-established procedure or recommended practice how to construct and equip these centers. Based on available standards for similar centers for military or civil application, reviewing the best practices and customizing those for industrial application, a comprehensive checklist of EOC necessities has been presented in this paper. By application of this checklist, the main factors necessary for functionality of an industrial EOC can be audited. As the weight of different factors is not equal in the functionality of the whole system, an evaluation method has also been developed for the very first time. This evaluation technique which has been constructed based on expert's opinion enables the analyst to highlight the weak points and even compare and rank different EOC design options against each other.

Appendix: EOC Assessment Checklist

D. Configuration and layout

Code	Item	Weight factor	Status				Weighted score
			Fully meets the criterion	Substantially meets the criterion	Partially meets the criterion	Does not meet the criterion	
D01	Safety of EOC against consequence of accidents						
D01.1	Safety against process hazards (i.e., fire, explosion)	7					
D01.2	Safety against toxic gas infiltration	7					
D01.3	Safety against natural and ecological hazards (i.e., flood, earthquake, …)	7					
D01.3.1	Resistance of EOC against the most violent flood in the past 500 years for newly constructed buildings and past 100 years for existing buildings (UFC 141-01, 3.5.1.3)	7					
D01.3.2	Resistance of EOC against the most violent earthquake in the past 100 years (ASTM E2668, Sect. 6.6.1)	7					
D01.3.3	Resistance of EOC against region-specific natural hazards (i.e., tsunami, volcano, …)	7					
D02	Proper accessibility to EOC for main members of team	5					
D03	Space and environment of EOC						
D03.1	Height of ceiling (at least 3 m according to UFC 141-01, 4.2.6.8)	3					
D03.2	Specification of proper space per person according to domestic regulations	3					
D04	Sound proof walls	1					
D05	Layout of center furniture						
D05.1	Assignment of specific area to each group (operation, planning, …)	3					
D05.2	Existence of enough public area for meetings and gathering	3					
D05.3	Proper location of screen display (so everyone can see it easily)	5					

(continued)

(continued)

Code	Item	Weight factor	Status				Weighted score
			Fully meets the criterion	Substantially meets the criterion	Partially meets the criterion	Does not meet the criterion	
D05.4	Proper location of each group (operation, planning, …)	5					
D05.5	Proper location of entrance	3					
D05.6	Specification of each person's place in the group with name plates	3					
D05.7	No physical/sonic interference between groups	3					
D05.8	Specific and accessible place for documents	5					
D05.9	Specific and accessible place for office supplies	3					
D06	Assignment of space for rest (sufficient area for 25% of involved persons)	1					
D07	Dedicated dining area	1					
D08	Easy access to toilets	1					
D09	Surveillance system for entrance and exit	3					
D10	Protection against internal fire	7					
D11	Firefighting facilities (detectors, portable extinguishers, …)	7					
Sum							

C. Communication facilities

Code	Item	Weight factor	Status				Weighted score
			Fully meets the criterion	Substantially meets the criterion	Partially meets the criterion	Does not meet the criterion	
C01	Phones						
C01.1	Availability of at least one landline for each group)	7					
C01.2	Satellite phone	5					
C01.3	Ring light phones	3					
C01.4	Headset for phones	3					
C02	Walkie-talkie with specified radio channel	5					
C03	Proper facility for paging and intercom	3					
C04	PCs/laptop						
C04.1	One computer system should be connected to display, and other one should be connected to Internet or intranet	7					
C04.2	Computers being used just in EOC (not dual-use)	3					
C04.3	Protected by antivirus software	3					
C04.4	Proper software installed	5					
C04.5	Availability of technical and process data about plant on computers	5					
C05	CCTV and recording system (video and voice recording)	1					
C06	Phone and radio communication recording system	1					
C07	Television	3					
C08	CD/DVD player	1					
C09	Video projection or LCD (at least three displays are needed for news, operation status, operation command)	5					

(continued)

(continued)

Code	Item	Weight factor	Status				Weighted score
			Fully meets the criterion	Substantially meets the criterion	Partially meets the criterion	Does not meet the criterion	
C10	Internet access	5					
C11	Fax (one for receiving and one for sending)	3					
C12	Photocopier	3					
C13	Online access to weather forecast	3					
C14	Display of plants CCTVs	3					
C15	Access to control and monitoring systems of plant	3					
C16	Emergency phone numbers (internal/external)	5					
C17	Megaphone	3					
C18	Regional maps	5					
C19	Plant layout drawings	7					
C20	Special location's maps (i.e., nuclear power plants, hospitals,…)	5					
Sum							

A. Stationery and office supplies

Code	Item	Weight factor	Status				Weighted score
			Fully meets the criterion	Substantially meets the criterion	Partially meets the criterion	Does not meet the criterion	
A01	Large table for group meetings	3					
A02	Specific desk for each group	5					
A03	Name plate for team members	1					
A04	Ergonomic chairs	3					
A05	Whiteboard (at least two), whiteboard markers	5					
A06	Files for documents	5					
A07	Large digital clock	5					
A08	Flashlight	3					
A09	Spare batteries	3					
A10	Diesel generator with fuel for at least two days	7					
A11	UPS for all critical apparatuses	7					
A12	Office supplies						
A12.01	Pencil	5					
A12.02	Pen	5					
A12.03	Paper	5					
A12.04	Glue	3					
A12.05	Highlight marker	3					
A12.06	Folders	3					
A12.07	Stapler	3					
A12.08	Sticker	3					
A12.09	Scissors	3					
A12.10	Punch	3					
Sum							

W. Welfare

Code	Item	Weight factor	Status				Weighted score
			Fully meets the criterion	Substantially meets the criterion	Partially meets the criterion	Does not meet the criterion	
W01	Lighting system						
W01.01	Lighting system with at least 500 lx luminous flux per unit area	5					
W01.02	Lighting system designed according to UFC 3 530 01	5					
W01.03	Lighting dimmer controller	1					
W02	Air ventilation system (20 CFM per person)	5					
W03	Proper heating system	5					
W04	Proper cooling system	5					
W05	Enough number of toilets	3					
W06	First aid kit	7					
W07	Proper PPE for all identified hazards	7					
W08	Drinking water storage	3					
W09	Dry food storage	3					
W10	Glass, plate, and cutlery	1					
W11	Trash bin	1					
W12	Proper facilities for rest	1					
W13	Tissues	1					
W14	Water dispenser	1					
W15	Supply for hygiene facilities						
W15.1	Toothpaste and toothbrush	1					
W15.2	Towels	1					
W16	Fridge, microwave heater	1					

I. Procedures and audits

Code	Item	Weight factor	Status				Weighted score
			Fully meets the criterion	Substantially meets the criterion	Partially meets the criterion	Does not meet the criterion	
I01	Procedures						
I01.1	Procedure for changing the configuration in case of EOC being dual-use	5					
I01.2	Layout of center equipment (in case of being dual-use)	5					
I01.3	Procedure for updating any document in EOC in case of any change or periodically	5					
I01.4	Procedure for updating emergency phone numbers in case of any change	5					
I01.5	Procedure for EOC activation	5					
I01.6	Procedure for EOC deactivation	5					
I02	Regular audit and inspection						
I02.1	Inspection of EOC facilities (every 3 months)	3					
I02.2	Inspection of the communication facilities functionality (every 3 months)	3					
I02.3	Inspection of computer and display systems (every 3 months)	3					
I02.4	Inspection of CCTVs and recording systems (every 3 months)	3					
I02.5	Inspection of CD/DVD player(every year)	3					
I02.6	Inspection of Internet connection (every month)	3					
I02.7	Inspection of photocopier and fax devices (every 3 years)	3					
I02.8	Inspection of weather forecast and display system (every month)	3					

(continued)

(continued)

Code	Item	Weight factor	Status				Weighted score
			Fully meets the criterion	Substantially meets the criterion	Partially meets the criterion	Does not meet the criterion	
I02.9	Checking maps and process documents whether updated or not (every year)	3					
I02.10	Checking internal and external emergency contacts whether updated or not (every year)	3					
I02.11	Checking office supplies, i.e., whiteboard markers, pens, …. (every 3 months)	3					
I02.12	Checking the digital clock (every 6 months)	3					
I02.13	Making sure that spare batteries are supplied	3					
I02.14	Checking the emergency generator (every 6 months)	3					
I02.15	Checking the UPS (every 6 months)	3					
I02.16	Making sure whether the luminous flux per unit area (LUX) is sufficient in EOC	3					
I02.17	Making sure that ventilation system (including heating and cooling system) is functional	3					
I02.18	Inspection first aid kit (every year)	3					
I02.19	Inspection water and dry food storages (every 6 months)	3					
I02.20	Inspection breathing apparatuses if there is any (every 6 months)	3					
I02.21	Regular training workshop to recall responsibilities, how to activate EOC and how to use the location of EOC facilities and equipment	3					

References

1. Abdolhamidzadeh, B., T. Abbasi, D. Rashtchian, and S. Abbasi. 2011. Domino effect in process-industry accidents: An inventory of past events and identification of some patterns. *Journal of Loss Prevention* 24: 575–593.
2. Canter, L.W. 1996. *Environmental Impact Assessment.* New York: McGraw-Hill.
3. Eldson, G. 2010. EOC facility checklist: Avoiding surprises. *Disaster Resource Guide Quarterly* 15 (2).
4. ENISA. 2013. Retrieved from European Union Agency for Network and Information Security. http://www.enisa.europa.eu/activities/risk-management/current-risk/bcm-resilience/glossary/c-d.
5. Fagel, M.J. 2011. *Principle of Emergency Management and Emergency Operations Centers.* NW: CRC Press.
6. Koo, Y.-H., Y.-S. Yang, and K.-W. Song. 2014. Radioavtivity release from the Fukushina accident and its consequences: A review. *Progress in Nuclear Energy* 74: 61–70.
7. Mariano, A., V. Kourafalou, A. Srinivasan, H. Kang, G. Halliwel, E. Ryan, et al. 2011. On the modeling of the 2010 Gulf of Mexico Oil Spill. *Dynamics of Atmospheres and Oceans* 52 (1–2): 322–340.
8. Meshkati, N. 1991. Human factor in large scale technological systems' accident: Three Mile Island, Bhopal, Chernobyl. *Industrial Quarterly* 5: 131–154.
9. Meshkati, N. 2006. Safety and human factors considerations in control rooms of oil and gas pipeline systems: Conceptual issues and practical observation. *International Journal of Occupational Safety and Ergonomics (JOSE)* 12 (1): 79–93.
10. Michigan State Police. 2003. *Design recommendations and criteria for emergency operations center.* Michigan: Michigan State Police Emergency Management Division.
11. Shrader, K. 2008. *Designing and effective operating center for city of Martinsville, Virginia.* Martinsville, Virginia: Fire EMS Department.
12. Stufflebeam, D.L. 2000. Guidelines for developing evaluation checklists: the checklists development checklist (CDC). *Evaluation Check list Project.*
13. Tseng, J., C. Shu, J. Horng, C. Kuan, and H. Hsu. 2007. Planning an emergency respone center in southern Taiwan science park. *IChemE.*
14. US Army, C. 2008. *Emergency Operation Center Planning and Design.* US: US Department of Defence.
15. Welker, T. 2003. *Design component of a modern emergency operation center (EOC) for city of clearwater.* Florida: National Fire Academy.

Development of Risk Acceptance Criteria for Indian Railways

Ashwin Malviya, Vishal Kumar Singh and Bikarama Prasad Yadav

Abstract Indian railways cater to the need of more than a billion people every year. Despite the undeniable fact that Indian railways form the backbone of Indian economy, its safety records are tainted. Over the past six months alone, three major railway accidents, resulting in the loss of 193 lives, have taken place. In order to improve this situation, it is required that proper quantitative risk assessment (QRA) is done for the railways. The prerequisite of doing a meaningful QRA is the availability of a well-defined Risk Acceptance Criteria (RAC). In this approach, the individual risk and societal risk are calculated with the help of statistical data of accidents eventuated from 1990 to 2016. After computation of individual risk and societal risk, the As Low As Reasonably Practicable (ALARP) zones are plotted. The ALARP zone gives a guideline of the region of risk acceptability. The ALARP zones have been calibrated by comparing criteria of various countries that can be used in Indian context. Risk acceptability region contributes in the formulation of RAC. With this risk-based approach of QRA, an RAC is developed for the Indian railways. Finally, the Indian Authorities may use formulated and uniform criteria from the RAC developed.

Keywords ALARP · Cumulative frequency F-N curve · Individual risk
Risk acceptance criteria · Societal risk

A. Malviya (✉) · V. K. Singh
Vedanta Limited, New Delhi, India
e-mail: ashu.malviya789@gmail.com

V. K. Singh
e-mail: vishalksingh3011@gmail.com

B. P. Yadav
University of Petroleum and Energy Studies, Bidholi, Dehradun 248007, India
e-mail: bpyadav@ddn.upes.ac.in

N. A. Siddiqui et al. (eds.), *Advances in Fire and Process Safety*,
Springer Transactions in Civil and Environmental Engineering,
https://doi.org/10.1007/978-981-10-7281-9_7

1 Introduction

Indian railways are really a microcosm of India and quite distinctive in character and the end product of safety is the cohesive fusion of myriad parts in Indian railways. The route has been distributed all over country in a criss-cross flow covering 65,436 km which constitutes in over 9,956 locomotives, 57,256 coaches and 2.56 lakh wagons which hauls nearly 2.8 millions of passengers every day. This mode of transport is adopt by every person as it one of the backbone of Indian economy system. But a wrong indication in one of the dot of thousand dots which completes the rail landscape, or an act of negligence by any staff member when the train is in motion, one of the millions who perform rush act on the road daily around the odd level across the crossing gates, or a person performing an irresponsible act by carrying flammable goods. Such events have potential to become a major accident and even cause fatality [1]. The Indian railways have accorded the safety in operation and security of public while travelling which has been the utmost requirement.

In this paper, we have discussed about the individual and societal RAC through As Low As Reasonably Practicable zones (ALARP). The RAC should reflect the importance of tolerable criteria of risk in Indian railway transportation is the purpose of this paper. This includes the probability of death per year of an individual due to exposure to particular hazard is individual risk while societal risk is measured for a group of people. This acceptance criteria is expressed in terms of F-N curve or multiple causality events of frequency distribution [2]. Vulnerability researches of hazard can be pursued back depending upon reliable statistics available. The tolerable criteria can be defined by the help of ALARP zone evaluation which appears out to be first safety concept which was developed in UK [3]. The ALARP approach is shown Fig. 1 which explains the range for upper and lower limit of risk. If the risk level lies within the lower level, then it is considered as tolerable risk and below which it is accepted with no practical interest in any circumstances. Thorough evaluation of ALARP is need to be performed to identify the safety is safe enough in all aspects or not.

2 Methodology

Quantitative Risk Assessment Principles and Estimation

It is mandatory to define risk measures and to which category the risk criteria can be applied before estimation of risk criteria. To obtain an equivalent perspective of the associated risk related to any sector, then it should be evaluated in two perspective which is mentioned below:

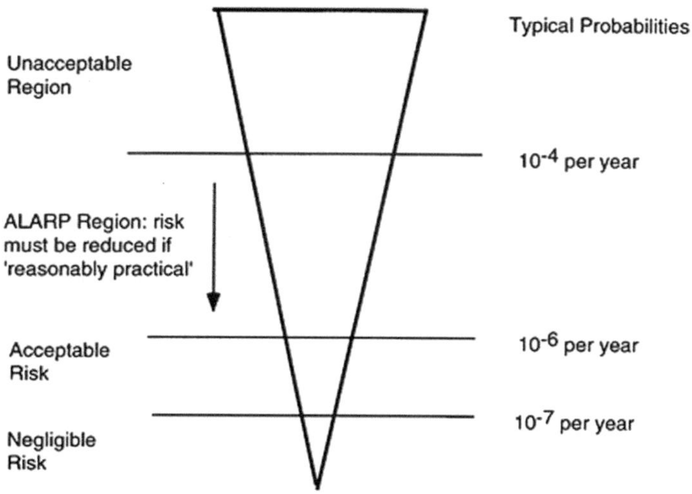

Fig. 1 Various level of risk and ALARP are defined based on the experience in UK [4]

1. risk involved with an individual is referred as individual risk.
2. a group of people involved in a risk is referred as societal risk.

A number of diversified measures, which address both individual and societal risks, can be applied for railways.

Individual Risk

Individual risk is when a single person is exposed to a hazard, i.e. the potential aftermath area of an event or set of events a person is exposed to. Therefore, if numbers of people are impacted by a single event in terms of scale, then it does not event the individual risk. Tables of numbers, unit numbers or various graphical representation can be calculated of individual risk which is based on formulated procedure which is mentioned as [5] individual risk per year (IR_y) is equal to the product of the frequency of occurrence of incident (F_y) and probability of casualty per year (P_y).

$$\mathrm{IR}_y = F_y \times P_y$$

where

$$F_y = \frac{\text{No. of incidents} \times 10^6}{\text{Total Passenger Travelled}}$$

and,

$$P_y = \frac{\text{No. of Casuality}}{\text{Passenger Travelled in Millions}}$$

All individual risk calculation methods in the paper are based on these relationships. Simplified techniques can reduce the amount of calculation, however, these techniques can also be used for more detailed analysis [5].

Compilation of Historical Data

Table 1 shows the estimated frequency and probability for passengers yearly in the period between 1990 and 2016. The data source is [6], representing the accidents. The constant decrease in the number of accidents has been seen but casualty is more or less same.

Table 1 Statistical database of accidents and calculation of individual risk

Year	Total accidents	Passenger travelled (million)	Frequency of occurrence (per million passenger travelled)	Casualty	Probability of occurrence of casualty (per million passenger travelled)	Individual risk (IR)
1990–91	532	3858	1.38E−01	220	5.70E−02	7.86E−03
1991–92	530	4049	1.31E−01	98	2.42E−02	3.17E−03
1992–93	524	3749	1.40E−01	96	2.56E−02	3.58E−03
1993–94	520	3708	1.40E−01	179	4.83E−02	6.77E−03
1994–95	501	3915	1.28E−01	84	2.15E−02	2.75E−03
1995–96	398	4018	9.91E−02	406	1.01E−01	1.00E−02
1996–97	381	4153	9.17E−02	83	2.00E−02	1.83E−03
1997–98	396	4348	9.11E−02	171	3.93E−02	3.58E−03
1998–99	397	4411	9.00E−02	280	6.35E−02	5.71E−03
1999–2000	463	4585	1.01E−01	341	7.44E−02	7.51E−03
2000–01	473	4833	9.79E−02	55	1.14E−02	1.11E−03
2001–02	415	5093	8.15E−02	144	2.83E−02	2.30E−03
2002–03	351	4971	7.06E−02	157	3.16E−02	2.23E−03
2003–04	325	5112	6.36E−02	135	2.64E−02	1.68E−03
2004–05	234	5378	4.35E−02	50	9.30E−03	4.05E−04
2005–06	234	5725	4.09E−02	315	5.50E−02	2.25E−03
2006–07	195	6219	3.14E−02	208	3.34E−02	1.05E−03
2007–08	194	6524	2.97E−02	191	2.93E−02	8.71E−04
2008–09	177	6920	2.56E−02	209	3.02E−02	7.73E−04

(continued)

Table 1 (continued)

Year	Total accidents	Passenger travelled (million)	Frequency of occurrence (per million passenger travelled)	Casualty	Probability of occurrence of casualty (per million passenger travelled)	Individual risk (IR)
2009–10	165	7246	2.28E−02	238	3.28E−02	7.48E−04
2010–11	141	7651	1.84E−02	235	3.07E−02	5.66E−04
2011–12	131	8224	1.59E−02	100	1.22E−02	1.94E−04
2012–13	121	8421	1.44E−02	60	7.13E−03	1.02E−04
2013–14	117	8397	1.39E−02	42	5.00E−03	6.97E−05
2014–15	131	8224	1.59E−02	118	1.43E−02	2.29E−04
2015–16	106	8107	1.31E−02	126	1.55E−02	2.03E−04

Societal Risk

Large number of people has potential to get affected from major incidents. The cumulative risk to which a group of people is exposed is expressed as societal risk, i.e. it is a measure of the specified level of harm to which a group of people will be exposed due to reliaziation of a hazard.

The tabular sets of numbers, graphical representations like frequency curve (F-N curve) can show societal risk estimates. The multiple casualty events show the frequency distribution is plotted in F-N curve, where N is number of casualty and F is the cumulative frequency of all events leading to casualty.

Historical data illustrations in the F-N curve are based on [7]. The frequency per year is calculated in the same way as in individual risk. To construct the F-N curve, a list of all yearly data and their associated frequencies and consequences, i.e. casualty is compiled and sorted by decreasing value of casualty. In the year 1995–96 is the most severe and has a casualty of 406, which occurs at a frequency of 9.91E−02 and so forth in the decreasing order. The cumulative frequency is calculated as shown in Table 2. Table 3 shows the estimated statistics for plotting of F-N curve.

Table 2 Calculation of cumulative frequency

Casualty in decreasing order	Year	Frequency	Cumulative frequency
C1	Y1	f1	F1 = f1
C2	Y2	f2	F2 = f1 + f2
C3	Y3	f3	F3 = f1 + f2 + f3
C4	Y4	f4	F4 = f1 + f2 + f3 + f4
C5	Y5	f5	F5 = f1 + f2 + f3 + f4 + f5
C6	Y6	f6	F6 = f1 + f2 + f3 + f4 + f5 + f6
C7	Y7	f7	F7 = f1 + f2 + f3 + f4 + f5 + f6 + f7

Table 3 Estimated statistics for the F-N curve

Year	Casualty	Frequency of occurrence (per million passenger travelled)	Cumulative frequency
1995–96	406	9.91E−02	9.91E−02
1999–2000	341	1.01E−01	2.00E−01
2005–06	315	4.09E−02	2.41E−01
1998–99	280	9.00E−02	3.31E−01
2009–10	238	2.28E−02	3.54E−01
2010–11	235	1.84E−02	3.72E−01
1990–91	220	1.38E−01	5.10E−01
2008–09	209	2.56E−02	5.36E−01
2006–07	208	3.14E−02	5.67E−01
2007–08	191	2.97E−02	5.97E−01
1993–94	179	1.40E−01	7.37E−01
1997–98	171	9.11E−02	8.28E−01
2002–03	157	7.06E−02	8.99E−01
2001–02	144	8.15E−02	9.80E−01
2003–04	135	6.36E−02	1.04E+00
2015–16	126	1.31E−02	1.06E+00
2014–15	118	1.59E−02	1.07E+00
2011–12	100	1.59E−02	1.09E+00
1991–92	98	1.31E−01	1.22E+00
1992–93	96	1.40E−01	1.36E+00
1994–95	84	1.28E−01	1.49E+00
1996–97	83	9.17E−02	1.58E+00
2012–13	60	1.44E−02	1.59E+00
2000–01	55	9.79E−02	1.69E+00
2004–05	50	4.35E−02	1.73E+00
2013–14	42	1.39E−02	1.75E+00

Comparison with RAC of Different Countries

From time to time, RAC is developed and presented to the national and international authorities of different countries since then it has been a subject for public debate across the board. After the discussion, a valuable judgement was raised on acceptable and unacceptable risks in terms of risks. Any railway safety standard is always implicit by an acceptable risk which has been possible by data analysis and the outcome of the result is known as RAC.

The individual risk criteria of different countries can be compared with the calculated individual risk of Indian railways as shown in Fig. 2. The maximum and minimum values of individual risk of Indian railways are been taken for the comparison with the individual RAC of different countries.

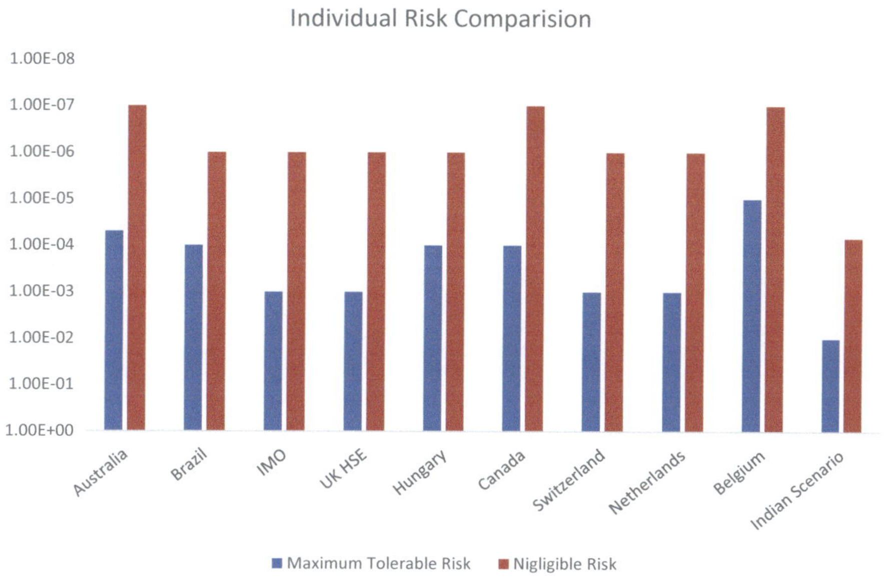

Fig. 2 Comparison of individual RAC of different countries to Indian scenario

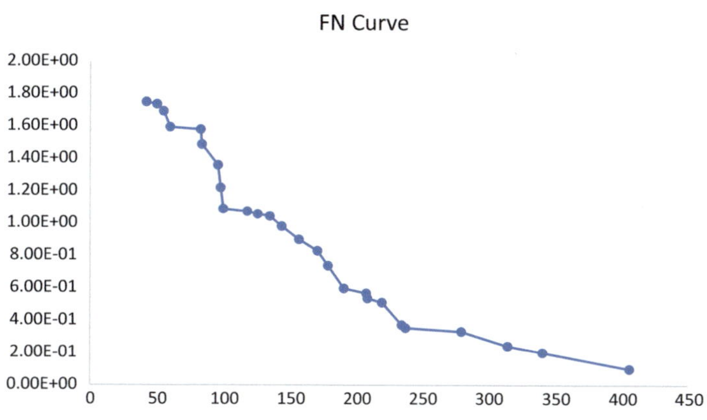

Fig. 3 F-N curve representing the societal risk calculated from the database [4]

Similarly, the societal risk criteria of different countries can also be compared with the calculated societal risk of Indian railways with the help of F-N curve. Fig. 3 shows the cumulative frequency which is been plotted in F-N curve. The societal RAC of different countries can also be seen as compared to Indian scenario (Fig. 4). All the criteria set by different countries is for the Railways and similar sectors. Only maximum tolerable risk of different countries is taken into consideration (Fig. 2).

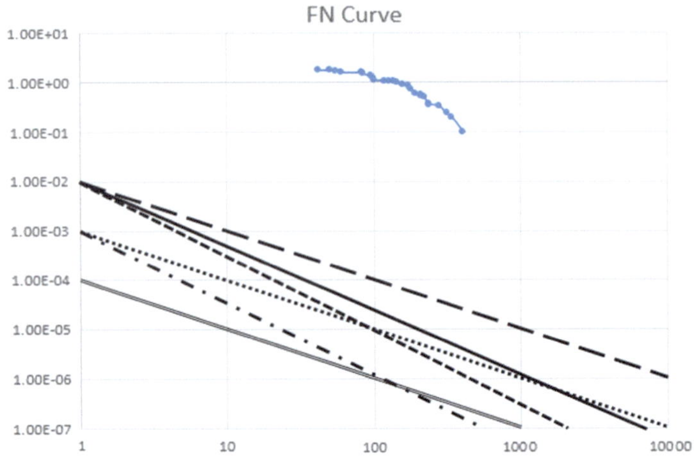

Fig. 4 Societal RAC comparison with Indian scenario

ALARP Principle

To reduce the risk to a practical state, the key principle is ALARP for achieving tolerable risk to reduce risks to a practical state. The tolerable risk [8, 9] is in arrange so the society can live with certain security and attain net benefits and the range of risk cannot be considered as risk which is negligible or risk that can be kept under review but which sometime might be ignored and to reduce it to its maximum or further if we can. Through this we can understand that it is an integral part of individual risk and societal risk procedures.

The ALARP method used in risk acceptance criteria that defines [8, 10]:

1. Intolerable risk—risk above the range which requires reduction.
2. Tolerable risk—risk is acceptable and risk reduction is optionable.
3. Negligible risk—broadly acceptable risk.

Risk should be reduced to lie in the area of ALARP region, i.e. between maximum tolerable risk and negligible risk and it should be reduced until it is not disproportionate to the costs.

The tolerable risk is not present at single universal level but different range is being developed for different sectors from different countries. If the risk is voluntary, ordinary and with delayed effects, then also risk can be acceptable. In some cases [11, 12], the acceptance of high-risk values can be understood considering these factors. But for the case of railways, higher risks cannot be accepted risks to which the people are exposed to which they have less or no control. The above comparison of the maximum tolerable risk with RAC of different countries for railways and similar sectors with calculated risks of current risk scenario of Indian railways gives a broad view of high-risk exposures to the people directly or indirectly involved.

3 Results and Discussions

Proposed Criteria

The proposed criteria are based on the comparison of RAC applicable to railways and similar sectors of different countries so as to frame RAC for Indian railways.

Individual RAC

Broadly acceptable maximum tolerable risk in most of the countries lie between 1.00E−03 and 1.00E−05, but our calculations show that the maximum tolerable risk for Indian railways is 1.00E−02. And similarly, negligible risk for most of the countries lies between 1.00E−06 and 1.00E−07, whereas for Indian railways it is 6.97E−05.
The proposed individual RAC for Indian railways can be seen in Table 4.

Societal RAC

The societal RAC is expressed with the help of F-N curve plotting ALARP region with the limit of casualty. The maximum tolerable risk for the same can be seen in the Fig. 3 with the involved limit of casualty for different countries. As plotted in the graph, the calculated societal risk for Indian railways lies above all the criteria of countries.

The proposed societal RAC for Indian railways can be seen in Table 4. The ALARP region and the limit of casualty are proposed in Table 5

Need of RAC for Indian railways

. The requirement for RAC is based upon the evolving tools which are needed for effective decision-making and risk assessment. Worldwide many countries in the parallel path are mandating the use of RAC so should Indian railways for:

1. Approval of plans for new facilities and operations.
2. Approval of plans for new offsite developments near vicinity of railway facilities.
3. Decreasing the risk exposure through effective risk reduction measures of existing facilities and operations.

Table 4 Proposed individual RAC for Indian railways

Maximum tolerable risk	Negligible risk
1.00E−03	1.00E−06

Table 5 Proposed societal RAC for Indian railways

Maximum tolerable risk	Negligible risk	Limit of casualty
1.00E−02	1.00E−05	10–1000

4. Decision-making in support of reissuance and renewal of approvals and licenses.
5. Proposed routes assessment for transportation of man and material through railways.
6. Mandating number of passenger to be transported through a particular network.

The current accident scenario shows, derailments constitute largest chunk of 50% of the total accidents followed by 36% at unmanned level crossing gates, 5% collisions, 4% at manned level crossing gates, 2% fire accidents and balance 3% accidents are due to miscellaneous reasons. From a study, a whopping 44% accidents where caused due to failure of railway staff, i.e. not having any knowledge of acceptable risk and risk

A systematic procedure is shown in Fig. 5 that can be followed by the Indian railways to reduce the risk to ALARP levels. This will help in the reduction of incidents as well as consequences of accidents by application of effective control measures to reduce the risk from intolerable to tolerable area. The training and

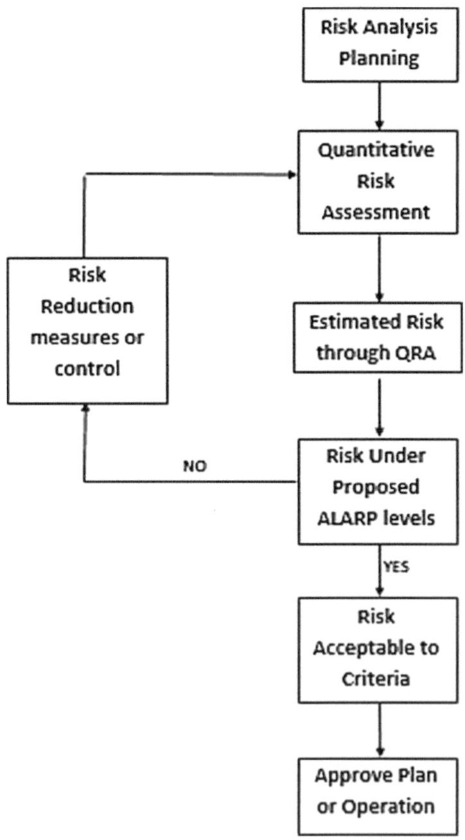

Fig. 5 Proposed procedure for risk evaluation

awareness about the risk reduction measures and control help the staff to bring the risk to ALARP level. The railways can use this criterion for an effective quantitative risk assessment so as to bring down the level of exposure to ALARP levels.

4 Conclusion

In this paper, risk-based approach is considered as a good concept for decision-making process. By carrying out risk analysis of activities, the Indian railways will get into a process of increased awareness of the risk generated from its own activity. This simplified method helps in providing conservative estimate of individual and societal risk through past accident analysis. The risks calculated can be termed as existing risks. The existing risks are classified and suggestions are made according to their tolerability to serve the purpose of ALARP principle. This principle gives a tolerable risk zone to decide criteria which are derived from suitable indicator, which allows international comparisons of RAC, used in railways and similar sectors. The RAC furnishes an increased focus towards risk reduction, both within the railways and by the authorities. As the calculated individual and societal risk levels of Indian railways are greater as compared to the RAC of different countries in the same sector, authorities should follow this risk-based decision-making process to bring down the levels to ALARP.

References

1. Sahni, J.N. 1953. *Indian railways: One hundred years, 1853 to 1953*. Ministry of Railways (Railway Board).
2. Faber, M.H., and M.G. Stewart. 2003. Risk assessment for civil engineering facilities: Critical overview and discussion. *Reliability Engineering & System Safety* 80 (2): 173–184.
3. Cullen, L.W.D. 1993. *The public inquiry into the Piper Alpha disaster* vol. 49(4). United States: Drilling Contractor.
4. Ale, B. 2005. Tolerable or acceptable: A comparison of risk regulation in the United Kingdom and in the Netherlands. *Risk analysis* 25 (2): 231–241.
5. Grossel, S.S., 2000. *Guidelines for chemical process quantitative risk analysis: By center for chemical process safety. American Institute of Chemical Engineers* pp. 750. New York, NY: Elsevier.
6. Mahalingam, A., and R.E. Levitt. 2007. Safety issues on global projects. *Journal of Construction Engineering and Management* 133 (7): 506–516.
7. Vrijling, J., W. Van Hengel, and R. Houben. 1998. Acceptable risk as a basis for design. *Reliability Engineering & System Safety*, 59 (1): 141–150.
8. Bowles, D.S. 2004. ALARP evaluation: Using cost effectiveness and disproportionality to justify risk reduction. *Ancold Bulletin* pp. 89–104.

9. Bray, D. 2004. *Risk tolerability and rail safety regulation*. NSW Transport and Population Data Centre.
10. Aven, T. 2004. *Risk management, in risk management and governance*. Springer. pp. 121–158.
11. Skjong, R. 2002. *Risk acceptance criteria: Current proposals and IMO position* pp. 4–6. Surface transport technologies for sustaninable development, Spain.
12. Litai, D. 1980. *A risk comparison methodology for the assessment of acceptable risk*. Massachusetts Institute of Technology.

Enhancing Safety Culture in Cement Industry Using Behavior-Based Safety Technique

Aswin Skaria Aliyachen, Bikarama Prasad Yadav and Soumyadeep Bhakshi

Abstract This paper presents implementation of behavior-based safety (BBS) approached for managing occupational risks and prevention of workplace injuries in cement industry. The study evaluates and reports the efficiency of implementation of behavior-based safety in cement industry. The efficiency of BBS is computed by comparing accident records before and after implementation of BBS. The result of the study shows that BBS can help in minimizing the accidents in workplace, specially the ones that occur because of human errors and negligence. BBS management technique can be applied to workforce coming from diverse cultural background, showing that it would be a good approach for improving the safety of frontline workers and that it has industry-wide application for any ongoing project or production process. The increase in safety performance after implementation of BBS will encourage more industries to adapt the approach for safeguarding their workers.

Keywords Behavior · Attitude · Steering committee · Safety performance
Accident triangle · Safety performance matrix

1 Introduction

Behavior of an individual or a group is defined by what they say or do. Psychologically, action or reaction of an individual or a group in response to external or internal stimuli is called behavior [1]. Behavior is determined as the behavioral intention or mental setup which results in a particular action [2]. Behavior is affected by behavioral intentions, which in turn are affected by attitudes toward the act and by subjective norms. A workplace observable action is direct and indirect causes of accidents.

A. S. Aliyachen · B. P. Yadav (✉) · S. Bhakshi
Department of Health Safety and Environment, University of Petroleum and Energy Studies, Dehradun, India
e-mail: bpyadav@ddn.upes.ac.in

N. A. Siddiqui et al. (eds.), *Advances in Fire and Process Safety*,
Springer Transactions in Civil and Environmental Engineering,
https://doi.org/10.1007/978-981-10-7281-9_8

Behavioral technique addresses on how people behave at workplace during working hours, and it also monitors once personal conduct within company premises [3]. By conducting observations, surveys and sampling, one can understand the behavior of a worker, and therefore, by applying BBS technique at workplace, safety performance can be improved [4].

The systematic process for the implementation of BBS system includes

- Study and analysis of safety management system.
- Constituting BBS committee.
- Providing education and training on BBS and its implementation technique.
- Review of the accident/incident record and measure current safety performance.
- Observation of employees measuring current safety performance using a checklist.
- Present and review the performance during each safety committee meetings.
- Improve the safety performance through training and supervision.

BBS is based on attitude and behavior of a person, therefore it is quite easy to identify the behavior or action of person at workplace since it remains visible to everyone. Behavior and attitude can be well explained by an iceberg structure where behavior, which is visible, accounts only for 10% and the attitude, what actually causes a person to exhibit a particular behavior is 90% and is below the surface in the iceberg structure [5]. It means attitude is what a person does when he is alone or when nobody is watching him. Attitude develops based on the situation and surrounding in which a person is born and brought up and is very difficult to change. Indeed if we want to change the attitude of a person, we need to change the behavior of a person first [3]. The only way to modify the attitude of a person is thus through behavioral change and this is what we aim to do through this study. The BBS technique can be implemented at the workplace where there is an existing safety culture with competent monitoring teams. Hence, it is important to have a safety management system in place [6].

1.1 Need for BBS

Research and experience indicate that

- 90% accidents are due to unsafe human acts or behaviors [7].
- At any time, 50% of actions in a workplace can be seen as unsafe acts [8].
- 30% employees lack safety awareness which gets reflected in their unsafe behaviors [9].

Unsafe behaviors are the cause of near misses, injury, or accidents. If we are able to control unsafe behaviors, we may not even have near misses.

Therefore, our focus should be on unsafe and safe behaviors in safety. BBS's key to success is the belief that safety is in hands of each person whether worker or an employee and they should feel empowered and responsible toward safety while performing any task.

1.2 History of BBS

H. W. Heinrich is the first person to introduce the concept of accident pyramid. He formulated an accident ratio in 1931 and published in his book 'Industrial Accident Prevention.' In his accident pyramid, relationship between major injury, minor injury, and near misses is represented as 1:29:300 [10] (Fig. 1). During 1969, a study of industrial accidents was done by Frank E. Bird, Jr. Bird, and a detailed analysis of 1,753,498 accidents reported by 297 cooperating companies in 21 different industrial groups employing 1,750,000 employees and who worked for 3 billion hours during the exposure period was done. The study revealed that the ratios of major injury, minor injury, property damage, and near misses are as 1:10:30:600 [10–12]. The 1:10:30:600 relationship in the ratio clearly indicates that accidents have happened due to lack of seriousness and ill-mannered approach from the management or employer. Most of the accidents could have been avoided if corrective actions were taken at the time of near misses. Near misses or property damage incidences have given enormous time to take action and to avoid major injuries [13] (Fig. 2).

In 2003, ConocoPhillips Marine Company conducted a study and revealed that for every single fatality there are at least 300,000 at-risk behaviors [14]. Behaviors are not only the non-compliances in safety programs, trainings, and parts of machinery (Fig. 3), but behaviors may also include bypassing safety components on machinery or eliminating a safety step in the production process that slows down the operations. With effective machine safeguarding and training, risk behaviors and near misses can be eliminated which can reduce the chances of the adverse situations [15], and therefore, controlled risk behavior may be present at workplace.

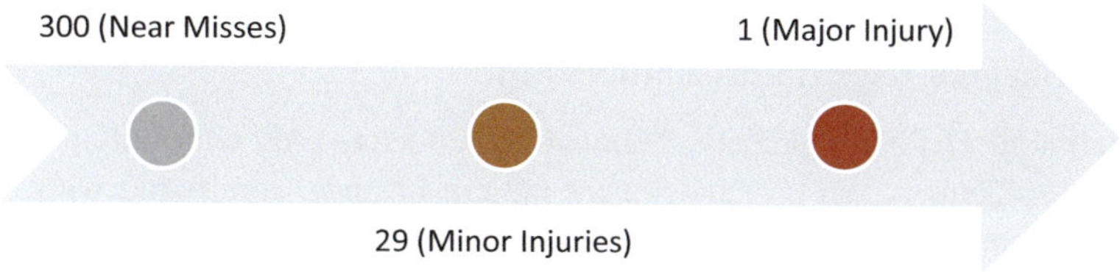

Fig. 1 Heinrich model of accident [2]

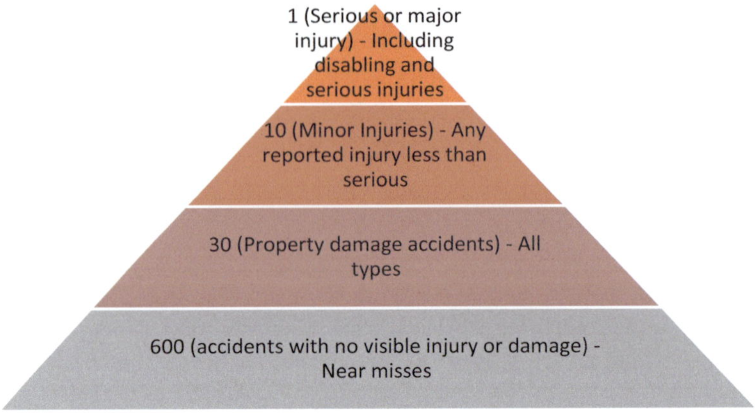

Fig. 2 Frank E. Bird accident pyramid [2]

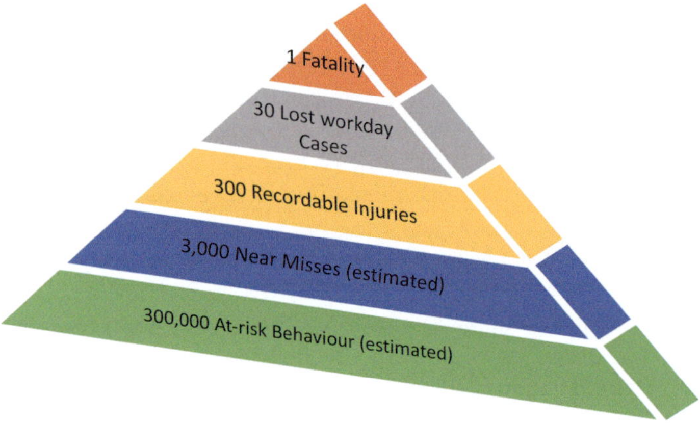

Fig. 3 Accident pyramid by ConocoPhilips Marine Company [2]

2 Methodology

Phase 1: Study of safety management system.

BBS technique can be implemented and will be effective only when an organization already has adequate safety management system in place complying with various national and international standards [16], for example ILO, ISO, and OHSAS as a part of process, implementing this technique improves overall safety performance, which requires effective participation from top management level to general workers. The management system works based on PDCA cycle its elements, e.g., policy, organizing, planning and implementing, evaluation, and action for improvement and audit. By applying and establishing, effective management system is the result of good BBS technique implementation [17].

Phase 2: Setting up of BBS committee.

BBS committee is the body which is responsible for the effective implementation of BBS [18]. The responsibility of this committee members covers to train workers on safe behavior, observation and inspection of workers, communicate and consult with workers on matters related to safety behavior, monitor performance, record the findings, and evaluate on a timely basis to monitor the safety performance. To evaluate this technique, the same approach is introduced in a reputed organization that manufactures cement. The whole concept was implemented and evaluated over a period of four months.

Phase 3: Training and education on BBS.

Based on requirement and technique, formal training about the BBS process was imparted among safety committee members, so that the same can be given to every worker by the line management showing the importance of safe behavior at workplace [19]. The aim of the training is to increase the effectiveness in safe behavior of employee, which plays an important role in eliminating major accidents. Based on BBS safety considerations, safe operating procedure for each operation was modified by including the concept of safe behavior consideration [20].

Phase 4: Determination of current safety performance.

Safety performance was taken into account by considering major accidents trend over a year. Six major performance indicators were identified based on overall safety evaluation with the help of BBS technique. The six major performance indicators were personal protective equipment (PPE) compliance, housekeeping, violation of operating procedure, non-routine work, material handling issues, and personal conduct issues. These performance indicators were quantified at each level in which baseline values, achievable goals, and attainable goals are fixed.

Phase 5: Observation/evaluation of workers and measurement of safety performance by using checklist.

Observation of workers is a key step in BBS, as it gives the status of safety performance at any instant. It can be done in many ways (i) by observing workers (ii) the way they perform a task (iii) inspection (iv) interviews and (v) consultation with workers. For this work, 30 workers from three prominent units (milling unit, packing unit, and mechanical workshop unit) were sampled as a part of this study. Every week performance evaluation was done and tabulated, based on the defined performance indicators. Checklists were used for counting each and individual unsafe behavior, which were recorded for analysis. Checklists were also used for safety committee meeting for enactment and for further action taken if any. Drastic changes were seen in some of the indicators (mainly personal conduct) for improvement in safe conduction of work.

Phase 6: Organizing safety committee meetings and developed measures for improvements.

Weekly safety performance was evaluated and discussed with the management and workers during safety committee meetings and action for the same was initiated for improvement. Special attention was given for those indicators whose performance has not shown considerable improvements [12, 21, 22]. As per recommendations, accident-reporting system was made simple and conveyed to the worker, and the importance of reporting accidents was impressed upon the workers. Accident investigation and their analysis were discussed with the workers. Workers were also approached for further improvement as a part of implementation of BBS. A decision was taken to display the lesson learnt after any incidence, and it was made sure that awareness is spread among all levels of employee to curb wrong practices. Workers or employee were consulted to identify some of the hidden issues which remain unidentified and were behind some of the unusual occurrences. Their suggestions were taken as added advantage to improve the culture of good performance. Various incentives, reward schemes were initiated for recognition, from top management, to boost morale and to encourage safety at work. Peer group influence was used for better output by infiltrating a group with highly safety conscious person, so that workers can be motivated more toward safety. Displaying safety posters, slogan contest, and celebrating safety day in organization were also helpful.

3 Developing Safety Performance Matrix

In this study, the safety performance was monitored by developing a safety matrix known as Eastman Kodak safety performance matrix [23, 24]. This is a simple yet powerful tool for monitoring the safety performance of any organization. For each safety performance indicator, there are 10 performance levels, 1 and 10 corresponds to minimum and maximum number of unsafe behaviors. We had set the baseline performance at level 7, goal at level 3, and stretch goal or minimum number at level 1 [23, 25]. The safety indicators are given weightage according to there importance out of 100. The basic performance matrix is shown in Fig. 4.

After the routine observation of 30 selected workers, the number of non-compliances corresponding to each performance indicators was tabulated. PPE compliance for January 22, 2016, as an example, the total number of non-compliances obtained from our observation was 28, which corresponds to the performance level 7. The weightage for PPE compliance is set as 20%. The score is obtained by multiplying level and weightage. Hence, score obtained is 140 (7 * 20). Our baseline score is 700, and our aim is to reduce the score to 300, which is our goal, and if possible reduce it furthermore toward our stretch goal, which is 100. Similarly, matrix was completed for each performance indicator for each week starting from January 22, 2016 (Figs. 5, 6, 7, and 8).

EASTMAN KODAK SAFETY PERFORMANCE MATRIX - 22 January 2016														
	Performance Level										Calculation			
Performance Indicator(PI)	1	2	3	4	5	6	7	8	9	10	Value	Level	Weight	Score
PPE Non-compliance	2	7	15	19	23	27	30	34	38	45	28	7	20	140
Housekeeping Issues	2	6	12	16	19	22	25	28	33	40	27	8	10	80
Violations of Operating Procedures	3	7	10	12	15	17	20	23	26	30	20	7	30	210
Number of Unusual Occurrences	2	4	8	9	11	13	15	17	22	25	18	9	20	180
Material Handling issues	1	5	10	12	15	17	20	23	26	30	21	8	10	80
Personal Conduct issues	5	10	20	26	31	35	40	46	54	60	38	7	10	70
													TOTAL SCORE	760

Fig. 4 Eastman Kodak performance matrix—22/01/2016

EASTMAN KODAK SAFETY PERFORMANCE MATRIX - 12 February 2016														
	Performance Level										Calculation			
Performance Indicator(PI)	1	2	3	4	5	6	7	8	9	10	Value	Level	Weight	Score
PPE Non-compliance	2	7	15	19	23	27	30	34	38	45	22	5	20	100
Housekeeping Issues	2	6	12	16	19	22	25	28	33	40	22	6	10	60
Violations of Operating Procedures	3	7	10	12	15	17	20	23	26	30	17	6	30	180
Number of Unusual Occurrences	2	4	8	9	11	13	15	17	22	25	15	7	20	140
Material Handling issues	1	5	10	12	15	17	20	23	26	30	17	6	10	60
Personal Conduct issues	5	10	20	26	31	35	40	46	54	60	29	5	10	50
													TOTAL SCORE	590

Fig. 5 Eastman Kodak performance matric—12/02/2016

EASTMAN KODAK SAFETY PERFORMANCE MATRIX - 26 February 2016														
	Performance Level										Calculation			
Performance Indicator(PI)	1	2	3	4	5	6	7	8	9	10	Value	Level	Weight	Score
PPE Non-compliance	2	7	15	19	23	27	30	34	38	45	19	4	20	80
Housekeeping Issues	2	6	12	16	19	22	25	28	33	40	20	6	10	60
Violations of Operating Procedures	3	7	10	12	15	17	20	23	26	30	15	5	30	150
Number of Unusual Occurrences	2	4	8	9	11	13	15	17	22	25	14	7	20	140
Material Handling issues	1	5	10	12	15	17	20	23	26	30	15	5	10	50
Personal Conduct issues	5	10	20	26	31	35	40	46	54	60	24	4	10	40
													TOTAL SCORE	520

Fig. 6 Eastman Kodak safety performance matrix—26/02/2017

EASTMAN KODAK SAFETY PERFORMANCE MATRIX - 11 March 2016														
	Performance Level										Calculation			
Performance Indicator(PI)	1	2	3	4	5	6	7	8	9	10	Value	Level	Weight	Score
PPE Non-compliance	2	7	15	19	23	27	30	34	38	45	18	4	20	80
Housekeeping Issues	2	6	12	16	19	22	25	28	33	40	17	5	10	50
Violations of Operating Procedures	3	7	10	12	15	17	20	23	26	30	14	5	30	150
Number of Unusual Occurrences	2	4	8	9	11	13	15	17	22	25	11	5	20	100
Material Handling issues	1	5	10	12	15	17	20	23	26	30	13	5	10	50
Personal Conduct issues	5	10	20	26	31	35	40	46	54	60	21	4	10	40
													TOTAL SCORE	470

Fig. 7 Eastman Kodak safety performance matrix—11/03/2016

	EASTMAN KODAK SAFETY PERFORMANCE MATRIX - 25 March 2016												
	Performance Level									Calculation			
Performance Indicator(PI)	1	2	3	4	5	6	7	8	9	10	Value	Level	Weight	Score
PPE Non-compliance	2	7	15	19	23	27	30	34	38	45	15	3	20	60
Housekeeping Issues	2	6	12	16	19	22	25	28	33	40	16	4	10	40
Violations of Operating Procedures	3	7	10	12	15	17	20	23	26	30	11	4	30	120
Number of Unusual Occurrences	2	4	8	9	11	13	15	17	22	25	9	4	20	80
Material Handling issues	1	5	10	12	15	17	20	23	26	30	11	4	10	40
Personal Conduct issues	5	10	20	26	31	35	40	46	54	60	16	3	10	30
													TOTAL SCORE	370

Fig. 8 Eastman Kodak safety performance matrix—25/03/2016

4 Result and Discussion

BBS program was implemented in an organization during January 2016. Implementing BBS is a time-consuming process, for complete accomplishment of set goals, since it deals with the attitude and behavior of employees. There was a sudden increase in accident reporting from employee side after BBS implementation, and this is because awareness among employee had increased, and due to this, gradual increase in safety performance was observed. For its continual improvements, program was thoroughly reviewed and corrected. The graphical representation of number of non-compliances from January 22, 2016, to March 25, 2016, is shown in Fig. 9. A steady decrease in number of non-compliances is seen, after the implementation of BBS. Another graph (Fig. 10) shows the performance score variation during the course of time.

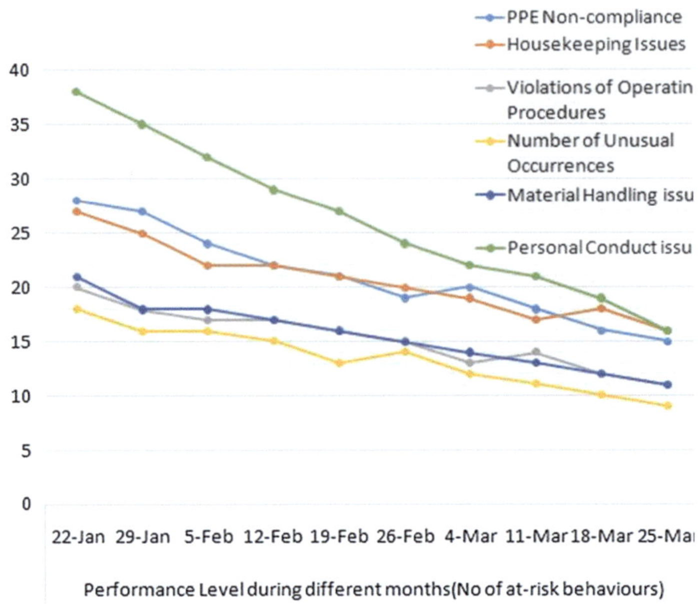

Fig. 9 Consolidated performance level variation for all indicators

Fig. 10 Performance score variation

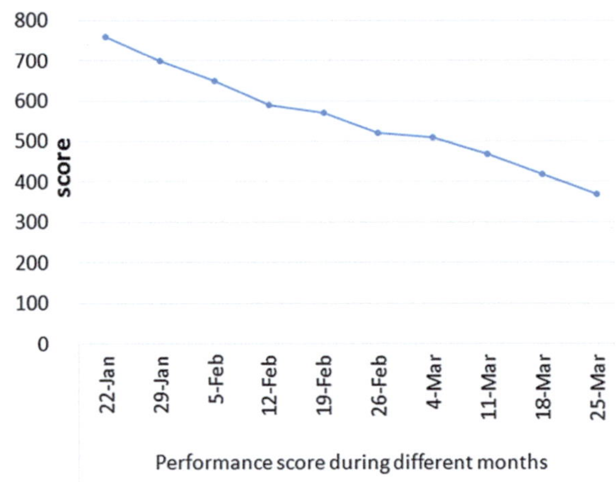

5 Conclusion

BBS technique was used to achieve higher standard of safety along with recommendation for control measures via various safety indicators, which are statistically valid, have ownership, trust, and unity among the team. BBS also develops empowerment and confidence toward employee safety at workplace. It also develops empowerment opportunities related to employee safety. BBS provides an organizational culture among top management, to line management, to workers to prove their core values toward safety at workplace. Safety always cost money, safety programs take time from manager to workers, accidents take time to investigate, but implementation of BBS doesn't require any cost, on the other hand, it builds trust, increases employability. Once an organization becomes an expert in leading the safety process through behavioral approach, it can further transfer that experience to other business sectors.

References

1. Sampson, J.M., S. DeArmond, and P.Y. Chen. 2014. Role of safety stressors and social support on safety performance. *Safety Science* 64: 137–145.
2. Petersen, D. 1989. *Safe behavior reinforcement*. Aloray.
3. Agrawal, N., and N. Jain. 2014. Identification of occupational diseases and injuries among the workers engaged in manual material handling of cement industry. *International Journal on Emerging Technologies* 5 (1): 78.
4. Li, H., et al. 2015. Proactive behavior-based safety management for construction safety improvement. *Safety Science* 75: 107–117.
5. Hurst, P.W., and W.L. Palya. 2003. Selecting an effective BBS process. *Professional safety* 48 (9): 39.
6. Chaib, R., et al. 2012. Promoting a culture of health and safety at work: Safety—A permanent priority. *WIT Transactions on Information and Communication Technologies* 44: 405–413.
7. Al-Hemoud, A.M., and M.M. Al-Asfoor. 2006. A behavior based safety approach at a Kuwait research institution. *Journal of safety research* 37 (2): 201–206.
8. Sertyesilisik, B., et al. Construction workers' personality: As a key for improving the occupational health and safety performance in construction. *Advances in Business and Management* 103.
9. Geller, E.S. 2004. Behavior-based safety: A solution to injury prevention: Behavior-based safety "empowers" employees and addresses the dynamics of injury prevention. *Risk and Insurance* 15 (12): 66.
10. Heinrich, H.W. 1941. *Industrial accident prevention; a scientific approach*, 488. New York, London: McGraw-Hill Book Company Inc.
11. Mearns, K., S.M. Whitaker, and R. Flin. 2003. Safety climate, safety management practice and safety performance in offshore environments. *Safety Science* 41 (8): 641–680.
12. Van Steen, J. 1996. *Safety performance measurement*. IChemE.
13. National Safety, C. 1946. *Accident prevention manual for industrial operations*. Chicago, Ill.: Printed by Wm. H. Pool co. x, 534 p.
14. Cole, H.P. 2012. Workplace injury and illness, safety engineering, economics and social capital. In *Handbook of occupational health and wellness*, 267–295. Springer.
15. Zain, M.F.M., et al. 2011. Production of rice husk ash for use in concrete as a supplementary cementitious material. *Construction and Building Materials* 25 (2): 798–805.
16. Chakraborty, J. 2016. Occupational safety and behavioural based safety in LPG processing unit a multidisciplinary approach.
17. Lodgaard, E., and K.E. Aasland. 2011. An examination of the application of plan-do-check-act cycle in product development. In *DS 68-10: Proceedings of the 18th International Conference on Engineering Design (ICED 11), Impacting Society through Engineering Design, vol. 10: Design Methods and Tools pt. 2, Lyngby/Copenhagen, Denmark, 15.–19.08.2011.*

18. Geller, E.S., S.R. Perdue, and A. French. 2004. Behavior-based safety coaching: 10 guidelines for successful application. *Professional Safety* 49 (7): 42.
19. Byrd, H. 2007. A comparison of three well known behavior based safety programs: DuPont STOP program, safety performance solutions and behavioral science technology.
20. DePasquale, J.P., and E.S. Geller. 2000. Critical success factors for behavior-based safety: A study of twenty industry-wide applications. *Journal of Safety Research* 30 (4): 237–249.
21. Levitt, R.E., and N.M. Samelson. 1993. *Construction safety management*. New York: Wiley.
22. Aksorn, T., and B. Hadikusumo. 2008. Critical success factors influencing safety program performance in Thai construction projects. *Safety Science* 46 (4): 709–727.
23. Geller, E.S. 2005. Behavior-based safety and occupational risk management. *Behavior Modification* 29 (3): 539–561.
24. Cooper, D. 2015. Effective safety leadership: Understanding types and styles that improve safety performance. *Professional Safety* 60 (2): 49.
25. Kaila, H. 2008. *Behaviour based safety in organizations: A practical guide*. IK International Pvt. Ltd.

Quantitative Estimation of Risk to Community Near an Ammonia Rail Wagon Loading Facility

Ankit Avasthy and Nihal A. Siddiqui

Abstract Quantitative risk assessment (QRA) is a key tool extensively used by process industries worldwide. The growth of process industries is crucial for the development of the economy of any country. With the advancement of process technologies and invention of new process and products, the risk associated with them is also increasing. This paper analyzes and quantifies the risk posed by an ammonia loading facility located north of a nearby residential area. The hazard identification was carried out in a systematic manner, and a number of maximum credible accident scenarios, with the potential to cause harm to surrounding residential population, were identified and analyzed for their consequences. The consequence analysis for the identified maximum credible accident scenarios was done with the help of a risk assessment software, and the forecasted consequences were compared with the accepted international criteria.

Keywords Incident scenarios · Risk analysis · Hazard identification
Risk criteria · Ammonia

1 Introduction

This research paper is an attempt to estimate the risk of fatality due to operational ammonia loading facility surrounded by residential population. To limit the example's complexity, representative set of incidents was selected based on various identified hazards and relatively simple models were used for consequence estimation. The number of incidents considered is kept small, and the weather conditions are limited to a single typical wind speed/stability class. The atmospheric dispersion model selected is Gaussian dispersion model. The level of treatment in this study might correspond to a project in the early stages of design.

A. Avasthy · N. A. Siddiqui (✉)
University of Petroleum and Energy Studies, Dehradun, Uttarakhand, India
e-mail: nihal@ddn.upes.ac.in

N. A. Siddiqui et al. (eds.), *Advances in Fire and Process Safety*,
Springer Transactions in Civil and Environmental Engineering,
https://doi.org/10.1007/978-981-10-7281-9_9

Process Description

The ammonia tank considered for the present study is mounted on a weighing scale, and liquid ammonia is transferred to a railcar using pressurized nitrogen. Liquid fill and vapor return lines are connected to the railcar with short sections of braided stainless-steel hose with threaded connections. Two remotely actuated emergency shutoff valves and an emergency vent ensure the safety of the ammonia tank. The ammonia supply tank and plant nitrogen systems are protected from back flow by check valves. The 10,000 gal (50 ton) ambient temperature rail tank car is fitted with a pressure relief valve and four connection points (two each for vapor and liquid) and is vented to the atmosphere. The liquid lines extend to the bottom of the tank car and are fitted with excess flow valves. The ammonia rail tank car is insulated with 4 in. of corkboard or urethane foam. The railcar, after inspection and required maintenance, is spotted on a weigh scale and properly positioned with brakes set and wheels chocked. The car and loading system are then electrically grounded [1].

The ammonia loading facility is located 100 m south of a populated area with a uniformly distributed population of 240 people. A nearby flammable-material pipe rack and flammable liquid railcars stored on neighboring rail lines have the potential to cause a pool fire in the ammonia loading area.

2 Risk Estimation Methodology

Knowledge of the probability of occurrence of any event and the severity of the consequence of the event in terms of loss of life, property, or damage to the environment are required for estimation of risk [2]. The target of risk assessment exercise is to identify potential hazards, estimate the probability of that hazard causing an accident, analyze the consequence of the accident, and evaluate the effects of the risk reduction measures [3].

A typical QRA study starts with hazard identification done with the help of qualitative tools and estimation of failure frequencies of various equipment, unit operations, and processes. Risk estimation is done with the help of quantitative tools. The selection of qualitative or quantitative or semi-quantitative risk assessment technique is decided based on the objective of the study. It depends on whether the risk assessment is done during design stage or during operation or expansion of the plant, and on the legal requirements and usage of the results [4]. The purpose of QRA is to assign numerical values to risk. Quantitative methods are used to identify the risk based on the qualifying hazards and assign risk level using various consequence modeling techniques [5]. Some of the most widely used QRA methodologies are based on the World Bank guideline, Dutch purple book, and CCPS guideline, etc. The QRA results are analyzed using individual risk graph or individual risk contours and F-N curve [6].

The outcome of QRA studies can help, for example, in land-use planning. A well-planned QRA is very important for risk management and in assessing risk of chemical transportation; it is widely used to improve safety [7].

2.1 Hazard Identification

Any QRA study starts with hazard identification. The Bureau of Indian Standards had published a standard where it is suggested that the fire hazard, attributable to ammonia is only moderate. As per Indian Explosives Act, 1884, ammonia is deemed to be an explosive, when contained in any metal container in a compressed or liquefied state. Even as ammonia is flammable and can burn explosively, the main hazard imposed by ammonia is its toxicity [8].

A structured method such as the hazard and operability (HAZOP) study is often used to enumerate a more complete list of incidents. The HAZOP procedure for this study has been reviewed from a number of sources including the AIChE/CCPS *Guidelines*. In this study, the HAZOP study has been conducted for the ammonia loading facility and various possible incidents have been identified.

The incident description (size and duration) is based on historical data and engineering judgment and is intended to represent a spectrum of possible real incidents. Pinhole leaks from any pipe or equipment item are not analyzed because they have been determined to be too small to cause public impact. Spontaneous catastrophic failure of the ammonia supply tank or rail tank car, although theoretically possible, has been judged to be too rare to contribute any significant risk for this study.

For this study, the incidents and incident outcomes are identical. Each ammonia release incident has only one incident outcome, a toxic cloud blowing downwind. The location of the toxic effect zone depends on the weather conditions and wind direction. Thus, each combination of incident, weather conditions, and wind direction results in a separate incident outcome case. Because only one weather condition is considered in this example, the incident outcome cases differ only in the direction of the wind. The effect zones have the same physical dimensions for all incident outcome cases.

2.2 Scenarios Selection

Out of various possible hazards identified through HAZOP study, representative set of incidents has been identified as scenarios. Various incident scenarios considered for the present study are listed in Table 1.

Table 1 Representative set of incident scenarios identified from HAZOP study

Scenario no.	Incident description	Duration (min)
1	Small liquid leakage (12.7 mm hole) due to valve/hose leak and impact failure	10
2	Small vapor leakage (12.7 mm hole) due to valve/hose leak, impact failure, and relief valve leak	10
3	Large vapor leakage due to external fire lifts relief valve	60

2.3 Consequence Analysis

For consequence analysis, the sources are identified and modeled. The source modeling is based on the source—such as from pipeline or tank—and nature of the discharge of materials and the type of failure. Then knowledge or informed decision on the probability and delay in ignition is used to model fire and or explosion.

The three representative Incidents 1, 2, and 3 (from Table 1) require three different discharge rate calculations. The calculations are straightforward for leaks of liquid or vapor from holes near a tank or large diameter pipe. Leak rates from longer lengths of piping will be reduced by pipe friction. However, for this case study, leaks are assumed to be unaffected by pipe length, and geometry and equations used are adopted from CCPS guidelines (Table 2).

Next step is to determine the toxicity relationship to be used for estimating fatalities from the exposure to ammonia vapor. The probit method is used to estimate fatal effects, and for this study, the lethal concentrations (LC) used are adopted from NIOSH published material. [9] (Table 3).

For determining the endpoint distances of ammonia vapor cloud, only one typical weather condition of wind speed of 4 m/s (13 ft/s) and neutral (D) atmospheric stability is considered for dispersion modeling. To further simplify calculations, the wind is assumed uniformly distributed in all directions. The magnitude of outcome for each of the studied cases was evaluated using ALOHA software (version 5.4.4). The dispersion model used for the study was Gaussian dispersion model (Figure 1; Table 4).

Table 2 Estimated ammonia release rate

Scenario	Description	Estimated release rate (kg/s)
1	Liquid leak	1.94
2	Vapor leak	0.14
3	Relief valve discharge	0.45

Table 3 Estimated LC for ammonia exposure

Exposure time (min)	Scenario	Estimated LC (ppm)
10	1 and 2	1105
60	3	197

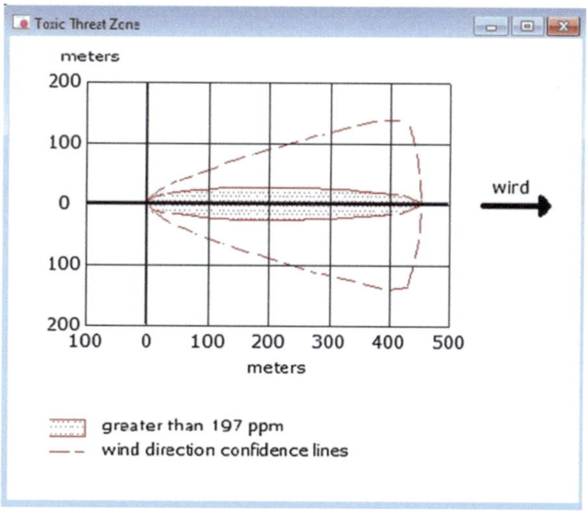

Fig. 1 Hazard footprint of Scenario 3

Table 4 Estimated endpoint distances for LC

Scenario	Duration (min)	LC (ppm)	Endpoint distance (m)
1	10	1105	292
2	10	1105	92
3	60	223	451

2.4 *Incident Frequency Estimation*

For simpler cases, the failure frequency data can be obtained directly from historical records or through past accident analysis. Failure frequency is estimated by dividing the number of recorded incidents by the total duration of operation (e.g., plant years, pipeline mile years). This straightforward technique directly provides the failure frequency for the top event without the need for detailed frequency modeling, and it is used for Scenarios 1 and 2 considered for the present study.

For Incident 3 (a large vapor leak caused by an external fire), historical data are not suitable for failure frequency estimation. External fire frequencies are strongly dependent on the features of each site. A simple fault tree model of the external fire scenario is developed to calculate the frequency from basic causative factors (Table 5).

Table 5 Estimated incident frequency

Scenario	Frequency of occurrence (year^{-1})
1. Liquid leak	5.8×10^{-4}
2. Vapor leak	6.6×10^{-4}
3. Relief valve discharge	3.0×10^{-6}

3 Results and Discussion

Risk counters are established and plotted on a map. Calculation of risk was done using the results of consequence analysis and data on failure frequencies. Individual risk contours were obtained by connecting the same risk level on a geographical map [10].

The risk to the residential population, called the societal risk, surrounding the installation was represented by F-N curves. The level of risk depends on factor such as input data, methodology, process data, operational data, meteorological conditions, failure frequency data, tools used for consequence assessment, simplifying assumptions made, etc. [11] (Figs. 2 and 3).

The computed risk levels were compared with the acceptable risk criteria and recommendations and methodologies for control measures and mitigations were developed. As per HSE-UK, the risk acceptance criterion for acceptable societal risk is 1×10^{-6}. The calculated risk levels indicate that the effects of considered

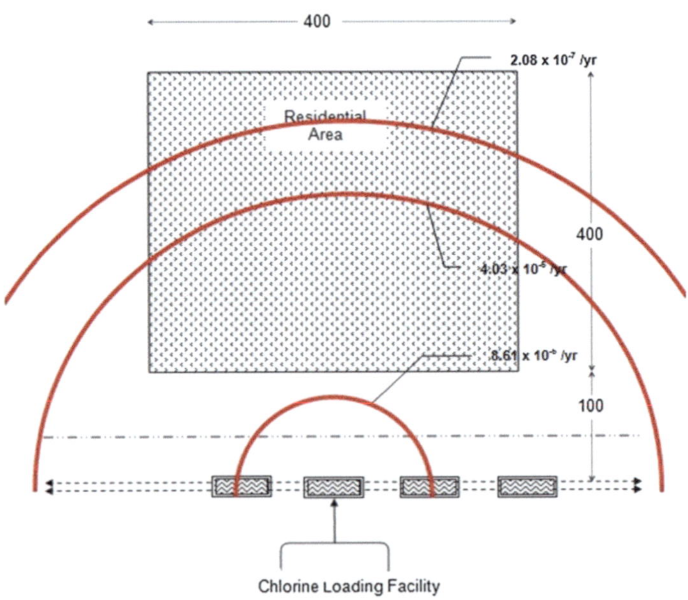

Fig. 2 Individual risk contours around ammonia loading facility

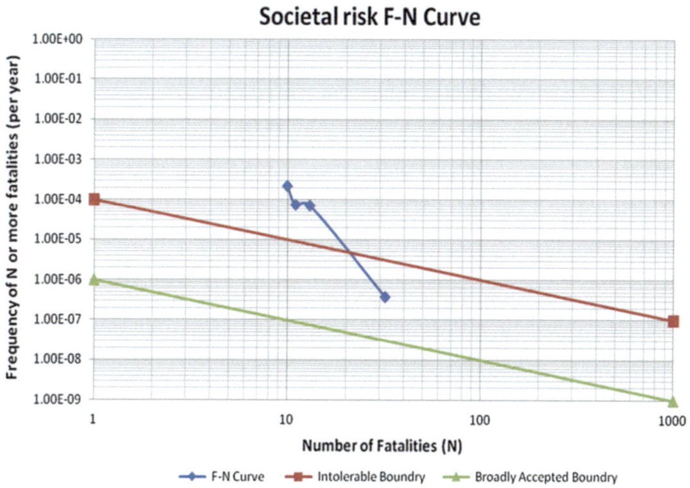

Fig. 3 Societal risk F-N curve for ammonia loading facility

scenarios will not be limited to within the plant premises and will affect the offsite residential population and property.

4 Conclusion

- Out of the identified three loss scenario, only Incident 1 (liquid ammonia leak for 10 min) and Incident 3 (relief valve discharge for 60 min) have the potential to affect the residential area north of ammonia loading facility.
- Incidents 1 and 2 are low-consequence, high-frequency hazards, whereas, Incident 3 is high-consequence, low-frequency hazard
- The maximum individual risk to which general public is exposed is estimated as 4.03×10^{-5}/year, which as per criteria falls in As Low As Reasonably Practicable (ALARP) region.
- The F-N curve drawn indicates that as per criteria, the societal risk levels in community fall mostly above and near intolerable region.

Risk levels as quantified in this study demonstrate that community is exposed to unacceptably high risk. The risk levels should be decreased at the cost of additional engineering and administrative controls, or through relocation of community or facility.

References

1. CCPS. 1989. Guidelines for process equipment reliability data, with data tables. In Center for Chemical Process Safety of the American Institute of Chemical Engineers, 345 East 47th Street, New York. New York 10017 (ISBN 0-8169-0422-7).
2. EPA of Western Australia. 2000. Guidance for risk assessment and management: Off-site individual risk from hazardous industrial plant, environmental protection authority. Western Australia, No. 2, July 2000.
3. CCPS. 2000. Guidelines for chemical process quantitative risk analysis, 2nd edn, Center for Chemical Process Safety of the American Institute of Chemical Engineers, 345 East 47th Street, New York. New York 10016 (ISBN: 0-8 169-0720).
4. Chiara Vianello, G.M. 2013. *Gas releases in urban area: Calculation of consequences through CFD modeling and comparison with standard software*. Padova: Italian Association Of Chemical Engineering.
5. Liu Liping, Y.L. 2013. The risk assessmennt research of hazardous chemicals rail and truck transportation. *Journal of Applied Sciences 13*: 2077–2081. (Pakistan: Asian Network for Scientific Information).
6. Li Jianfeng, Z.B. 2011. A typical small-scale leak and dispersion simulation. *International Journal of Energy and Environment* 1039–1052.
7. Tseng, J.M., T.S. Su, and C.Y. Kuo. 2012. Consequence evaluation of toxic chemical releases by ALOHA. In *International Symposium on Safety Science and Technology*, 384–389. Taiwan: Elsevier Ltd.
8. Hanna, S.R., D.G. Strimaitis, and J.C. Chang. 1991. Hazard response modeling uncertainty (a quantitative method). In *Evaluation of Commonly-Used Hazardous Gas Dispersion Models,* vol. II.
9. Centers for Disease Control and Prevention, Retrieved 4 Oct 2016. http://www.cdc.gov/niosh/idlh/7664417.Html.
10. International Labour Organization. 2004. Retrieved 14 Sept 2014, from International Labour organization: http://www.ilo.org/legacy/english/protection/safework/cis/products/safetytm/mah.htm.
11. ERPG. 2008. *Emergency response planning guidelines*. Akron, Ohio: American Industrial Hygiene Association. http://www.aiha.org/1documents/Committees/ERP-erpglevels.pdf.

Network Guided Robot

Rohit Smkaria, Rajesh Singh, Anita Gehlot, Sushaban Choudhary and Mayank Sharma

Abstract This paper describes a technique for a robot to be guided to a position, in indoor environments with the use of wireless network nodes. The system comprises of number of wireless static nodes with capability to change the guiding parameters for robot using onboard switches. The static nodes are placed at strategic locations in environment. The receiver node would be carried by robot. The communication between robot and wireless nodes will be carried out using ZigBee compliant RF modem. ZigBee which is 2.4 GHz IEEE 802.15.4 compliant transceiver-based wireless protocol is used to communicate between robot and wireless nodes through a master controller. The wireless nodes will send data packet to robot containing information about the movement of robot in desired direction for safe navigation in a given environment. This capability renders network node as intelligent. This technique is of great importance to logistic warehouse where robots are involved and robots move on specified path given by the central network controller. With this technique, wireless network can figure out an alternate path for another robot if one of them fails in order to prevent system coming to a standstill

Keywords Robot · Wireless nodes · Data packet · ZigBee

1 Introduction

Luna et al. narrate about the difficulties encountered in guiding a number of robots having multiple path through wireless nodes which could avoid any deadlocks in the system despite having only local knowledge of its environment thereby reducing computation loads on robot to calculate goal path [1]. Ye et al. describe different wireless communication strategies between multiple robots as an important

R. Smkaria · R. Singh (✉) · A. Gehlot · S. Choudhary · M. Sharma
University of Petroleum and Energy Studies, Dehradun, India
e-mail: rsingh@ddn.upes.ac.in

R. Smkaria
e-mail: rohit.samkaria93@gmail.com.com

N. A. Siddiqui et al. (eds.), *Advances in Fire and Process Safety*,
Springer Transactions in Civil and Environmental Engineering,
https://doi.org/10.1007/978-981-10-7281-9_10

factor for enabling efficient communication between wireless nodes and robot [2]. Shenoy et al. describe movement of mobile robot in an un-localized network with inbuilt capability for navigating to the desired location without the use of GPS but only with wireless sensor nodes [3]. Ladd et al. proposed a system using RF signal strength to approximate the position of the robot in a given environment [4]. Allaoua et al. proposed a control strategy for controlling speed of DC motor; in this, PID parameters are optimized using particle swarm optimization algorithm which makes the system much more optimal toward reaching goal or desired speed [5]. Singh et al. describe successful use of particle swarm optimization for tuning PID parameters of a system [6]. Kennedy et al. proposed a radical technique named as particle swarm optimization for optimization of nonlinear functions, this algorithm has its roots in behavior of flocking birds. This algorithm has shown its superiority over other algorithms such as ACO (Ant Colony Optimization), Genetic algorithm [7]. Kim et al. discuss about robust tuning of PID using constrained particle swarm optimization [8]. Sultana et al. narrate about the use of ZigBee-based wireless nodes as solution to multi-mobile robot communication among each other to solve complex tasks requiring multiple agents [9]. Huasong et al. describe importance of wireless communication of multi robot docking platform through ZigBee protocol [10]. Yeh et al. explains the use of WSN (Wireless sensor networks) integrated with ZigBee protocol for sharing information such as in case of rescue robot and other area such as RoboCup which connects operating person to robot through long distances [11]. Wang et al. narrate the innovative use of networked nodes for use in environment monitoring and precision agriculture by monitoring all desired parameters wirelessly through a central node for gathering relevant data [12].

2 Hardware Developments

2.1 Proposed System

The experimental system consists of two wireless nodes being placed at appropriate distance from each other for system to work effectively. In the system, individual nodes possess the capability to make robot go in desired direction with desired speed to avoid any deadlocks created due to another robot working in the grid. The robot before moving first sends out a data packet to the nearest node which then responds to the request of the robot and gives moving instructions to the robot. The robot also possesses the capability to avoid obstacles with use of ultrasonic sensors-placed robot chassis. To achieve desired speed, the robot runs a PID algorithm that compensates for any deviation in desired speed. Particle swarm optimization is being used to calculate optimum values for PID gain constants. For robot to know its speed, the robot is fitted with encoded motors. The wireless nodes possess dynamic behavior, which allows it to be placed at different location. Once robot movement information is being programmed into nodes, system can run

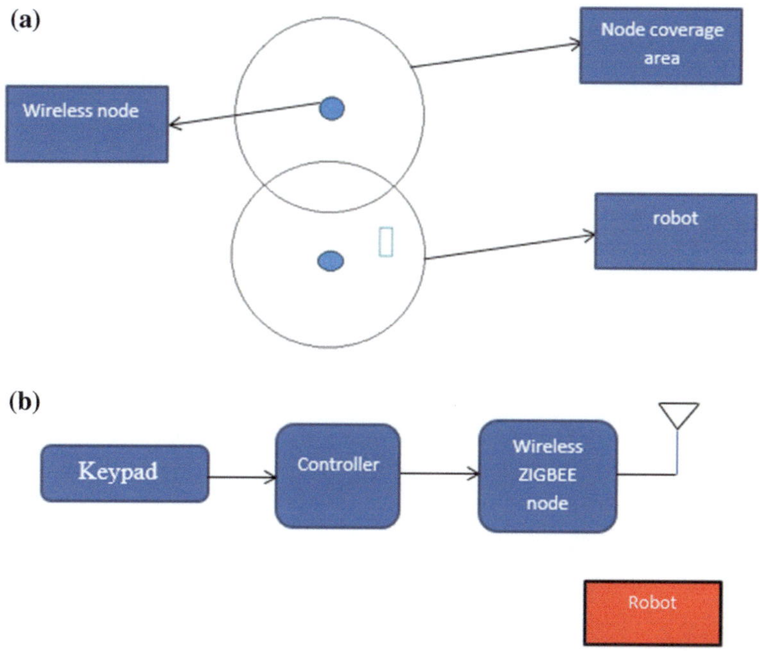

Fig. 1 **a** Shows the basic layout of the system, **b** graphical representation of the node and robot interaction

anywhere in the desired location. This combination of capabilities of wireless nodes and robust movements of robot which can communicate with each other to achieve goal makes the entire system intelligent. Figure 1a, b shows the basic layout of the system.

2.2 *Wireless Node*

Wireless node consists of a LCD display (16 × 2), keypad for passing information to the node, battery for providing on board power to the node, and ZigBee (2.4 Ghz). The required distance to be travelled by the robot is given input to the node using keypad. This information is transmitted to robot with ZigBee RF modem (Fig. 2).

S. No.	Device/ module	Make/model no.	Specifications and working
1	LCD	Sunrom model no. 3013 [11]	16 × 2 LCD is used in wireless nodes to display settings
2	Atmega16	Atmel	It is a low-power CMOS 8-bit microcontroller based on the AVR RISC architecture for robot, wireless nodes
3	ZigBee	Sunrom model no. 1195 [11]	It is a wireless module, having two way data communication availability. It operates on license free 2.4 GHz band
4	Keypad		4 × 4 matrix keypad is used
5	Ultrasonic sensor	Sunrom model no. 3001 [11]	Ultrasonic sensor is used to measure the distance between two objects. Sensor gives pwm output which can be easily read
6	Magnetic compass	Sunrom model no. 3932	This product is 3-Axis Digital Compass IC. It gives direction data with respect to earth magnetic field. It operates on I2C mode
7	L293d	Sunrom model no. 1289 [11]	It is a high voltage and high current four channel drive IC used to drive
8	DC motors	Sunrom model no. 3213	It is high-efficiency and low-cost geared motor
9	Power supply		For robots and receiving device, all the devices operates on 5 V accept dc motors which operate on 9 V, so 9 V battery is used with 7805 regulator to achieve 5 V

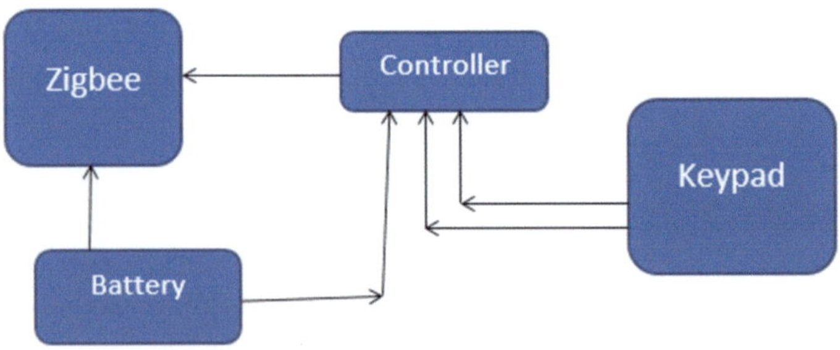

Fig. 2 Wireless node architecture

3 Software Development

3.1 Algorithm for PID Control of Robot with PSO Optimization

1. Specify all the constraints (overshoot, rising time, and settling time), robot transfer function, and PSO constants along with number of particles.
2. Initiate particle position and velocities.
3. Evaluate the system with unit step response.
4. Compute constraint for each and every particle in the system followed by total error.
5. Check correctness for each and every particle by comparing every value of particle with its previous one, respectively.
6. Particle having lowest error becomes global best value for the system.
7. Update all particle's position and velocities using step (3).
8. Check if all desired constraints of the system are met if yes then stop else start with step (1).

3.2 Controller Architecture for PID Control of Robot

Based on system transfer function, once the PSO algorithm has given out optimized values for Kp, Ki, Kd these values can directly be applied to the controller for controlling desired parameters of the robot such as speed and direction as given by node (Fig. 3).

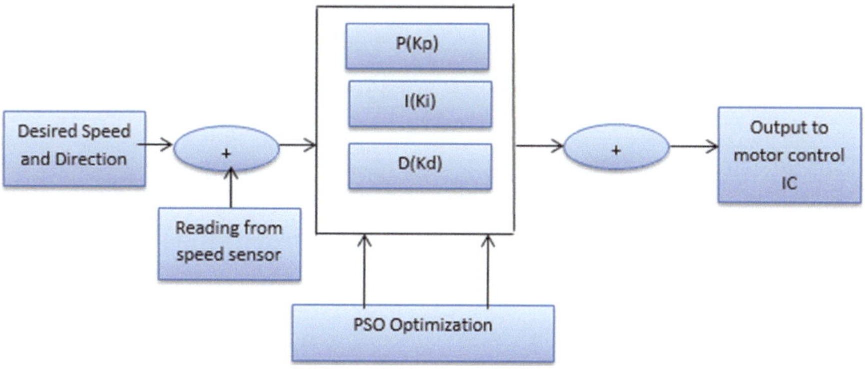

Fig. 3 Controller architecture for PID control of a robot

3.3 Implementation of PID with PSO Optimization

The tuning of PID parameters and optimization with PSO algorithm has been done with MATLAB and Simulink environment. Initially, the PID is programmed with MATLAB and all the parameters such as overshoot, rising time, and settling time, Kp, Kd, Ki were calculated. Next, the PSO optimization was done on the calculated values of PID controller. Hence, these new optimized values of Kp, Ki, Kd were imbedded into robot controller for precise control of robot speed (Figs. 4 and 5).

4 Circuit and Simulation

Figure 6 shows the Proteus design of circuit for the hardware used. The robot includes an ultrasonic sensor for avoiding obstacles through its path. The bot contains two 12 V dc motor for its movement, and the motors are controlled by L293d IC which helps robot to move its motors forward or backward simultaneously. The robot runs on Atmega16 microcontroller platform, which provides all the necessary signals for accomplishing the task of robot movement.

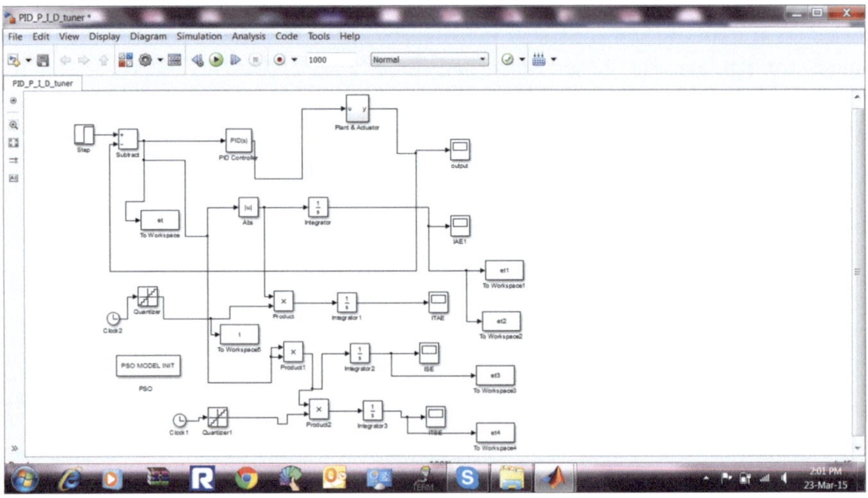

Fig. 4 PID tuner with PSO optimization

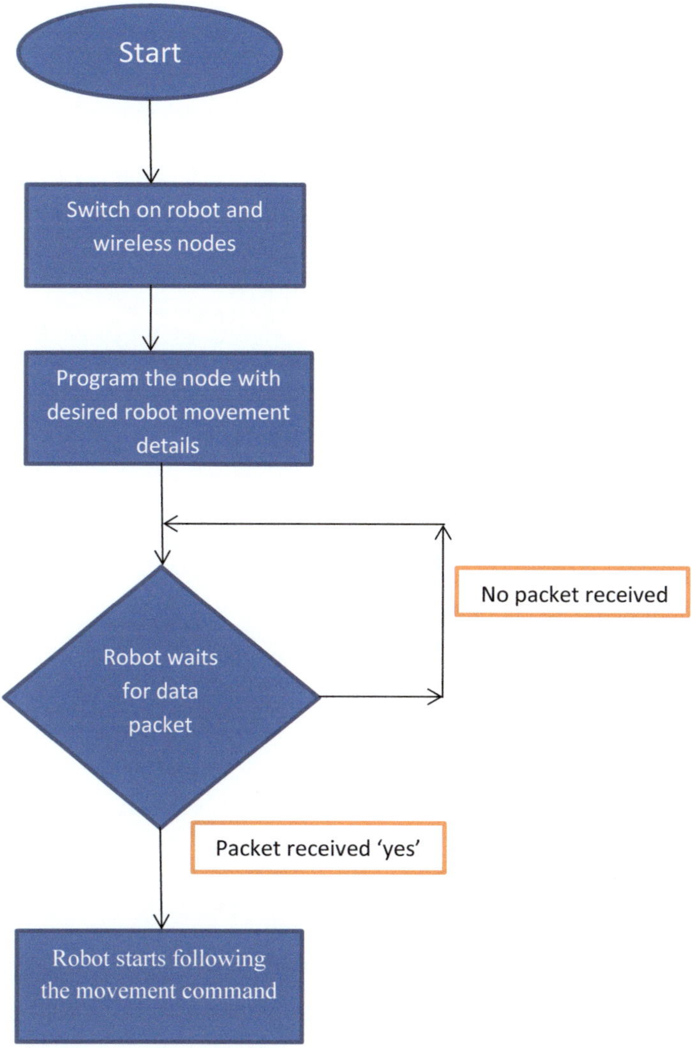

Fig. 5 Flow chart of system working

5 Experimental Setup

The setup consists of robot and wireless node. The robot consists of ultrasonic sensor, ZigBee RF modem, motors, and Atmega16 microcontroller. The wireless nodes consist of a ZigBee module and Atmega16 microcontroller. Figure 7a, b shows the experimental setup of robotic and nodes platform, respectively.

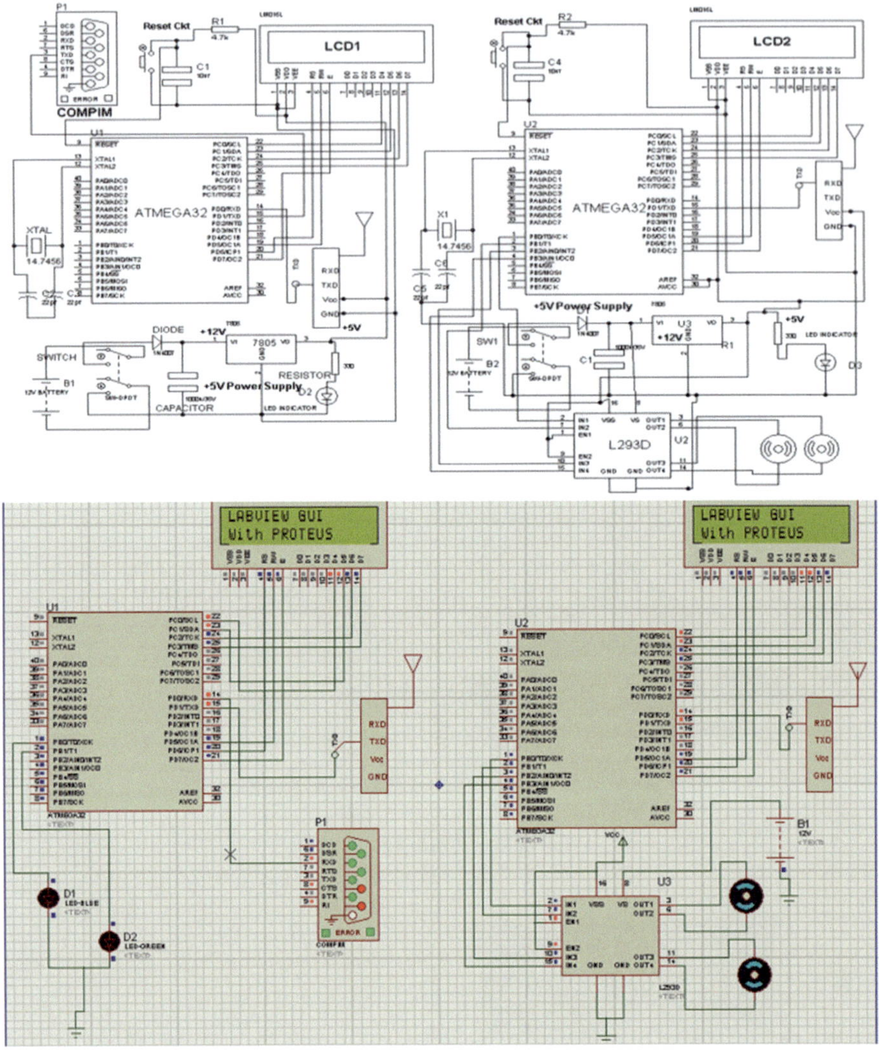

Fig. 6 Proteus model of robot section

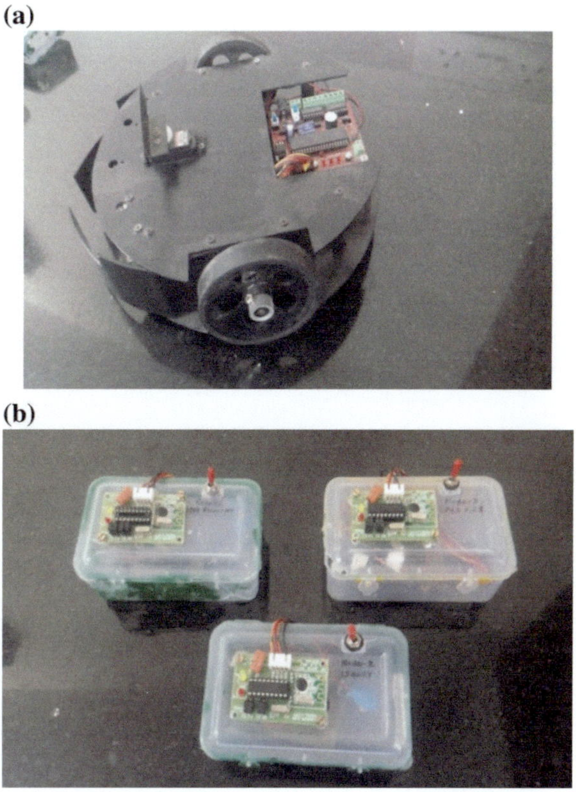

Fig. 7 **a** Experimental setup of the robotic platform, **b** experimental setup of the nodes platform

6 Results and Conclusion

Figure 8a, b shows the different curves of the control parameters obtained with the use of PSO optimized values for tuning of PID gain parameters, respectively.

Hence, the proposed system worked as per demand of the system. The algorithm discussed in this paper was applied to carry out the task of controlling the robot over network having ability to guide robot through a terrain or specified area which is within the range of wireless nodes.

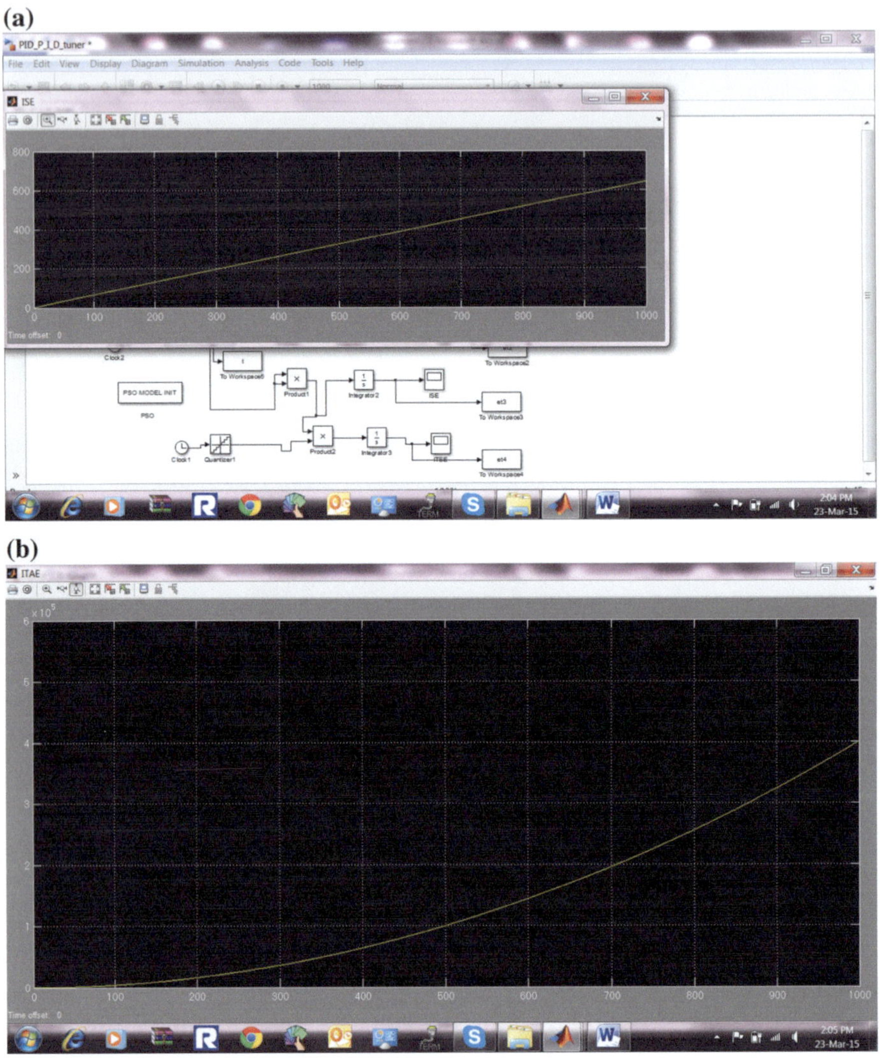

Fig. 8 Snapshot for the PID gain graph

References

1. Luna, Ryan, and Kostas E. Bekris. 2010. Network-guided multi-robot path planning in discrete representations. In *IEEE/RSJ International Conference on Intelligent Robots and Systems (IROS)*, 18–22 Oct 2010, pp. 4596–4602.
2. Ye, Wei, et al. 2001. Evaluating control strategies for wireless-networked robots using an integrated robot and network simulation. In *IEEE International Conference on Robotics and Automation, 2001. Proceedings 2001 ICRA*, vol. 3, pp. 2941–2947.

3. Shenoy, Suresh, and Jindong Tan. 2005. Simultaneous localization and mobile robot navigation in a hybrid sensor network. In *IEEE/RSJ International Conference on Intelligent Robots and Systems, 2005. (IROS 2005)*, 2–6 Aug 2005, pp. 1636–1641.
4. Ladd, Andrew M., Kostas E. Bekris, Algis Rudys, Lydia E. Kavraki, and Dan S. Wallach. Robotics-based location sensing using wireless ethernet.
5. Allaoua, Boumediène, Brahim Gasbaoui, and Brahim Mebarki. 2009. Setting up PID DC motor speed control alteration parameters using particle swarm optimization strategy. *Leonardo Electronic Journal of Practices and Technologies*.
6. Singh, Rajesh, et al. 2014. Design and experimental evaluation of PSO and PID controller based wireless room heating system. *International Journal of Computer Applications* 107.
7. Kennedy, James, and Russell Eberhart. 2010. Particle swarm optimization. In *Encyclopedia of machine learning*, 760–766. Berlin: Springer.
8. Kim Tae-Hyoung, Ichiro Maruta, and Toshiharu Sugie. 2008. "Robust PID controller tuning based on the constrained particle swarm optimization." *Automatica* 44.4: 1104–1110.
9. Sultana, Taskeen, and Zeenath. 2014. Communication of multi mobile-robots' based on ZigBee network. *International Journal of Science and Research (IJSR)* 3 (8).
10. Huasong, Min, Liu Zhenglin, Wei Hongxing, and Wang Tianmiao, Control of a self-assembly modular robot system over a wireless ZigBee network. In *World Congress on Intelligent Control and Automation*, 6–9 Jul 2010.
11. Yeh, Syh-Shiuh, Cheng-Chung Hsu, Tzu-Chiang Shih, Jen-Pin Hsiao, and Pau-Lo Hsu. 2008. Remote control realization of distributed rescue robots via the wireless network. In *SICE Annual Conference*, 20–22 Aug 2008.
12. Wang, Ning, Naiqian Zhang, and Maohua Wang, Wireless sensors in agriculture and food industry—Recent development and future perspective. http://www.ncbi.nlm.nih.gov.

Case Study on Vapour Cloud Explosions (Buncefield and Jaipur Explosion)—A Review

R. Shristi Vidusha, Dipti Sharath, Bikrama Prasad Yadav and Abhishek Nandan

Abstract Fire being an essential element in man's life has not only helped in building up lives but also have taken up a lot of lives. Since 1966 and before that, there had been a lot of fire accidents reported till date. Many of those fires have had engulfed a lot of peoples' lives. Two such accidents that happened in the oil terminals four years apart gained greater significance worldwide. First is the Buncefield incident in 2005, followed by IOC Jaipur fire in 2009. Huge monetary losses were incurred, with complete destruction of major areas of the installations. The similarity in both these incidents is that they resulted from vapour cloud explosions. In Buncefield, the vapour cloud formed due to the spillage of 300 tonnes of winter grade MS followed by the overfilling of the storage tank 912, which found an ignition source in a nearby car park leading to the first explosion, followed by a series of explosions that engulfed over 20 storage tanks. This giant and the biggest storage tank explosion surprisingly did not kill anyone but injured 43 employees [1]. In Jaipur, accidental spillage of MS during a hammer blind reversal job created a vapour cloud which exploded, resulting in an uncontrollable fire engulfing 12 tanks. Unfortunately, the explosion killed 12 people and injured over 200 workers. The MET department had recorded a tremor of 2.3 on Richter scale while the explosion caused the shattering of glass window around 3 km from the terminal. Both terminals had burned for days, making it a daunting task for the fire fighters to extinguish the intense fire. As such accidents are on a rise, globally, it shows the lack of implementation of existing safety standards. This paper aims at discussing the potential causes and consequences resulting from both these accidents and further talks about the personal views in preventing and mitigating similar accidents in future.

R. Shristi Vidusha (✉)
Weatherford Drilling Services, Dubai, UAE
e-mail: shriztee@gmail.com

D. Sharath · B. P. Yadav · A. Nandan
Department of Health Safety and Environment Engineering, University of Petroleum and Energy Studies, Bidholi, Dehradun 248007, India

N. A. Siddiqui et al. (eds.), *Advances in Fire and Process Safety*,
Springer Transactions in Civil and Environmental Engineering,
https://doi.org/10.1007/978-981-10-7281-9_11

Keywords Fire · Vapour cloud explosion · Motor spirit · Oil terminal
Buncefield · Jaipur

1 Introduction

As technology grew so did the associated risks and hence evolved the concept of safety. Safety is not a priority, but is a value considered to protect human lives and prevent destruction of infrastructure and environment. Since the first known explosion in the storage tanks, it has become every company's responsibility to prevent the enormous amount of losses in dollars and lives of people. Yet, the disasters continue inspite of all the layers of protection in a plant. Tank farms are the temporary industrial facility storage regions where petroleum and other oil products along with wide range of other fluids are stored. If they store petroleum, they are called as oil depots. From these depots or farms, these stored chemical products are transported to the customers (storage sites or industrial users). These storage facilities are the areas of high risk, as they store large amounts of petroleum products. Over 242 accidents have occurred in the last four decades worldwide [2]. Many of these explosions at the storage tank terminals are found to be the vapour cloud explosions which are focused mainly in this paper. The vapour cloud explosions occur when a vapour cloud forms because of loss of containment and when it finds an ignition source. Such explosions are still not under man's control [3]. Let it be the 1966 explosion at Feyzin, France, or be the 2013 HPCL blast at VIZAG, India, both were caused by the human negligence inspite of all the safety measures.

In this paper, the two major catastrophic tank fires are being discussed viz., Buncefield (2005) and Jaipur (2009). Both the accidents share the similarity that they are vapour cloud explosions. When Buncefield accident happened on 11 December 2005 the loss incurred was very high not only for the HOSL Corporation but also for its stake holders. It has not taken any lives but has cost heavily [4]. After the Buncefield incident, the effect on environment was also quite adverse. The fire water that got absorbed by the ground polluted the ground water table and exceeded the threshold of 1 hectare causing an environment impact to the Seveso directive. The pump heads up to 3 km from the depot had to be shut down [5]. On 29 October 2009 another accident of same kind took place at Jaipur. The terminal was engulfed by the fire that broke out from the storage tank at Indian Oil Corporation Ltd. terminal (IOCL). The disaster resulted in the major effects on the environment around 2 km radius from the terminal. Every buildings and equipments in the vicinity were destroyed by the fire. This paper focuses about the actual events that led to the explosion and the major recommendations that have been posed by the regulatory boards.

2 Storage Tank Fires and VCEs

The storage tanks at the oil contain large volumes of the flammable and sometimes highly flammable liquids which are stored for transporting the fuel to various refineries through pipelines [6]. These above ground storage tanks are prone to fire accidents because of many errors and failures like human and maintenance error. Vapour cloud explosion being the common type of explosion occurring at the storage tanks has taken many tolls at the human and the business losses. Top depots in world have also faced such devastating fate because of minor errors [7]. Some of the storage tank fires and vapour cloud explosions that have occurred in the past have been reviewed for its cause and effects and are tabulated in Tables 1 and 2.

Table 1 List of vapour cloud explosions reviewed

S. No.	Year	Place	Name of the company	Gas in vapour cloud	No. of fatalities/ injured	Error
1	1966	Lyon, France	Total refineries	Propane	18/81	Human error (valves failed to closed)
2	1970	Port Hudson, US	Phillips Pipeline Company	Propane	NA	Pipeline break
3	1974	Flixborough, England	Nypro UK	Cyclohexane	28/36	Leak from the reactor
4	1975	Beek, Netherlands	Dutch State Mines (DSM)	NA	14/107	Fracture in depropanizer
5	1984	San Juanico, Mexico city	Pemex State Oil Company	LPG	600/7000	Pipe rupture and pressure drop
6	1997	Vishakapattinam, India	HPCL	LPG	22/11	Leakage of gas while pumping
7	2000	New Mexico	El Paso Natural Gas Company	Natural gas	12/NA	Pipeline rupture
8	2005	Texas, USA	British Petroleum	Hydrocarbon	15/170	Leak and human error
9	2009	Puerto Rico	Caribbean Petroleum Corporation	Gasoline	0/3	Overflow

Table 2 Notable storage tank failures

Date	Place	Contents of the tank	No. of injured/killed	Reason of accident
21st December 1985	Naples, Italy	Gasoline	170/5	Overflow
16th October 1995	Pennsylvania, USA	Hydrocarbon and water	1/5	Spark from hot work entered through corroded holes
17th July 2001	Delaware, USA	Sulphuric acid	8/1	Ignition of flammable vapours
5th June 2006	Mississippi, U.S.A.	Hydrocarbons, toluene, xylene, ethyl benzene and naphthalene fumes	1/3	Ignition of flammable vapours

3 Buncefield Fire, 11 December, 2005

3.1 About the Terminal

The Hertfordshire Oil Storage Terminal, London, known as the Buncefield depot experienced on such accident that incurred huge loss fortunately at the cost of zero lives. The 60 million gallons capacity depots dealt with every year, around 2.37 MT of oil products. The depot stored and distributed the fuels for London and the South East of England. The operation at the Buncefield depot was very simple that it used to pipe down the petroleum products from three refineries viz., at Coryton, Stanlow and Lindsey and store in large storage tanks [8]. The terminal was operated by the owner and operator of the UK Oil Pipeline Network (UKOP), British Pipelines Agency (BPA) [9]. Since the withdrawal of two shell refineries from the depot by 2005, Hertfordshire Oil Storage Ltd (HOSL) along with British Petroleum (BP) acted as terminal's major operator. BPA stored aviation kerosene at the depot while receiving and distributing the fuels from UKOP to HOSL [10]. In its main area, the HOSL terminal had the loading gantry and the tanks 910 through 915 which were large and were not able to be viewed directly from the control room and their overview was provided by the security CCTV cameras installed at the site. The depot that had been operated successfully without any major accidents since 1960 met a major disaster on early morning 6.01 A.M. of 11th December 2005 due to the explosion of nearly 20 tanks [11].

3.2 How It Started?

The witnesses of the accident heard a sharp sound of a jet engine which was also interpreted as a huge pressure sound, and it was mistaken as blast wind or violent quaking of a lorry since the noise lasted for 500–1000 ms [12]. At approximately 05.38 A.M. on the D-Day, the CCTV footage showed that the mist started flowing from the north-western edge of the HOSL part of the depot slowly through the north-eastern parts filling the car parks and finally reaching BPA Bund A at an average speed of 0.6 m/s [13]. The mist spreading was initially having a shallow depth which was increased to 2 m at the car parking and 4–5 m in Bund A. On seeing a weak spray coming out of a tank, the worker reported which suggested that the tank 912 may be the source of the vapour cloud because of over filling [14]. The notable fact was that there were only two workers working at the site then.

3.3 The Safety Features of the Tank

The instrumentation provided in the tank 912 helped in measuring and monitoring the temperature and the liquid level inside the tank. The instrumentation included the servo level gage and temperature sensor and was connected with the automatic tank gauging (ATG) system that was connected with all tanks from the entire depot in common [15]. An independent safety switch was also fitted in the tank which provides flashing light as a visual alarm and loud buzzing sound as an audible alarm at the event of any overflowing. The switch was also connected in a way to trip the valves down to prevent filling the tank over the allowable level. The ATG system was capable of trending the data with an event logging system and an integrated alarm system [16].

3.4 The Timeline of the Disaster

The tank 912 was scheduled for filling with winter grade motor spirit in the eve of the disaster at around 7.00 P.M. and was about to continue till the next morning. The tank was expected to get filled around 5.20 A.M., and the ATG system was supposed to shut the filling process down [17]. Since the protection system did not operate, for the reasons still not known, the tank 912 was overfilling past the ATG high-level alarm and the level gauge was stuck because of which the gauge was showing no flow in the tank since 3.00 A.M. [18]. Hence, the alarms were not triggered. Hence, no staff-on-duty noticed that the tank was over filling, and the time to divert the filling to the next tank has come. This overfilling of tank 912 was continuing for more than 25 min and reached the flow rate of 960 $m^{3/}h$ when it was

8 min to the explosion. There were around 300 tonnes of gasoline that was overflowing, and the vapour cloud covered an area of around 120,000 m^2 [19].

A spark from the pump house of HOSL site ignited the vapour cloud causing the first major explosion at around 6:01 A.M. This spark led to a huge fire that swallowed up 20 tanks at the depot. In the direct neighbourhood of the terminal, the magnitude of the explosion was recorded to be 1000 mbar. [9]. The explosion was recorded 2.4 on Richter scale causing structural damages to the surrounding areas around 2 km away from the depot. [8]. It was 14th December 2005 when the last firing fuel tank was extinguished. The fire took almost 59 h to get extinguished by over 32 fire brigades and 650 fire fighters [20]. The fire fighters used 68 ML of water and 786,000 L of foam concentrate having perfluoro octane sulphonate (PFOS) [5]. Fortunately, the disastrous fire killed no one but injured 43 workers at the depot.

3.5 Control Measures Recommended

1. Overfill indicating devices could be placed completely separated from the tank gauges so that even if the gauge fails the alarm do not fail, and hence, the operators would get alerted at any such circumstances.
2. Shift handover must be carried out properly and the safety communication must be enhanced among the workers for better handover of the shift [21].
3. Better operations must be carried on at the control room by allotting clearer roles and improving the interfacing with front line staffs. This would create a potential to detect the emergencies early and respond to it before anything goes out of the hand.
4. The secondary containment (bunds) and the tertiary containments (drainage) have to be built strong enough to hold the loss of containment, if happens in the future.
   - The secondary containments have to be inspected periodically for the joints and the parts were pipes penetrate them.
   - The tertiary containments must be large enough to hold the liquids enough to confine the pollution inside the site only.
5. The previous reports on near misses or any incidents must be reviewed, and immediate and long-term corrective actions must be taken, and these actions are to be reviewed for changes periodically to continually improve the safety management system.
6. The safety critical equipments must be designed strong enough to serve their purposes and safety studies like hazard and operability (HAZOP) and safety integrity levels (SIL) must be carried on effectively.

3.6 *Legal Impacts of the Incident*

Control of Major Accident Hazards Regulations 1999 ("COMAH Regulations") is the driving regulations existing within UK which was framed after the famous Seveso disaster. The main focus of the regulations is to reduce the risk level existing inside any industry, and it facilitates the regulation board to undertake the complete investigation on any incident in the industries which are covered by this regulation. This regulation came up with certain amendments after the Buncefield accident. The amendments focused on the role of the operators and the importance of the risk analysis studies that must be taken in an industry. Some of the recommendations are as follows:

1. The operators must review the emergency plans that would include the potential procedures for the vapour cloud explosions on-site and off-site also [22].
2. The emergency plans must include the public health implications that might come out of vapour cloud explosions.
3. As a part of risk assessment, all the dangerous substances and normal substances used under normal/abnormal operations must be included while defining the scope of any operations.
4. Risk assessment must be carried out for the moving vapour clouds and the contents of the vapour clouds to foresee the consequences of such occurrences.

4 Jaipur Fire, 29 October, 2009

During the late hours of 29th of October, 2009, a gasoline vapour cloud exploded after a major containment loss during line up for operation in the IOCL terminal in Jaipur, India. In the accident, 11 people died and many were injured, with villages having to be evacuated. The fire in the storage terminal burned away all the flammable material lasting for more than a week. The entire facility was destroyed, including the administrative building, fire water pumping stations and the emergency station [23].

4.1 *Description of the Accident*

During the late shift at IOCL terminal, the operators were lining up kerosene and motor spirit (gasoline) tanks for a pipeline transfer operation to BPCL terminal. At the time of the operation, there was one shift operator and three operators, who were present at the terminal. During the lining up process, gasoline leaked out of the hammer blind and splashed onto the worker. The worker was unable to stop the leak, and so was the other operator present in the vicinity of the MS tank 401 A.

The operators could not perform the emergency spill response procedures and lost their lives [24]. The gasoline formed a pool in the tank farm, which quickly evaporated into a vapour cloud which spread over an area of 180,000 m^2. The vapour cloud ignited, after obtaining an ignition source nearby, resulting in a massive pressure wave. The pressure wave from the blast destroyed nearby structures. The fire spread to the other tanks, including kerosene and diesel (HSD), resulting in further blasts [25]. Approximately, over 2000 tones of gasoline were released; out of which, 4% formed a vapour cloud which exploded. As the wind speed was low, it prevented dispersion of the vapour. The burning tanks could not be extinguished, as a total of 11 tanks were on fire. The officials decided against any fire extinguishing proceedings and emergency procedures, as that would be very dangerous, risking many lives [26]. Hence, the fire burned till the fuel was exhausted. Approximately, 60,000 kL of petroleum products was lost in the fire. The emergency operations never took place, as the operators involved had been overcome by the fire. By the time the senior staff and the civil authorities arrived at the terminal, the fire had engulfed a major portion of the installation, making their entry dangerous. The losses that were incurred by IOCL were multifold. Along with a loss of over 280 crores, there was a lot of public outrage, for the loss of the lives. It seemed obvious that it was a case of negligence that resulted in such a huge amount of damage. After the accident, there were major changes recommended in the operating procedures and fire fighting systems for a POL terminal [27].

4.2 Lessons Learnt from the Accident

(a) The safe operating procedure must take into account all the hazards and must be reviewed periodically.
(b) The safe operating procedures must be intimated to all employees, including senior management, operators and workers.
(c) Management should make sure that unfailing systems are in place to give timely pointers on the current practices, along with provision for monitoring of the current system management systems
(d) The emergency procedures must be formulated, and mock drills are to be performed periodically, and the state of readiness is to be assessed.
(e) The safety policy of the organization must reflect in the attitude of the higher management, which is to be inculcated by the workers.
(f) Facilities which store hazardous material must ensure strict compliance to safety procedures.
(g) Automation of operations is to be considered to reduce human error.

Ministry of Petroleum and Natural Gas under the Govt. of India constituted an independent seven-member committee led by Mr. M.B. Lal, Ex. Chairman, HPCL, to prepare a report to understand the cause of the accident. The committee made 118 recommendations to be implemented at oil installations which store petroleum products. Currently, modifications are in progress in all installations under the public sector. The recommendations were divided into categories, which fell into the different layers of safety protection systems. They were

- Engineering related
- Operation related
- Procurement related
- Training related
- Policy related
- OISD related
- Ministry related.

There are more than hundred recommendations as per the committee guidelines. These few were important from the perspective that, had these been followed earlier, the Jaipur incident might not have occurred, and many lives could have been saved.

- The hammer blind valves should be replaced with other types of valves such as pressure balancing type plug valves/ball valves as the accident was caused due to the malfunctioning hammer valve.
- The body valve which is closest to the tank must be a remote operated shut off valve (ROSOV) inside the dyke, on the tank nozzle which can be remotely operated from either outside the tank dyke or from the control room. This ROSOV must be fail and fire proof. This helps in shutting down the dispatch and filling lines, in case of loss of containment scenarios.
- Adequate lighting should be provided in operating areas. Minimum lighting lux level should be as: tank farm area/roads—20, main operating area/pipe manifolds—60 and pump house areas—100.
- For floating roof tanks, leak protection system must be present, so that there is no leakage during roof water/water draining operations. The swivel joint might also malfunction resulting in leakage. Leakage of oil can lead to loss of containment and vapour cloud explosions.
- The piping within the tank farm must not be haphazardly placed. The manifold must be placed in such a way that it can be accessed easily. This allows for quicker access for the shutting down of valves and also helps in preventing injuries during attempts to access the valves.
- In the recommendations related to instrumentation engineering, an important one was that there must be an indicator present in the control room, which shows whether the tank dyke valve is opened or closed. And a mandatory check must be done to ensure it is in working condition. The indicator must also show to what degree the valve is open or closed.
- Buildings not related to terminal operation including canteen should be located outside the plant area. These areas are more populated and are severely affected during accidents. As the safety systems in these areas are not as stringent, they can also be the cause for accidents.

- The main emergency response centres like the emergency control room, fire water tank farm and the fire water pump house must be located at a safe distance from the oil tank farm. It must be damaged as a result of any accident in the terminal. Automation of tank farms and terminals is to be done with sophisticated systems both in hardware like devices and software programs for reducing human error. Another important recommendation is that the total quantity of fire water stored must be calculated for a scenario where two major fires are to be extinguished/controlled simultaneously.
- The emergency exit gate should be located at a considerable distance from the main entrance gate and must not be obstructed in any way, to provide easy access to the emergency response and rescue personnel. Management of change procedure should be immediately implemented.
- The full shutdown system activation also should be included in all the regular mock drills. The mock drills are to be conducted on a regular basis, and roles and responsibilities must not only be assigned, but also understood. The time taken for the response to an accident must be recorded, and efforts must be taken to improve it.
- Emergency procedures should be written to incorporate all major and minor accident scenarios. It should be made available to all personnel in the installation outlining the actions to be taken by each during a major incident.
- Quantitative and qualitative risk assessment like hazard and operability studies are to be carried out by a suitable consultancy and regularly updated.
- Long-range foam monitors (more than 1000 GPM) which are operated remotely with a provision of variable flow must be provided for fighting tank fires. These enable in curbing the fire at the initial stage itself.
- The rim seal fire detection and protection system are to be installed in all Class “A” products in the terminal. Class A products as specified in the OISD, which refer to the highly flammable materials like motor spirit. Rim seal fire protection system helps in the early detection and quick extinguishing action of fire.
- Medium expansion foam generators are to be provided, so as to arrest the formation of ignitable vapour clouds, from pools in case of loss of containment scenario.
- Closed circuit television is to be installed in high-risk areas like the tank farm areas and pumping stations for video surveillance. To detect any deviation from the routine, an alarm can be provided along with the CCTV. It must be made mandatory to ensure the cameras are in working condition.
- Hydrocarbon detectors should be installed in all areas susceptible to a containment of loss of highly flammable petroleum products, such as Class A products like gasoline. These quick action detectors can also be connected to alarm systems in the control room. Sometimes, the leak continues for long periods and is left undetected, as it happened in the case of both Buncefield and Jaipur tank fire.
- VHF (wireless handsets) handsets should be provided to each of the operating crew. It should be made mandatory for the operating personnel to enter the battery limit area only with VHF sets. The communication between the

Table 3 Comparison of the two disasters

Facts	Buncefield disaster	Jaipur disaster
Type of explosion	VCE	VCE
Fuel leaked	Winter grade MS	MS
No. of people killed	0	11
No. of people injured	43	45
Error	Overfilling of tank	Improper line up procedure
Cause of error	Failure of safety systems	Lack of knowledge of SOP
No. of tanks engulfed	20	11
Richter scale impact	2.4	2.3
After-math regulations	The COMAH regulation	The M.B. Lal Committee

operation personnel and the control room is vital, and any other system can be used so that any emergency situation can be communicated to the control room quickly.

It is vital to understand that though safety systems can help in reducing accidents substantially, the role of the humans cannot be neglected. A right attitude towards understanding and implementing the safety protocols must be present to have an accident free work environment. The above given recommendations can bring down the rate of accidents, only when it is diligently followed.

5 Comparison Between the Incidents

These notable and unforgettable disasters have some similarities and differences that are tabulated in Table 3.

6 Summary and Conclusion

The paper starts with the small discussion about certain vapour cloud explosions and details the incidents happened at Buncefield Oil Depot, 2005, and Jaipur Oil Terminal, 2009. The losses incurred after the disasters were enormous in all financial, social and environmental aspects. Like how the Bhopal gas tragedy gave birth to regulations like OHSAS and amendments in Factories Act of India, these two disasters also have brought some regulations. After the Buncefield incident, Control of Major Accident Hazards Regulations 1999 (The "COMAH Regulations") was amended and the recommendations in specific to those industries similar to that of the damaged depot were proposed. Similarly, after the Jaipur fire accident, which was a pure man fault induced accident, MB Lal committee was setup which investigates every accident and gives strong recommendations in specific to the industry and the accidents. The committee has given such recommendations to the IOCL terminal after studying the entire scenario. All these recommendations have a greater impact and also give the industries the better way to operate safely so that the accidents in near future will have lesser likelihood of such occurrences.

Acknowledgements The authors would like to thank the department of HSE at UPES, Dehradun, for the successful completion of this paper.

References

1. Capelle-Blancard, G., and M.-A. Laguna. 2008. The Buncefield oil depot explosion: Where there's smoke, there's (stock market) fire? *Applied Financial Economics Letters* 4 (2): 103–107.
2. Chang, J.I., and C.-C. Lin. 2006. A study of storage tank accidents. *Journal of Loss Prevention in the Process Industries* 19 (1): 51–59.
3. Atherton, W., and J. Ash. 2007. Review of failures, causes & consequences in the bulk storage industry. In *University research conference.* UK: Liverpool John Moores University.
4. Chlimintza, E.M. 2008. Resource management in fire fighting organizations: Lessons from the Buncefield oil depot fire and the Greek 2007 Forest Fires. In *the 12th Biennial Conference of the International Association for the Study of Commons, Governing Shared Resources: Connecting Local Experience to Global Challenges*, Cheltenham.
5. Nicholas, M. and A. Whitfieldb. 2013. The Buncefield accident and the environmental consequences for fuel storage sites and other sites in the UK, Regulated under the Seveso directive. *Chemical Engineering* 31.
6. Glavan, C., and E. Palaneeswaranb. 2012. The price and complications of safety compliance–An Australian perspective. *Chemical Engineering* 26.
7. Eckhoff, R.K. 2014. Boiling liquid expanding vapour explosions (BLEVEs): A brief review. *Journal of Loss Prevention in the Process Industries* 32: 30–43.
8. Tam, V.H. 2011. The Buncefield accident—Why was the explosion so severe? *Loss Prevention Bulletin* 222.
9. Howard, C. 2013. The Buncefield incident—7 years on: A Review. *Measurement and Control* 46 (3): 76–82.
10. Bunch, J.M., R. Beresford, and P.P. Engineer. *The Buncefield Enquiry Findings and Costain's Approach to Best Practice in Integrity Level Assessment.*
11. Batista-Abreu, J.C., and L.A. Godoy. 2013. Thermal buckling behavior of open cylindrical oil storage tanks under fire. *Journal of Performance of Constructed Facilities* 27 (1): 89–97.
12. Atkinson, G., and L. Cusco. 2011. Buncefield: A violent, episodic vapour cloud explosion. *Process Safety and Environmental Protection* 89 (6): 360–370.
13. Gant, S.E., and G.T. Atkinson. 2011. Dispersion of the vapour cloud in the Buncefield Incident. *Process Safety and Environmental Protection* 89 (6): 391–403.
14. Atkinson, G., et al. 2015. Flammable vapor cloud generation from overfilling tanks: Learning the lessons from Buncefield. *Journal of Loss Prevention in the Process Industries* 35: 329–338.
15. Nivolianitou, Z., et al. 2012. A methodology for the hazard assessment in large hydrocarbon fuel tanks. *Chemical Engineering* 26.
16. COmAH, C. 2011. *Buncefield: Why did it happen.* HSE Books.

17. Mannan, M.S., et al. 2007. Our inability to learn from the past: is Buncefield another example. In *12th International symposium on loss prevention and safety promotion in the process industries, IChemE symposium series*.
18. Board, B.M.I.I. 2005. *Buncefield major incident investigation*. Initial Report to the Health and Safety Commission and the Environment Agency of the investigation into the explosions and fires at the Buncefield oil storage and transfer depot, Hemel Hempstead, on, vol. 11.
19. Johnson, D.M. 2010. The potential for vapour cloud explosions—Lessons from the Buncefield accident. *Journal of Loss Prevention in the Process Industries* 23 (6): 921–927.
20. Murray, V., C. Keshishian, and G. Urquhart. 2010. Chemical hazards and poisons report: From the centre for radiation, chemical and environmental hazards, June 2010 Issue 17. *Chemical Hazards and Poisons Report* 17: 3–72.
21. Wilkinson, J., and R. Lardnerb. 2013. Shift Handover After Buncefield. *Chemical Engineering* 31.
22. Page, L., and S. Wessely. 2006. Health complaints following an air disaster: A "second disaster". *Occupational and Environmental Medicine* 63 (10): 647–648.
23. Sharma, R.K., et al. 2013. Assessment of an accidental vapour cloud explosion: Lessons from the Indian Oil Corporation Ltd. accident at Jaipur, India. *Journal of Loss Prevention in the Process Industries* 26 (1): 82–90.
24. Chauhan, A., et al. 2010. Jaipur Indian Oil fire of 29 September 2009 and associated atmospheric and meteorological changes using multi sensor data. In *38th COSPAR Scientific Assembly*.
25. Giri, R., S. Prakash, and P. Rani. *Plume Generated by Oil Depot Fire and Associated AOD Changes*.
26. Abbasi, M.H., E. Benhelal, and A. Ahmad. 2014. Designing an Optimal safe layout for a fuel storage tanks farm: Case Study of Jaipur Oil Depot. *World Academy of Science, Engineering and Technology, International Journal of Chemical, Molecular, Nuclear, Materials and Metallurgical Engineering* 8 (2): 147–155.
27. Girdhar, M. 2012. Jaipur Fire and its Environmental effects. *Fire Engineer* 37 (3): 21–22, 27–29.

Broadband Stacked Microstrip Antenna with Genetically Designed Patches

Raj Gaurav Mishra and Jeevani Jayasinghe

Abstract This paper presents the broadband performance and broadside radiation of a stacked microstrip antenna design. This antenna consists of two substrates stacked one over the other incorporating an air gap in between. The substrates have $\epsilon_r = 4.4$ (dielectric constant) and $h = 1.58$ mm (height). One patch of the antenna has been designed to resonate at a frequency of 5 GHz. Genetic algorithms (GA) have been used to select the dimensions of the other patch in order to have broadband performance. Moreover, two slots have been cut on patches to improve performance further. The dimensions and the positions of the slots, the position of the feed, and the size of the air gap have also been optimized by using GA. The novelty of this antenna design is synchronized GA-based optimization of useful parameters. This optimized design has a −10 dB bandwidth of 3.4 GHz and broadside radiation with gain about 5 dB throughout the bandwidth. Its performance has been compared with stacked microstrip antennas that have not been optimized.

Keywords Antenna radiation patterns · Genetic algorithms · Microstrip antennas Optimization

1 Introduction

Microstrip antennas with conventional patch shapes have limited bandwidth along with lower gain/directivity and lower efficiencies [1]. Therefore, researchers have proposed various techniques to design improved antennas. One such technique is to

R. G. Mishra (✉)
Department of Electronics, Instrumentation and Control, CoES,
University of Petroleum and Energy Studies, Dehradun, India
e-mail: rgmishra@ddn.upes.ac.in

J. Jayasinghe
Department of Electronics, Wayamba University of Sri Lanka,
Kuliyapitiya, Sri Lanka
e-mail: jeevani@wyb.ac.lk

N. A. Siddiqui et al. (eds.), *Advances in Fire and Process Safety*,
Springer Transactions in Civil and Environmental Engineering,
https://doi.org/10.1007/978-981-10-7281-9_12

increase the distance in between the radiating patch and the ground plane. This can be done by suspending the patch on air [2], by using thick substrates [3] or by using multiple substrates [4]. Therefore, microstrip antennas with multiple substrates are famous among researchers in the field of antennas to obtain broadband, multiband, miniature, and high-directivity characteristics.

For dual-frequency and broadband performance, a microstrip antenna designed by using two stacked patches is presented in [4]. It operates at 1.8 and 3.5 GHz bands. Research evaluation of a U-slot imposed patch stacked on H-shaped parasitic patches is presented in [5]. The bandwidths at upper and lower resonances are approx. 10.2 and 3.7%, respectively. A rectangular shaped dual-layer single-feed microstrip antenna having one short pin is discussed in [6]. This quad-band antenna is suitable for four wireless communication applications (e.g., GSM 900, 1800, IMT 2000, Wi-Fi, Bluetooth) with maximum gain (4.4 dB at 2.42 GHz). The design is optimized for the location of single short pin. The design of a stacked microstrip antenna in [7] shows two closely spaced resonant frequencies to achieve broadband performance. Research paper [8] suggested the expansion of the antenna bandwidth using an E-shaped patch along with second stacked dielectric substrate on the top of a U-slot patch. It achieves an impedance bandwidth of about 59.7% (for range in between 3.28 and 6.07 GHz). Throughout the entire frequency band, radiation patterns are relatively constant. Another broadband E-shaped stacked microstrip antenna is proposed in [9]. In this design, the impedance bandwidth achieved is about 38.4%. By using a washer on the probe, a much wider input impedance bandwidth of about 44.9% is achieved. Throughout the entire frequency band, radiation patterns are relatively constant. The stacked microstrip antenna proposed and presented in [10] has a bandwidth of about 35%. In this publication, effects on bandwidth for various antenna parameters were explored. A broadband microstrip antenna design with slot is presented in [12–14]. The significance of selection of size and position of the slot on the antenna design is also discussed.

Review of the publications based on stacked microstrip antennas shows that two or three substrates have been used most commonly to improve performance. Some designs contain air gaps between substrates to improve performance further. However, use of very thick substrates or air gaps increases the total height of the antennas and may not be suitable for some applications.

Therefore as proposed in this paper, genetic algorithms (GA) is used to design a broadband microstrip antenna. GA is used for tuning of several antenna parameters altogether, while maintaining the air gap size as smaller as possible. Another objective is to maintain broadside radiation throughout the entire bandwidth. The antenna design procedure is explained in Sects. 2, 3 and presents the results, and Sect. 4 concludes the research outcomes.

2 Design Procedure

This section presents the design procedure of the broadband stacked microstrip antenna. Two substrates having a dielectric constant $\epsilon_r = 4.4$ and height $h = 1.58$ mm are used to etch patch radiators (Fig. 1). The upper patch dimensions have been selected as $18.6 \times 14\ \text{mm}^2$ in order to resonate around 5 GHz at the fundamental mode. GA is used to select the dimensions of lower patch. There is an air gap between two substrates, and the most suitable air gap size is selected by using GA. Each patch contains a slot, and their properties such as dimensions and positions are also selected by using GA. A 50 Ω coaxial feed is connected to the lower patch. GA selects the best feeding position out of 32 possible points spread on the patch area.

Hence, eight antenna parameters are optimized by using GA (Table 1). Possible ranges have been assigned to each antenna parameter. All possible values are encoded and are represented as "0"s and "1"s. They represent genes of chromosomes in the GA optimization procedure. The chromosome size has been defined based on the number of antenna parameters, possible range, and number of candidates (Table 2). The feed position is defined by using five genes as there are 2^5 candidates. Each other antenna parameter is defined by using three genes.

Therefore, a chromosome with 26 numbers of genes has been used in the GA optimization procedure. The optimization procedure has started with a randomly generated 20 chromosomes. All the chromosomes have gone through crossover operation and mutation probability was 2.5%. The generation size has been kept 20 throughout the optimization process.

The objective of the optimization process is to enhance the bandwidth and to obtain a radiation pattern in the broadside direction. Therefore, the fitness has been defined as:

$$F = \sum_{f_i=f_1}^{f_2} G(f_i) - \sum_{f_i=f_1}^{f_2} L(f_i) \tag{1}$$

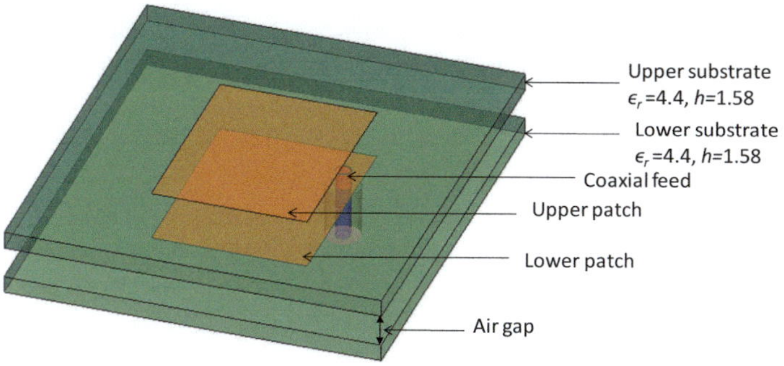

Fig. 1 Antenna configuration

Table 1 Possible ranges of antenna parameters

Antenna parameter	Number of candidates	Possible range
Length of the lower patch (L_l)	8	8–15 mm
Width of the lower patch (W_l)	8	15.0–18.5 mm
Length of the lower slot (L_{ls})	8	1–8 mm
Width of the lower slot (W_{ls})	8	1–8 mm
Length of the upper slot (L_{us})	8	1–8 mm
Width of the upper slot (W_{us})	8	1–8 mm
Feed position	32	All over the patch
Air gap size	8	3.0–6.5 mm

Table 2 Structure of the chromosome

Antenna parameter	L_l	W_l	Air gap size	L_{ls}	W_{ls}	L_{us}	W_{us}	Feed position
Number of genes	3	3	3	3	3	3	3	5
Assignment of genes	0, 1, 2	3, 4, 5	6, 7, 8	9, 10, 11	12, 13, 14	15, 16, 17	18, 19, 20	21,..., 25

where $G(f_i)$ is the gain along the broadside direction and $L(f_i)$ is defined as [11]

$$L(f_i) = \begin{cases} \rho(f_i) & \rho(f_i) \geq -10\ \text{dB} \\ -15\ \text{dB} & -15\ \text{dB} \leq \rho(f_i) < -10\ \text{dB} \\ -30\ \text{dB} - \rho(f_i) & -30\ \text{dB} \leq \rho(f_i) < -15\ \text{dB} \\ 0 & \rho(f_i) < -30\ \text{dB} \end{cases} \tag{2}$$

where $\rho(f_i)$ is the reflection coefficient at frequency f_i in the frequency range of f_1 = 4 GHz to f_2 = 7 GHz at 100 MHz intervals. This fitness function encourages having reflection coefficients between −10 and −30 dB, because the major objective is to increase the bandwidth as much as possible instead of having very low reflection coefficients in a narrowband.

A Visual Basic script was written to perform all GA operations and then imported to the HFSS. Numerous antennas have been simulated over several generations in the HFSS environment. The best fitness in the generation increases gradually as this problem is a maximization problem. Simulations have been carried out until the fitness converges, and the design with maximum fitness has been selected as the optimized antenna.

3 Results and Discussion

The optimized design has a smaller lower patch with a size of 16 × 12 mm². The substrates are having an air gap of 4.5 mm. The nonconducting slots have different sizes. The sizes of the lower slot and the upper slot are 4 × 2 mm² and 8 × 1 mm², respectively (Fig. 2).

The optimized stacked antenna has a $S_{11} < -10$ dB bandwidth of 3.4 GHz in the frequency range of 3.9–7.3 GHz (Fig. 3). Therefore, the fractional bandwidth is 60.7% showing broadband performance. The optimized antenna resonates at four frequencies (i.e., 4, 4.9, 5.9 and 6.6 GHz) having the following reflection coefficients, respectively (i.e., −14, −15, −27 and −25 dB). Due to multiple resonances at frequencies close to each other, the optimized antenna shows wideband performance. The current patterns on patches are different from each other at different resonant frequencies (Fig. 4).

The radiation characteristics of the optimized antenna have also been explored. The antenna radiates along the broadside direction throughout the bandwidth (Fig. 5). The radiation pattern at each 1 GHz interval is shown in Fig. 5. Gain along the broadside direction is about 5 dB within the bandwidth.

The resonant behavior of the optimized microstrip antenna is compared with that of a conventional stacked antenna, which has two patches with similar sizes. Further, it contains no slots. It resonates at two frequencies with narrowband performance (Fig. 6). When a lower patch with the optimized size is used, the upper band shows broadband performance. The lower patch with both the optimized size and the optimized slot improves the resonant behavior slightly. However, still the antenna does not have a continuous frequency band. In contrast, the optimized design has no discontinuities in the operating bandwidth. Hence, simultaneous optimization of several antenna parameters has helped to design a broadband stacked antenna.

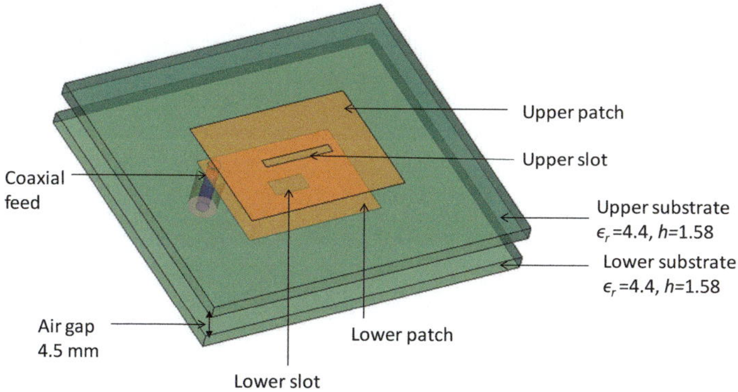

Fig. 2 Optimized antenna

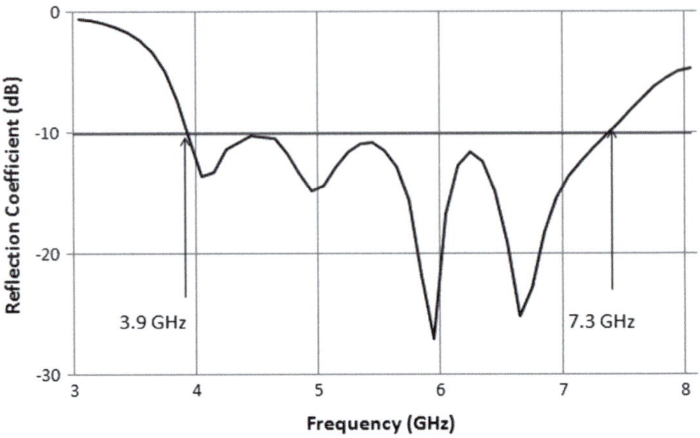

Fig. 3 S_{11} plots

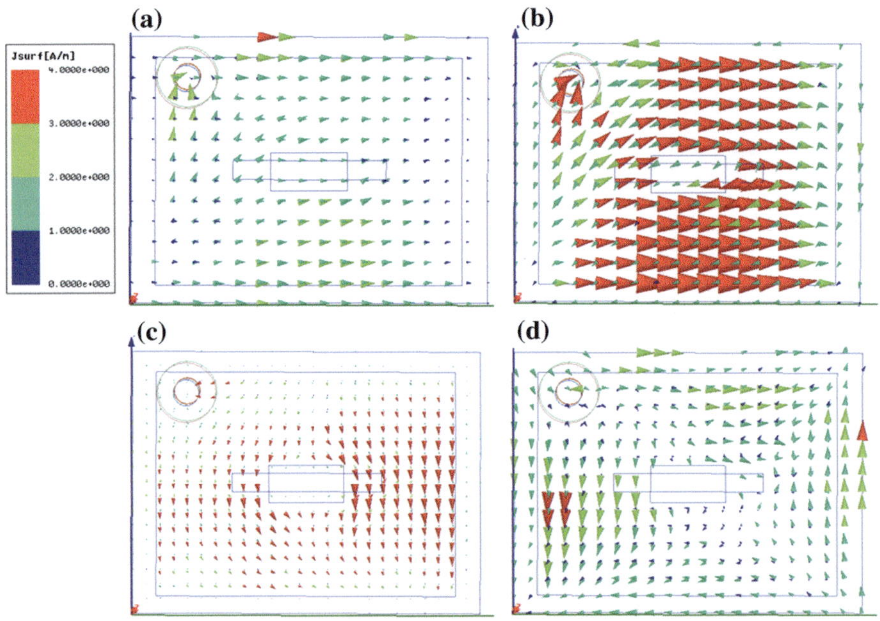

Fig. 4 Current patterns at resonant frequencies **a** 4 GHz, **b** 4.9 GHz, **c** 5.9 GHz, **d** 6.6 GHz

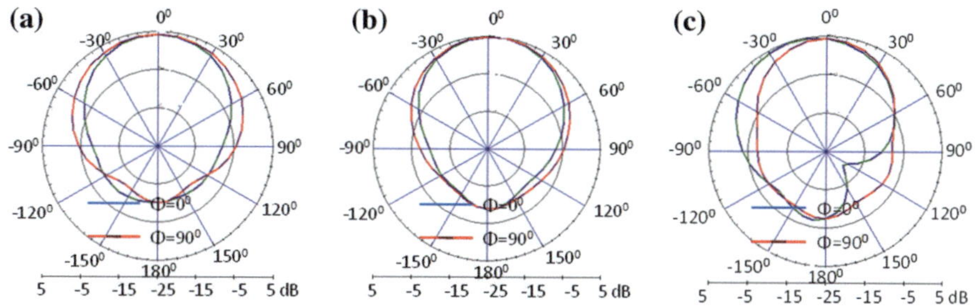

Fig. 5 Radiation behavior at different frequencies **a** 4 GHz, **b** 5 GHz, **c** 6 GHz

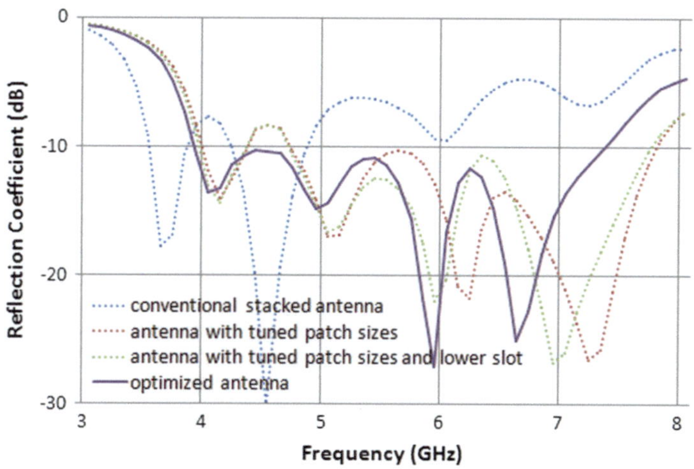

Fig. 6 Comparison of S_{11} plots of antennas with different configurations

4 Conclusion

This paper presents a novel stacked microstrip antenna with two dielectric substrates separated by an air gap. Size of only one patch is fixed, while several other parameters such as the lower patch size, slot dimensions, air gap size, and feed position have been optimized. The optimized antenna has a fractional bandwidth of 60.7% and broadside gain of about 5 dB throughout the band.

References

1. Balanis C.A. 2005. *Antenna Theory and Design*. 3rd ed. New York: Wiley.
2. Jayasinghe, J.W., J. Anguera, and D.N. Uduwawala. 2012. A simple design of multi band microstrip patch antennas robust to fabrication tolerances for GSM, UMTS, LTE, and

Bluetooth applications by using genetic algorithm optimization. *Progress In Electromagnetics Research M* 27: 255–269.

3. Jayasinghe, J.M.J.W., and D.N. Uduwawala. 2013. Optimization of the performance of patch antennas using genetic algorithms. *Journal of National Science Foundation* 41 (2): 115–122.
4. Anguera, J., L. Boada, C. Puente, C. Borja, and J. Soler. 2004. Stacked H-shaped microstrip patch antenna. *IEEE Transactions on Antennas and Propagation* 52 (4): 983–993.
5. Ansari, J.A., P. Singh, S.K. Dubey, R.U. Khan, and B.R. Vishvakarma. 2008. H-shaped stacked patch antenna for dual band operation. *Progress In Electromagnetics Research B* 5: 291–302.
6. Faisal, G.M., K.M. Quboa, and D.M. Ali. 2014. Quad-band dual-layer microstrip antenna design for mobile handset. *American Journal of Electrical and Electronic Engineering* 2 (2): 51–56.
7. Ansari, J.A., and B.R. Ram. 2008. Broadband stacked U-slot microstrip patch antenna. *Progress In Electromagnetics Research Letters* 4: 17–24.
8. Matin, M.A., B.S. Sharif, and C.C. Tsimenidis. 2007. Probe fed stacked patch antenna for wideband applications. *IEEE Transactions on Antennas and Propagation* 55 (8): 2385–2388.
9. Ooi, B., S. Qin, and M. Leong. 2002. Novel design of broad-band stacked patch antenna. *IEEE Transactions on Antennas and Propagation* 50 (10): 1391–1395.
10. Khodier, M., and C.G. Christodoulou. 2000. A technique to further increase the bandwidth of stacked microstrip antennas. *Antennas and Propagation Society International Symposium, IEEE* 3: 1394–1397.
11. Jayasinghe, J.M.J.W., J. Anguera, and D.N. Uduwawala. 2015. On the behavior of several fitness functions for genetically optimized microstrip antennas. *International Journal of Scientific World* 3 (1): 53–58.
12. Mishra, Raj Gaurav, Ranjan Mishra, and Piyush Kuchhal. 2017. Design of broadband monopole microstrip antenna using rectangular slot. In *Proceeding of International Conference on Intelligent Communication, Control and Devices, Volume 479 of the series Advances in Intelligent Systems and Computing,* 683–688. ISBN: 978-981-10-1707-0. http://dx.doi.org/10.1007/978-981-10-1708-7_78.
13. Mishra, Ranjan, Jeevani Jayasinghe, Raj Gaurav Mishra, and Piyush Kuchhal. 2016. Design and performance analysis of a rectangular microstrip line feed ultra-wide band antenna. *International Journal of Signal Processing, Image Processing and Pattern Recognition* 9 (6): 419–426. https://doi.org/10.14257/ijsip.2016.9.6.36.
14. Mishra, Ranjan, Raj Gaurav Mishra, and Piyush Kuchhal. 2016. Analytical study on the effect of dimension and position of slot for the designing of ultra wide band (UWB) microstrip antenna. In *5th IEEE International Conference on Advances in Computing, Communications and Informatics (ICACCI)*, 978-1-5090-2028-7, Sept 2016.

A Novel Distance Authentication Mechanism to Prevent the Online Transaction Fraud

Vipin Khattri and Deepak Kumar Singh

Abstract Online money transaction or online banking is an electronic transfer system that enables customers of a financial institution to perform financial transactions on a Website or application operated by the same or any other institution. After the growth of financial institution and also in their consumer base, online transaction has gained lots of transaction. Customer now can use the details of their credit card, debit card, or any other secure credentials for online transaction. The frauds related to online transactions are also growing at the same pace as the online transaction itself. Besides all security measures, various types of frauds have been reported for online payment. To prevent these frauds, researchers are constantly working to enhance the security measures for online transactions. Plethora of the literature is available regarding the same. The purpose of this paper is to propose a new authentication mechanism for committing a valid and secure online transaction. This mechanism will be applicable for credit card/debit card both whether it is used for internet shopping, point of sale, or money transfer. The basic idea is that precaution is better than cure. It means that a strong authentication mechanism for online transaction should be implemented so that fraud could not be committed.

Keywords Online transaction · Online transaction fraud detection
Two-factor authentication · Multifactor authentication · OTP (one-time password)
Credit card fraud · GPS (global positioning system) authentication

---

V. Khattri
Faculty of Computer Science & Engineering, Shri Ramswaroop Memorial University, Barabanki, India
e-mail: vipinkhattri@gmail.com

D. K. Singh (✉)
Department of Computer Science, Integral University, Lucknow, India
e-mail: deepak.iiita@gmail.com

N. A. Siddiqui et al. (eds.), *Advances in Fire and Process Safety*,
Springer Transactions in Civil and Environmental Engineering,
https://doi.org/10.1007/978-981-10-7281-9_13

1 Introduction

The main motive of introducing the online transaction is to perform transaction from anywhere, i.e., without going to point of sale or financial institution and anytime at the customer's convenience using their credential. This comfort has a heavy cost in terms of online transaction frauds. Consequences of these frauds are loss of money, trust, and effort [1]. Fraudster always tries to commit fraud in different ways to escape being nabbed. To prevent the online transaction frauds, various authentication techniques [2–4] and fraud prevention techniques [5–23] have already been developed. Although many security measures exist and different systems adapt different technique, committing online transaction fraud are increasing at alarming rate. In view of this, the present work proposes a novel authentication technique before committing the valid online transaction. This paper is divided into four sections. Section 2 contains different frameworks and issues related to authentication for online transaction. In Sect. 3, author proposes a new global positioning system (GPS)-based authentication technique using smart phone. Section 4 contains the case studies of online transaction fraud and impact of propose mechanism on these case studies. Section 5 contains the conclusion and future work.

2 Different Frameworks and Issues

2.1 Different Frameworks

There are various techniques implemented to prevent the online transaction fraud. These techniques are broadly classified in two ways:

Authentication before the Commit of the Transaction

Two-factor authentication or multifactor authentication is a prevalent mechanism that adds some additional steps to the current procedure of authentication [24–26]. The main aim of multifactor authentication is to confirm that only authenticated user could perform transaction and prevent the unauthorized user. Earlier, one factor uses the static information of user credential such as password which is provided by the user during each transaction and this static information can be hacked. As a result, an unauthorized user can commit the fraud. For making transaction more secure, two and multifactors of different type have been introduced. The information for second or third factor is provided by the user, and the value of these factors is dynamic in nature, i.e., on every transaction it gets changed. Second or multifactor authentication such as one-time password, visual cryptography, fingerprint, graphical password, SMS, security tokens, and smart phone authentication is used in various online transaction applications.

Kulat et al. [2] implemented a novel approach for preventing online transaction fraud, i.e., dual-level security mechanism using OTP keeping in mind the comfort of user. In this mechanism, authentication is implemented based on cookies on a user machine and OTP generation. The study of Jamdar et al. [3] implemented a secure authentication mechanism of using fingerprint to detect online transaction fraud. As per research, during authentication of fingerprint phase, fingerprint will be matched with the previously stored pattern; if match not found then transaction is considered as suspicious, and this suspicious fingerprint is recorded to identify the fraud in future. The paper of Akole et al. [4] implemented a secure transaction using visual cryptography for detection of fraud during online transaction. As per research, the authentication image is broken into number of shares. Separately shared image reveals no data, but when you combine these separate images, by stacking the share, then secret image is reconstructed and reveals data, and this data is used for authentication.

Authentication during the Commit of the Transaction

To validate the transaction to be genuine during the commit of the transaction, various techniques are used such as artificial neural network [5–11], hidden Markov model [12–15], genetic [16–18], Bayesian [19, 20] artificial immune system [21, 22], data mining [23]. The study of Agrawal et al. [12] implemented fraud detection by analyzing the spending behavior and shopping behavior of the customer and classify the pattern of behavior of the customer using hidden Markov model and then using genetic algorithm to find the fraudulent transaction. The study of Behera et al. [5] analyzed the transaction pattern and using fuzzy c-means clustering algorithm, and classified it. These patterns compare the current transaction with previously stored transaction to find the fraudulent transaction. Singh et al. [19] implemented Bayesian learning to classify the current transaction into fraudulent or non-fraudulent transaction by considering the parameter such as location and spending behavior. This spending behavior is classified using k-means clustering algorithm and detected the fraudulent transaction. The study of Renuga Devi et al. [27] implemented the Naïve Bayesian classifiers and random forest approach. In this approach, author analyzed the behavior of customer in terms of transaction for fraudulent pattern.

2.2 *Issues with Existing Techniques*

There are some flaws or gray areas in existing techniques, in which improvements are required. Therefore, researchers have to find new techniques or upgrade existing techniques. Research quoted some cases which have some issue in existing techniques.

Case I: One-Time Password: This technique is used by all the financial institutions for authenticating the transaction. In this, fraudster can acquire the user credential using phone fraud then fraud will be committed.
Case II: Fingerprint: Biometric authentication is very useful and very secure technique, but it has serious drawback that any fraudster can steal fingerprint using glass and form a fake fingerprint. In this case, if your password is hacked then you can change it, but if your fingerprint is hacked then how will you change your fingerprint.

3 Proposed Mechanism

The main aim of this research is to implement strong authentication technique to stop most of the fraudulent online transactions. The authentication system proposed will not only increase the security but will also be easy to use. In this mechanism, smart phone is used as a second/third factor, as it has become an integral part of human life. The use of smart phones has doubled in last two years with total users are in the tune of 250 million (Fig. 1) [28].

The use of smart phone has penetrated deeply in almost all sections from less privilege to privilege class and all geographical regions (from urban to suburban and rural India). By using smart phones, user generally navigates number of important activities which are ranging from education, health, entertainment, job to finance. More than 50% user uses their smart phone for online baking.

In this mechanism, smart phone and transaction devices are required; in the absence of any device, transaction will not be committed. This mechanism can be seen as one lock with two keys. Availability of both the keys at same place and time is mandatory to open the lock.

3.1 Integral Part of the Mechanism

- **Smart Phone with GPS activation application**:
  Every user who is registered with financial institution and wants to commit online transaction should have smart phone with GPS activation application. This application will be designed for the users of the financial institutions. The application should be installed in the smart phone of the registered user and must be used before any online transaction, using some authentication code. The function of this application is to enable (on) GPS of user's smart phone before initiating online transaction, and user could use this facility during online transaction for authentication.

- **Transaction Device**:
  This device can be a smart mobile phone, desktop/laptop, or point of sale device, etc., this device is used to make online transaction.
- **ATM/Credit/Debit Card**:
  The card should be issued by the financial institution as per requirement of the user so that user could perform online transaction using mobile phone, desktop/laptop, or point of sale device.

3.2 Minimum Requirements

- User should have an account in the financial institution.
- User should be registered to acquire the facility of secure authentication through GPS for online transaction
- Mobile phone number should be registered with the financial institution.
- GPS activation mobile application should be installed in the user's mobile.
- ATM/debit/credit card should be issued to user by financial institution.

3.3 Authentication Procedure

The basic idea behind the proposed methodology is to get the location of user using his or her smart phone and match it with the location of point of sale (PoS) used for transaction. The key assumption is that the user is keeping his mobile phone near him at the time of transaction. To stop fraud transaction after smart phone theft, several options can be used like blocking the transaction or taking preventive measures like securing the phone. The smart phone application proposed may consist of fingerprint verification, as most of the high-end phones coming today are equipped with fingerprint sensor.

Fig. 1 Number of smart phone users

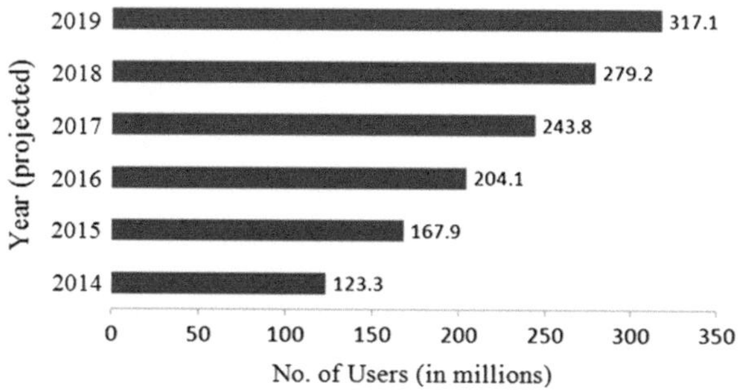

Step 1: Using GPS activation mobile application, activate GPS with some security code.
Step 2: Initiate online transaction by providing some credential as per financial institution requirement.
Step 3: Confirm all the details and found ok.
Step 4: Authentication system of financial institution will find the location of mobile phone and location of transaction device.
Step 5: Find the distance between the mobile phone and transaction device using latitude and longitude received by both mobile device and device used for transaction
Step 6: If distance is less then critical value (decided by financial Institution: Fig. 2) then authentication is approved and then go for review transaction **else** Transaction will be stopped and consider as fraudulent transaction (Fig. 3)
Step 7: STOP

The flow chart of the process is depicted in Fig. 4. This system will have remarkable advantages against almost all the type of fraud during online transaction.

Stolen Card:

In case of only stolen card, online transaction fraud will not commit because GPS authentication is required, and in the absence of GPS application, the transaction process will not be completed.

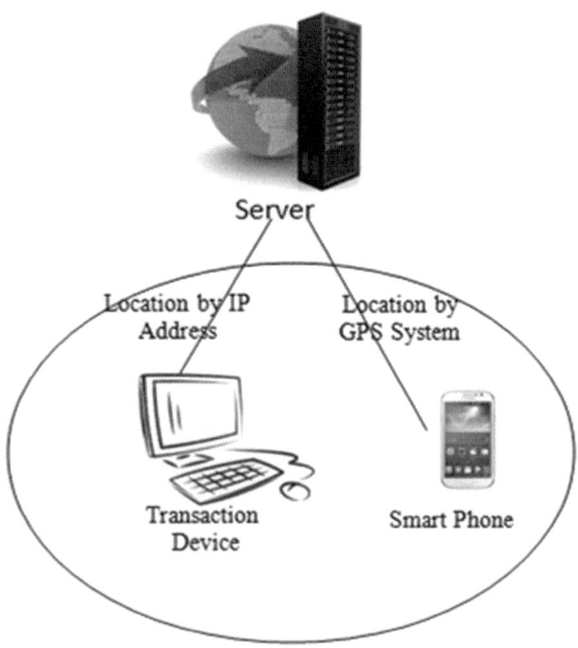

Fig. 2 Valid authentication distance between transaction device and smart phone within the range

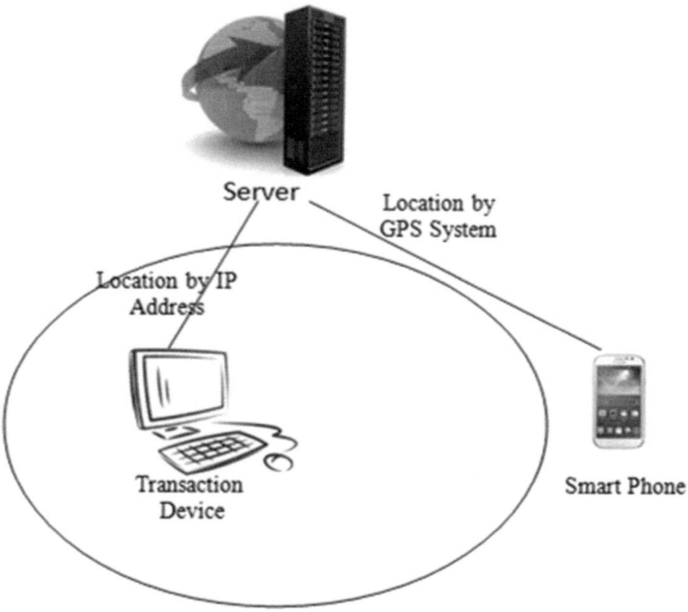

Fig. 3 Not valid authentication distance between transaction device and smart phone not within the range

Skimming Card:

In case of skimming card, online transaction fraud will not commit because GPS authentication is required, and in the absence of GPS application, the transaction process will not be completed.

Card Number Generator:

In case of creating fraudulent valid credit card, online transaction fraud will not commit because GPS authentication is required, and in the absence of GPS application, the transaction process will not be completed.

Phone Fraud:

In case of providing all the details of your credentials (including OTP and PIN) to the fraudster, online transaction fraud will not commit because GPS authentication is required, and in the absence of GPS application, the transaction process will not be completed.

If GPS activation mobile application is on in the mobile phone of the registered user, transaction will not be committed because distance between location of transaction device and location of mobile phone will be mismatched.

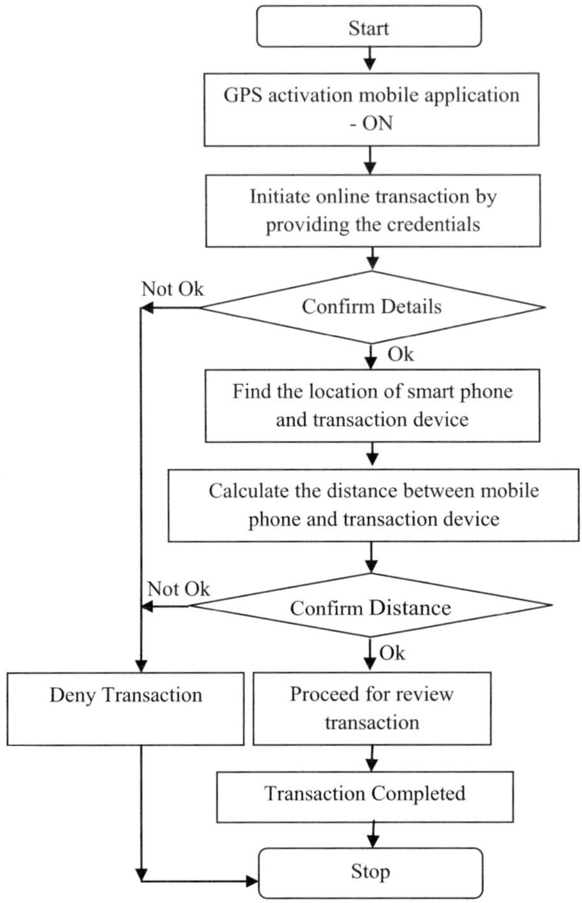

Fig. 4 Authentication procedure

In this system location of mobile phone and transaction, device plays a major role. This system will not require any additional cost, time, and effort of financial institution and user. Just require GPS activation mobile application. Any user can handle these things.

4 Case Studies of Online Transaction Fraud and Impact of Propose Mechanism on Online Transaction Fraud

In the following section, case studies have been discussed which shows the benefits which a user may get after the implementation of the proposed mechanism.

4.1 Case Study: Phone Fraud

This is a real-life incidence which happened with a citizen of small city of Uttar Pradesh (India).

A phone call was made by the fraudster in the morning at 7:30 AM to Mr. XYZ who was in hurry for the office. Conversation between fraudster and Mr. XYZ was as follows:

Fraudster: I am from Bank ABC and your ATM card has been blocked and Rs. 5000/- penalty has been imposed. To validate kindly check SMS on your mobile.

XYZ: What should I do?

Fraudster: Please do not worry. Do you agree to unblock the ATM Card?

XYZ: Yes

Fraudster: Tell me your ATM Card Number

XYZ: 5xxxxxxxx3819 (Fig. 5)

Fraudster: Tell me your ATM Card Expiry Date

XYZ: 05/35 (Fig. 5)

Fraudster: Tell me your CVC Number

XYZ: Please tell me your identity how can I believe on you because you are asking confidential details.

Fraudster: Sir this is part of validation. For confirming please check your mobile to receive a message from Bank ABC

XYZ: A message is received from Bank ABC
CVC Number: xx9

Fraudster: One more message you will receive in which OTP will be mentioned

XYZ: Yes I have received

Fraudster: Tell me your OTP Number

XYZ: 145512 (Fig. 6)

Fraudster: Your ATM card has been unblocked Thank you.

Fig. 5 ATM debit card

Fig. 6 SMS with OTP

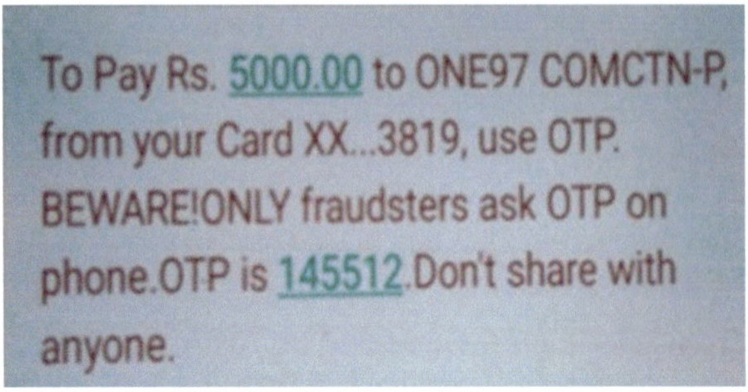

Result:

Online fraud transaction has been committed and Rs. 5000/- should be deducted. But due to less amount in the account, Rs. 5000/- could not be deducted and bank charges (Fig. 7) is imposed on the account holder XYZ.

Discussion:

In case of sufficient amount in the account, the amount will be deducted, and in case of insufficient amount in the account, the bank charges will be deducted.

4.2 Case Study: Skimming Card

This is a real-life incidence which happened with a citizen of small city of Uttar Pradesh (India)—published in newspaper (November 2015).

In a City XYZ, one family went to the restaurant for dinner. They had their meal and the payment was made by the ATM card but they did not know that ATM card was skimmed by the fraudster (restaurant owner), i.e., all the details (ATM card number, expiry date, CVC number, PIN number) was acquired by the fraudster. After payment, family went away. Fraudster used this skimmed card for online transaction.

18.06.16	10000053250297061661001371 7CANC710.00		710.00	892.83Cr
25.06.16	INTEREST CREDIT		74.00	966.83Cr
05.07.16	INSUF BAL POS DECLINE CHARGE-050716 TRANSFER TO 098353032238	17.00		949.83Cr

Fig. 7 Bank charges for decline payment ATM debit card

Result:

Online fraud transaction has been committed and loss of money from the account holder.

Discussion:

In case of sufficient amount in the account the amount will be deducted, and in case of insufficient amount in the account, then bank charges will be deducted.

4.3 Implementation of Proposed Mechanism on Both the Above-Mentioned Cases

Using multifactor authentication mechanism, i.e., smart phone with GPS Application, the distance between the transaction device of fraudster and user's smart phone will not be within the range (set by bank) and authentication will be failed before committing the transaction and bank will deny the online transaction. As a result, fraud will not be committed although all the details (ATM card number, expiry date, CVC number, PIN number, and OTP number) of user were acquired by the fraudster. Finally, no loss will be imposed on account holder.

5 Conclusion and Future Work

Huge amount of online transaction frauds are reported. This is due to either the user's mistake or intelligence of fraudster. This fraud gives the result in terms of loss of money, trust, and time. To prevent this type of online transaction fraud, various new techniques are evolved. These techniques have been implemented before commit of the transaction such as two/multifactor authentication technique and during the commit of the transaction such as analyzing the shopping behavior of the customer using hidden Markov model. Researchers are trying to reduce this type of fraud by evolving new techniques or upgrade the existing techniques.

This paper focuses on new technique by using multifactor authentication, i.e., smart phone as a third factor before the committing of the transaction. Benefit of new proposed authentication technique will increase the security level and prevents the fraud during online transaction without hampering or modification in existing system.

Future research includes the implementation of this authentication technique which is applicable for multifactor authentication with financial institution and analyzes the impact and accuracy. It also includes the development of new techniques to prevent the online transaction fraud during the commit of online transaction.

References

1. Bhatla, T.P., V. Prabhu, and A. Dua, 2003. Understanding credit card frauds. Cards business review. 1 (6).
2. Kulat, A., R. Kulkarni, N. Bhagwat, K. Desai, and M.P. Kulkarni. 2016. Prevention of online transaction frauds using OTP generation based on dual layer security mechanism. *International Research Journal of Engineering and Technology (IRJET).* 3 (4): 1058–1060.
3. Jamdar, P.A., S.K. Bendale, and D.A. Durgawale. 2016. Secure mechanism for credit/debit card transaction fraud detection by using fingerprint authentication system. *International Journal of Computer Science and Network Security (IJCSNS).* 16 (3): 35–38.
4. Akole, P., N. Mane, K. Shinde, and A. Swati Khodke. 2016. Secure transactio: An credit card fraud detection system using visual cryptography. *International Research Journal of Engineering and Technology (IRJET).* 3 (4): 1719–1724.
5. Behera, T.K., and S. Panigrahi. 2015. Credit card fraud detection: A hybrid approach using fuzzy clustering & amp; neural network. In *Advances in Computing and Communication Engineering (ICACCE), Second International Conference on IEEE,* 494–499.
6. Bekirev, A.S., V.V. Klimov, M.V. Kuzin, and B.A. Shchukin. 2015. Payment card fraud detection using neural network committee and clustering. *Optical Memory and Neural Networks.* 24 (3): 193–200.
7. Carneiro, E.M., L.A.V. Dias, A.M. da Cunha, and L.F.S. Mialaret. 2015. Cluster analysis and artificial neural networks: A case study in credit card fraud detection. In *2015 12th International Conference on Information Technology-New Generations (ITNG) IEEE,* 122–126.
8. Van Vlasselaer, V., C. Bravo, O. Caelen, T. Eliassi-Rad, L. Akoglu, M. Snoeck, and B. Baesens. 2015. APATE: A novel approach for automated credit card transaction fraud detection using network-based extensions. *Decision Support Systems* 75: 38–48.
9. Khan, A.U.S., N. Akhtar, and M.N. Qureshi. 2014. *Real-time credit-card fraud detection using artificial neural network tuned by simulated annealing algorithm*, 113–121. Telecommunication and Computing, ITC: In Conf. on Recent Trends in Information.
10. Brause, R.W., T.S. Langsdorf, and H.M. Hepp. 1999. Credit card fraud detection by adaptive neural data mining. In *Proceedings of the 11th IEEE International Conference on Tools with Artificial Intelligence*, 103–106.
11. Aleskerov, E., B. Freisleben, and B. Rao. 1997. Cardwatch: A neural network based database mining system for credit card fraud detection. In *Computational Intelligence for Financial Engineering (CIFEr), Proceedings of the IEEE/IAFE, 1997,* 220–226.
12. Agrawal, A., S. Kumar, A.K. Mishra. 2015. Implementation of novel approach for credit card fraud detection. In *Computing for Sustainable Global Development (INDIACom), 2nd International Conference on IEEE, 2015,* 1–4.
13. Prakash, A., and C. Chandrasekar. 2015. An optimized multiple semi-hidden Markov model for credit card fraud detection. *Indian Journal of Science and Technology.* 8 (2): 165–171.
14. Prakash, A., and C. Chandrasekar. 2012. A novel hidden Markov model for credit card fraud detection. *International Journal of Computer Applications.* 59 (3): 35–40.
15. Khan, M.Z., J.D. Pathan, and A.H.E. Ahmed. 2014. Credit card fraud detection system using hidden Markov model and K-clustering. *International Journal of Advanced Research in Computer and Communication Engineering.* 3 (2): 5458–5461.
16. Assis, C.A., A. Pereira, M.A. Pereira, and E.G. Carrano. 2014. A genetic programming approach for fraud detection in electronic transactions: In *Computational Intelligence in Cyber Security (CICS), 2014 IEEE Symposium on IEEE,* 1–8.
17. RamaKalyani, K., and D. UmaDevi. 2012. Fraud detection of credit card payment system by genetic algorithm. *International Journal of Scientific & Engineering Research.* 3 (7): 1–6.
18. Duman, E., and M.H. Ozcelik. 2011. Detecting credit card fraud by genetic algorithm and scatter search. *Expert Systems with Applications* 38 (10): 13057–13063.

19. Singh, P., and M. Singh. 2015. Fraud detection by monitoring customer behavior and activities. *International Journal of Computer Applications* 111 (11): 23–32.
20. Panigrahi, S., A. Kundu, S. Sural, and A.K. Majumdar. 2009. Credit card fraud detection: A fusion approach using Dempster-Shafer theory and Bayesian learning. *Information Fusion.* 10 (4): 354–363.
21. Halvaiee, N.S., and M.K. Akbari. 2014. A novel model for credit card fraud detection using artificial immune systems. *Applied Soft Computing* 24: 40–49.
22. Soltani, N., M.K. Akbari, and M.S. Javan. 2012. A new user-based model for credit card fraud detection based on artificial immune system. In *Artificial Intelligence and Signal Processing (AISP), 16*th *CSI International Symposium on IEEE, 2012*, 029–033.
23. Philip, N., and K.K. Sherly. 2012. Credit card fraud detection based on behavior mining. *TIST International Journal for Science, Technology Research* 1: 7–12.
24. Kennedy, E., and C. Millard. 2016. Data security and multi-factor authentication: Analysis of requirements under EU law and in selected EU Member States. *Computer Law & Security Review.* 29,32 (1): 91–110.
25. Aloul, F.A., S. Zahidi, and W. El-Hajj. 2009. Two factor authentication using mobile phones. In *AICCSA*, 641–644.
26. Zhang, F., A. Kondoro, and S. Muftic. 2012. Location-based authentication and authorization using smart phones. In *IEEE 11th International Conference on Trust, Security and Privacy in Computing and Communications*, 1285–1292.
27. PATTERN, T.B.O.B. 2014. Fraud detection in card not present transactions based on behavioral pattern. Journal of Theoretical and Applied Information Technology 61(3).
28. www.eMarketer.com: https://apsalar.com/2016/03/mobile-and-smartphone-usage-statistics-for-india/. 2015.

Fire Safety Management in India: A Review

Ayush Vidyadharan, Joji John, Cherish Thomas and Bikarama Prasad Yadav

Abstract Fire is something that can be expected at any structure, may be at your home or at your workplace, or in a hospital, or in public places, almost anywhere. Fire in any kind of occupancy would certainly have the potential to cause harm to its occupants and severe damage to property. Due to fire and related causes in India itself, almost 25,000 fatalities are occurring every year. On an average, around 21 males and 42 females die each day because of such fire and similar causes and thereby made the concept of fire safety management an area of concern for Indian scenario. Fire safety management in building basically is a day-to-day management concept for aligning a building's fire safety procedures in place so that they can be used at the time of need. This is moreover a responsibility of each and every individual to restrict any flaws in the fire safety procedures and to prevent any fire occurrences from happening and as being a human being to protect the lives in such situations. No matter how good active and passive fire protection systems you are having in place, it all becomes useless once the people who are supposed to use it are unaware of how to use them in need and are unaware of the significance of the use of such systems and their role in fire prevention and suppression. For solving all the above-stated problems, a proper fire safety management program should be established in the building in order to get all the things right in the case of a fire outbreak. For the purpose of ensuring a 100% safety from fire in the building, we would say a healthy fire safety management system is essential. Various views on fire safety management are reviewed in this paper. The applicability of this management system is different in case of existing buildings and new buildings. Engineering strategy is applied at the design level of fire safety provisions. A fire manual with clearly stated objectives should be given. In case of existing buildings, a ranking system should be set in place in order to check the present fire safety provisions, and at the same time, new fire codes must be tallied for active fire protection system.

A. Vidyadharan · J. John · C. Thomas · B. P. Yadav (✉)
Department of Health, Safety and Environment Engineering, University of Petroleum and Energy Studies, Dehradun, India
e-mail: bpyadav@ddn.upes.ac.in

N. A. Siddiqui et al. (eds.), *Advances in Fire and Process Safety*, Springer Transactions in Civil and Environmental Engineering, https://doi.org/10.1007/978-981-10-7281-9_14

Along with this, a safety management system for fire is to be brought into the picture. A fire safety plan should be made with fire safety design methodology.

Keywords Safety management system · Fire manual · Fire safety plan

1 Introduction

The importance of fire safety management has been playing a vital role in fire safety in building [1]. A fire grading report (Post-War Building Studies No. 29) prepared in 1952 in UK [2] gave us an overview by providing us with the first relevant document on fire safety management. The following were described:

- Regular inspection of all doors, passages and staircases.
- All provisions for means of escape to be effectively maintained.
- A major concern should be given to the external facilities that can create an unsafe environment due to external exposures.

Findings and suggestions to the management, and evacuation procedures in case of fire are given in the appendices of fire precautions in the design and construction of buildings, Part 2: Code of practice for shops, British Standards Institution, London, UK (1985) and [3]. The key points, as described in the two standards, are summarized below [3, 4]:

- Provide effective guidance and support to management on deployment of personnel for various domains such as fire safety, training and education, record keeping, notice preparation.
- Undertaking suitable measures in case of fire, evacuation procedures, etc.
- To train the mind in such a way that it should react rationally when it confronts with a dangerous fire event or situation.

A fire safety plan has to be developed by keeping in mind the occupancy and size of the building.

It is important to involve fire safety engineers in fire safety management [5]. And at the same time, they should play their part in designing the fire safety plan and implementing engineering performance-based fire codes [6] as they are the competent (authorized) persons. The familiarization of the fire safety management with fire safety design philosophy is necessary. It is also necessary to ensure that everyone is evacuated on time; this can be made possible by having a fully effective and efficient fire emergency plan in place. This should give a cover to everyone: let it be employees, any service users and even the visitors who are around. And this should not just be it, it should also consider people having physical disabilities, mobility, sensory or some learning and hearing impairments and all such issues which can affect their ability to move on stairs or else hinder their evacuation. The emergency plan should be so that evacuation should not totally depend on fire and rescue department but can be done even without them around.

2 Recent Views

Among some of the most important aspects of fire safety and protection in building is a fire safety management. "Clear listing of Management" in BS5588: Part 11 on "Codes of practice in design offices, industrial, storage and other similar type of buildings" in 1997 [7]. Effective management having appropriate staff training and knowledge is important in taking right actions and having the occupants evacuate safely. The management will be totally responsible for the maintenance of all the fire protective passive building designs like means of escape, and similarly, the active fire protecting systems also like the fire extinguishers which are portable the warning systems, emergency lighting facilities, instruction manual to staffs as per Indian norms, Basically, the following were discussed [2]: for carrying out these things in the right manner, it is to be ensured that a safety management plan is in place for the building. This would include matters related to:

- Requirement of testing, review and regular maintenance for fire safety equipment.
- Inside the premises, the contractors who are working.
- If any alterations have been brought in the building, the procedures to be followed in that case.
- Emergency procedures (including evacuation management).
- Routine precautions/testing.
- Methods by which the fire alarm usage can be minimized.
- Commissioning and handover of fire safety installation, fire safety manual
- Staff training.
- Evacuation procedures and responsibilities of each staff.
- Extension and alterations.
- Monitoring and review of fire safety manual.
- Fire risk assessments carrying out responsibilities and
- Special events to be carried out by specific procedures.

Following points are also highlighted for consideration [2]:

(a) The goal of safety management program for fire is to maintain and also to improve the fire safety measures and precautions within the structures. The preventive measures against fire and the activities to be taken care of, in an outbreak of fire event, should be properly made aware to the inmates of the building through awareness programs and mock drills. In order to prevent the loss of life and minimize the damage to property, as a result of accidents occurring due to fire, the role of a fire safety manager is vital. As the fire safety manager is the right person who is in charge to guide the firefighting team to the zone which needs attention, he should have adequate fire-fighting training. For special buildings with higher occupancy, fire safety manager having work experience in that area should be preferred.

(b) The provision of a competent person is to be made mandate, to keep and maintain a fire safety manual, at the areas of work. The following should be included in the manual:

- fire safety planning measures, design of system and constructions—their role in the overall evacuation process, importance of safety management and its applications;
- Records of documents submitted during the design stage, and changes if any, for employing several kinds of fire protection system, for various scenarios;
- The design stage documentation or any alterations that are approved after designing, for using different types of protection system for various circumstances;
- Duties and roles assigned to management and staff;
- Building alteration work records;
- Fire routine;
- Updated drawings of the buildings and identification and classification of different smoke control zones, fire detection zones; updated drawings of all fire precaution measures.

Recent reviews [8] stated that various areas of hazards that the fire safety management is associated to are, people in and around a building, the components within and the building itself. It is very important to life safety as confirmed by many, multi-fatality fires [9] are result of failure in adopting the right decisions.

SMS ensures in utilizing the opportunity to the core so that overall safety can be assured. As mentioned in [10], fire safety management is considered critical and integral for a successful fire safety engineering design and development. Qualitative design review (QDR) team should encompass personnel answerable to fire safety management. Timely audits are to be performed in order to check the efficiency of the system. It is the responsibility of QDR team to prepare a fire safety manual. Annex D (Information) of a guide [11] contains the list of things that should be included in the fire safety manual.

The strategies laid down in the safety management regulation are able to create various levels of SMS to be established in an organization. Discussions on fire safety management were done on application of fire performance concepts to design objectives [12]. It was mentioned that fire safety management procedures play an important function in the areas of fire prevention, fire control, occupant's evacuation and safety system's maintenance. On periodic basis, audits which are independent in nature, based on the procedures of management in the field of fire protection and prevention, are to be carried out. Guidance on key aspects is given in Annex C of that guide [12]. Dailey has given a guide for the people responsible for managing fire in the work areas [13]. An approach which is systematic to fire safety was integrated in the national fire protection association (NFPA) and fire safety concept tree [14].

American Society of Safety Engineers, USA published a handbook on fire safety management [15], in which the fire safety concepts are followed. Implementation and development of an efficient fire safety management will:

- Lower the premiums dealing with property insurance.
- Realize quality gains.
- Enhance customer service and goodwill.
- Prevent business interruptions.
- Hugely affect the profitability of an organization.
- Develop an efficient work environment.

We should see to it that a safety management program for fire should be done in a well-organized way. Action plan's sequence for developing an effective program is:

- Access requirements and capabilities.
- Effectiveness evaluation.
- Identify, evaluate and enforce life safety, controls for prevention and protection against fire.
- Careful analysis of facilities.
- Careful analysis of fire hazards.

It should be divided into eight elements:

- Timely Inspection and maintenance.
- Personnel training and education.
- Mechanisms for suppressing fire.
- Services for emergency.
- Fire possibility evaluation.
- Fire prevention.
- Keeping of records.
- Effective Communication.

All details above mentioned are based on performance. If the organization has met that, it ensures the accomplishment of objectives by the organization.

3 Objectives of Fire Safety Management

The following are the main objectives that have to be ensured from a safety management for fire [1]:

- To make all the fire safety measures provided available at all times.
- Ease with which the occupants can use the fire safety measures.
- Assistance to occupants will be provided to escape to a safer place.

If the above conditions are not met [1], it would imply that the management had not done enough to ensure the proper evacuation of the occupants, which would result in loss of lives. This has been the case with all the big fire occurrence in both indoors [9, 16].

There are three main roles that fire safety management has to play:

- The fire safety measures and devices provided are kept in good condition, and this should be ensured.
- In case of fire, actions are to be initiated that will save occupants and would help to shift them to safer place.
- Whenever we come across a change of building, a change in use or new technology in fire services installations, the adequacy of existing safety measures should be reviewed and modified.

All fire safety requirements and provisions should be maintained properly and are to be ensured by fire safety management. Assistance from fire brigade is another area that needs to be maintained. Fire fighters are to be informed of the existing fire protection system and should be provided with guidance on the site. The maintenance of a safety management system would also include measuring and checking or auditing; this is the way a contractor will get to know if the safety management of industry of his relevance is correctly working or not, and at the same time, he can know the scope of improvements also. All of this would make the system more effective and efficient.

4 The Fire Safety Plan

A fire safety plan is to be prepared in fire safety management program [1]. A fire safety plan is a written procedure that includes plan to enhance safety in workplace which covers all the actions that the employers and employees must practice to ensure safety of every person in the workplace during a fire incident. The fire safety plan should mainly consist of:

- Maintenance strategies—for proper inspection and maintenance of fire safety and firefighting systems.
- A fire action plan that is well-defined and necessary actions to be carried out in case of fire.
- A training plan to train the staffs on various fire scenarios.

A good maintenance plan should provide with necessary services for maintaining a well-functioning and safe working of firefighting appliances at the time of fire. A maintenance plan should include:

- Maintenance of an active fire protection system like sprinkler systems, detectors, fire hydrants and extinguishers.
- Maintenance of a passive fire protection system like fire doors and escape routes.

- Proper inspection of the system performance and effectiveness in regular intervals.
- Occupants should be accessible to all information and drawing on warning signs, layout and escape routes.
- Good housekeeping like ensuring zero obstacles in emergency escape routes and proper disposal of garbage.

A staff training plan should contain:

- Training regarding the proper use of fire firefighting equipment.
- Training on general fire dynamics for effective evacuations.
- Description on roles and responsibilities of each employee.
- Fire warden.
- Guiding of occupants to safe zones during fire.

An effective training, which includes fire drill, provides employees with information on their surroundings, different safety systems installed in the building, providing awareness of all hazardous conditions that can lead to fire and how to react to such outbreak of fire.

An effective fire action plan should include:

- Assembling and leading the occupants to safe zones.
- Attacking the fire.
- Assisting the fire marshal.
- Reporting to the person in charge.
- Fire evacuation strategy.
- Roll call.
- Access to the firefighting equipment.

In addition to the safety plans mentioned above, a "fire prevention plan" is also to be considered [8]. The objective of an effective fire prevention plan is to prevent the occurrence of fire in a workplace. It can point out hazardous substances or materials on the site that can be a source to the occurrence of a fire and also the different firefighting necessities that are to be installed in the workplace to ensure control on fire from spreading. The written plan should be made available to all employees in the workplace for reviewing. However, an employer in charge of less than 10 or equal to 10 numbers of employees can communicate to the employees regarding the safety plan orally.

A fire prevention plan must necessarily include:

- Data on all the major hazards that can lead to fire, procedures for safe storage and handling of hazardous substances, potential sources of ignition, and fire protection and fighting equipment necessary to mitigate each and every major hazard [29 CFR 1910.39(c)(1)].
- Methodologies for controlling the accumulation of combustible- and flammable-natured hazardous material wastes [29 CFR 1910.39(c)(2)].

- Regular interval procedures for maintenance and inspection of safeguards installed on equipment producing heat that can lead to fire accident [29 CFR 1910.39(c)(3)].
- The details of employees maintaining the equipment to control or prevent fire sources [29 CFR 1910.39(c)(4)].
- The details of employees who are responsible for controlling the hazards related to fuel source [29 CFR 1910.39(c)(5)].

This prevention plan can help in identifying and maintenance of potential substances that could be a source of ignition or can help in restricting the use of combustible materials which may lead to rapid spreading of fire upon ignition. This fire prevention plan can be good for bigger organizations like airport authorities, railway organizations and tunnel managements in preventing occurrence of fire and resulting disasters.

Fire safety plan should have two modes of operation:

- Normal mode which includes maintenance plan, training plans and fire prevention plans.
- Emergency mode which includes fire action plan that helps in effective firefighting.

5 New Buildings and Integration with the "Engineering" Approach

During the concept and design stages of buildings, issues were raised regarding the importance of integrating fire safety management into fire safety engineering, an engineering approach on safety design [17]. Using advanced technique such as fire models in the design stage was an effective approach [18, 19]. It is very important that a fire safety engineer should understand how the effectiveness of the safety system design is being influenced by fire safety management as the fire safety provisions might not be maintained. The main duties of fire engineer are to take care of the active protection systems, passive protection systems, building designs and to determine the effectiveness of fire safety management while designing safety systems. In case of a fire outbreak, a fire safety manager has to carry out the safety tasks in normal and emergency mode. A fire safety manager has the power to delegate intermediate staffs for carrying out the duties if the building management team is too big.

In practice, the fire safety engineer determines the work of the manager for fire safety. The fire safety engineer and the fire safety manager should have good knowledge of fire dynamics. The main duties or works are determined in three aspects [5]:

- Regular inspection regarding the effectiveness of the passive building safety designs [20].
- Proper testing and maintenance of active fire protection system.
- Managing of the building in conformation of fire safety designs like controlling fire and occupancy load.

6 Existing Buildings

Buildings that are constructed at different times followed the fire safety regulations of that time, respectively. Internationally, certain fire codes are accepted and implemented in the passive designs of building and systems for active fire protection include the means of access code and of escape code, fire resisting construction code and fire service installations code.

In an existing building, the assessment of their fire safety aspects, both passive and active, was done by fire safety ranking system based on the current fire codes. These rankings were useful in identifying deviations of the fire safety provisions from the codes and hence application of appropriate fire safety management scheme in case of errors. However, the applications of these codes are still difficult to satisfy all the requirements. Hence, the fire codes on basis of engineering performance for designing perfect fire safety provisions [21] were being proposed as an alternative solution. But the development of these new codes will be time consuming, and the cost will be high [22].

7 Conclusion

An effectively implemented fire safety management ensures total safety to buildings. However, these safety managements are not under local building regulation's control. The fire codes used for the safety provisions for the new buildings are yet to be implemented. A reasonable standard for fire safety in buildings can be incorporated by the integration of fire safety management in the safety regulations followed by the existing buildings [23, 24]. The reason that leads to such changes was to reduce the fire brigade's workload in important situations and to also increase the involvement of building management's participation in fire safety. The fire precaution regulations implemented in 1997 in UK included the fire safety management.

Local government's interest in building management and fire safety has gradually increased. The Home Affairs Department can provide with the concept of effective building management. But regarding the fire safety management schemes, further studies are required.

In fact, the total fire safety system that includes designs, construction and management of active and passive fire safety provisions [25] should be properly worked out. During this designing period, a fire safety manager should be appointed with responsibility in taking necessary actions in the two modes, normal and emergency, of operation. The main areas of concern for fire safety designs are:

- Old high rise buildings.
- Higher education systems.
- Airport and railway terminals.
- Tunnels: both for trains and vehicles.
- Shopping malls.

Appointing an authority to take care of fire safety management will be appreciable, despite the hardship of training such persons for carrying out the duties.

A firm legislative control and regular inspection on fire safety management can ensure an effective implementation of these safety plans.

References

1. Malhotra, H. 1987. Fire safety in buildings–building research establishment report. Borehamwood, Herts, UK. Department of the Environment, Building Research Establishment, Fire Research Station.
2. Thompson, P.A., and E.W. Marchant. 1995. A computer model for the evacuation of large building populations. *Fire Safety Journal* 24 (2): 131–148.
3. British Standard, B. 1983. *5588: Fire precautions in the design and construction of buildings. Part 3 Code of practice for office buildings*. British Standards Institution.
4. Chow, W. 2005. Building fire safety in the Far East. *Architectural Science Review* 48 (4): 285–294.
5. Chow, W.K., and G.C. Lui. 2002. A proposed fire safety ranking system for karaoke establishments and its comparison with the NFPA-fire safety evaluation system. *Building and Environment* 37 (6): 647–656.
6. Babrauskas, V. 2016. *Heat release rates*, 799–904. SFPE handbook of fire protection engineering: Springer.
7. Thompson, P.A., and E.W. Marchant. 1995. Testing and application of the computer model 'SIMULEX'. *Fire Safety Journal* 24 (2): 149–166.
8. Chow, W. 2001. Discussion on applying the American fire safety evaluation system for business occupancies in Hong Kong. *International Journal on Engineering Performance Based Fire codes* 2 (4): 92–97.
9. football stadium bibliography.pdf.
10. DD240, B. 1997. *Fire Safety Engineering in Building, Part 1: Guide to the Application of Fire Safety Engineering Principles*. British Standard Institution.
11. Richardson, J.K., and M. Seaton. 1997. Performance-based fire codes: Why we need them… How we make them work. *NFPA Journal* 91 (1): 72–77.
12. ISO, T. 1999. 13387: Part 1: Fire safety engineering-the application of fire performance concepts to design objectives. International Organization for Standardization (ISO).
13. Dailey, W. 2000. A guide to fire safety management. Perpetuity Press.
14. Chow, W. 2001. Fire safety requirements in karaokes: Comments on the new karaoke establishment bills. *International Journal on Engineering Performance-Based Fire Codes* 3 (2): 59–66.

15. Chow, W. 2001. Instant responses–On the attack fire at World Trade Centre. *International Journal on Engineering Performance-Based Fire Codes* 3 (3): 128–129.
16. Report of the Tribunal of Inquiry on the fire at the Stardust, Artane on the 14th February, 1981.
17. Buchanan, A. 1999. Implementation of performance-based fire codes. *Fire Safety Journal* 32 (4): 377–383.
18. Forney, G.P., and W.F. Moss. 1994. Analyzing and exploiting numerical characteristics of zone fire models. *Fire Science and Technology* 14 (1/2): 49–60.
19. Yang, K. 1999. Role of fire field models as a design tool for performance-based fire-code implementation. *International Journal on Engineering Performance-Based Fire Codes* 1 (1): 11–17.
20. Chow, W. 1998. Numerical studies on recent large high-rise building fire. *Journal of architectural engineering* 4 (2): 65–74.
21. Hadjisophocleous, G.V., et al. 1998. Literature review of performance-based fire codes and design environment. *Journal of Fire Protection Engineering* 9 (1): 12–40.
22. Tubbs, B. 1999. How far have we come? The ICC's perspective. In *Proceedings of the Second Conference on Fire Safety Design in the 21st Century.*
23. UK Public general acts, fire protection act 1971.
24. Building Regulations and Fire Safety—Procedural Guidance.pdf.
25. Stollard, P. 1994. *Design against fire: An introduction to fire safety engineering design.* London: Spon Press.

Analysis of Fire Protection Facilities in Hospital Buildings

B. Abhishek Shastri, Y. Sivaji Raghav, R. Sahadev and Bikarama Prasad Yadav

Abstract Hospitals are considered to be safer places where injuries are treated rather than generated. Even though hospitals follow safety measures and monitor the reports of accidents happened, there is always a boundless chance for a new accident to happen because hospital environment is charged with compound inherent risks. Apart from risks, hospitals are a combination of numerous hazards in different parts of the building like biological, chemical, physical, and fire hazards [1]. There have been major incidents in hospitals reported in the past, including the fire accident that occurred in AMRI, a hospital located in Kolkata, which killed almost 90 people in the year 2012 [2]. It is one of the recent accidents happened in hospitals in India. In the morning of May 15th, 1929 Cleveland hospital in USA encountered a massive fire accident followed by poisonous gas release from burning nitrocellulose X-ray film ignited by a nearby light bulb [3]. Around 125 people died in this disaster since there was no proper evacuation strategy and lack of fire protection systems. Fire accidents taking place in hospitals are not easily controllable, and casualties are more because the evacuating techniques are typically diverse as most of the patients cannot move out by themselves. And if the accident happens nearby anesthesia care [4] or intensive care, patients inside the wards may not be aware of fire and could not respond properly to call for help. This is why a fire accident is always very precarious in a hospital. This paper emphasizes about the hazards associated with fire in the hospital environment, proper evacuation in case of fire, and appropriate fire fighting equipment. Evacuation plan and fire extinguishing equipment should be easily approachable, readily usable, and maintained properly in regular intervals, and competent persons in evacuation techniques should be available. The constructions of buildings also play a significant role in case of evacuation. Proper guidelines followed during the construction of a hospital building can also help in avoiding various hazards which may turn into

B. Abhishek Shastri · Y. Sivaji Raghav · R. Sahadev · B. P. Yadav (✉)
Department of Health, Safety and Environment Engineering, University of Petroleum and Energy Studies, Bidholi, Dehradun 248007, India
e-mail: bpyadav@ddn.upes.ac.in

N. A. Siddiqui et al. (eds.), *Advances in Fire and Process Safety*,
Springer Transactions in Civil and Environmental Engineering,
https://doi.org/10.1007/978-981-10-7281-9_15

disasters. The aim of this paper is to identify hazards and provide long-term and short-term control measures to deliver safety for the staff members and the patients who have the right to be treated with dignity and also protect them during any unexpected situations.

Keywords Hospital fires · Evacuation · Emergency · Fire safety
Fire hazards · Healthcare

1 Introduction

The very first requirement in a hospital is that it should do the sick no harm. In order to fulfill this, administration in hospitals needs to conceptualize and regularize certain set of criterions to function and should not, in any situation, deviate from safety standards depicted by governing bodies. Starting from the construction of hospitals to their functioning, several procedures are to be followed to avoid hazards, specifically fire. Hospitals have variety of materials that can contribute toward a happening of fire. And the persons inside a hospital also are of various physical and mental characteristics that have dissimilar reactions toward fire. Considering the general population inside a hospital, it constitutes of patients ongoing treatment under supervision. While some patients can rationalize a situation, there are some patients who cannot [5]. Evacuation during fire accidents becomes difficult for these patients. In the past, it has been witnessed that hospitals are susceptible to fire hazards and are proven to have tedious evacuations with high fatalities.

This paper is a quantitative and qualitative approach toward possible fire scenarios in hospitals, incidents, and mitigating techniques. The purpose of this piece is to deliver long-term as well as short-term solutions to fight fire hazards in hospitals.

Effective safety management system requires implementation of meticulously planned logistics and readiness in case of an emergency [6]. The past accidents happened in hospitals reflect poorly on the implementation of proper safety protocols and inadequacy of emergency response (Table 1). Safety must be made as a value for the organization.

2 Methodology

This investigation focuses on minor data significant to fire accidents in hospitals globally. A survey over mishaps in hospital fires till the present scenario was done through rereading the internet, articles, newspapers, investigation, and other reports. In the history of fire building, heterogeneous fire incidents were perceived. A detailed study on fire incidents of 13 hospital buildings were used here. All the incident cases were constructed on some norms to be designated. For example, one

Table 1 Prominent hospital fires

Date	Hospital	Place	Injuries and deaths
February 18, 1923	Manhattan State Hospital [7]	New York City	25 deaths
1948	Highland Hospital [8]	Asheville, North Carolina	9 deaths
April 4, 1949	St. Anthony's Hospital [9]	Effingham, Illinois	70 deaths
January 7, 1950	Mercy Hospital [10]	Davenport, Iowa	41 deaths
December 23, 1956	Doctor's Memorial Hospital	Minnesota	8 deaths
July 14, 1960	Guatemala Mental Hospital [11]	Guatemala	225 deaths
1945	Hartford Hospital [12]	Connecticut	20 deaths
February 12, 1968	Shelton Hospital [13]	Shrewsbury	21 deaths, 14 injured
July 5, 1972	Coldharbour Hospital [13]	Sherborn, England	30 deaths
December 9, 2006	Moscow Hospital [14]	Moscow, Russia	46 deaths
April 26, 2013	Moscow Psychiatric Hospital [11]	Moscow, Russia	36 deaths
2013	Psychiatric Hospital in Luka [11]	Novgorod Region, Russia	37 deaths
2014	A hospital in Jangseong Country	South Korea	22 deaths

of the norms included higher rate of injuries and casualties. As a result, number of people were killed and injured during fire incidents happened. Also, the hospitals while handling fire accidents have to meet the faults and glitches. In Fig. 1, the blue area denotes three fire accidents before 1950; the red area denotes six fire accidents which occurred between 1950 and 2000; and the green area denotes four accidents chosen after 2000.

3 Evaluation and Outcomes

Considering all the 13 incidents, administrations of these hospitals took similar erroneous actions in fire safety management. A total of eight mistakes were excerpted by reviewing these fire accidents [11]. The mistakes include the following:

1. Absence of mechanized fire fighting systems.
2. Non-compliance of law enforcement.

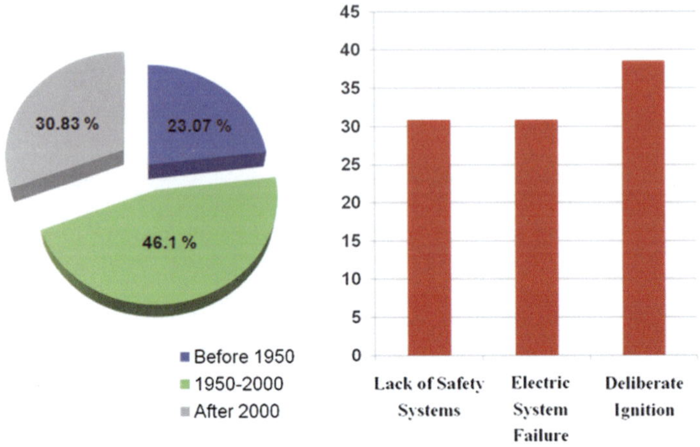

Fig. 1 Accidents considered for study

3. Myopic planning.
4. Maintenance and management of fire fighting appliances.
5. Incompetency of hospital staff regarding safety issues.
6. Combustible materials used and stored in the building.
7. Poor accident management methodology.
8. Legal and administrative bottlenecks regarding security and safety systems.
9. Insufficient or non-availability of mechanical aids for patients during emergency evacuation.
10. Inability of patients to evacuate the building [15].

Prior to the year 1950, it was found that the foremost reason of higher number of victims in hospital fire incidents which had occurred was due to the lack of installation of automatic fire fighting appliances. Fire at early stages is sensed by automatic sprinklers, and if occurred, smothering is done with extinguishing media. In simple arguments, automatic systems alarm the inhabitants in the building or public, once they sense the source of fire. They offer a longer timescale for inhabitants to evacuate the building. Also, the extinguishing media captivates nitrous fumes. Hence, water spray by automatic sprinklers is efficient in decreasing the degrees of risk from this danger. Water spray cool down combustible substances below their ignition points and provides high level of moisture, although only an insignificant proportion of the fresh gases are captivated in water. Similarly, there was no provision of suppression systems, fire and smoke detectors, and alarm systems. Hospitals equipped with automatic fire system always have a good chance of escaping damage. Unfortunately, there were problems in planning maintenance and management of fire safety systems [16].

Health and Safety Management System Management Assessment Tool:

Management leadership: Demonstrate commitment to improve safety and health.
Employee participation: Encouraged to communicate openly with management and report safety and health concerns.
Hazard identification and assessment: Processes and procedures are in place to continually identify workplace hazards and evaluate risks.
Hazard prevention and control: Processes, procedures, and programs are implemented to eliminate or control workplace hazards.
Fire Safety and emergency awareness using education and training: On hazard recognition and control.
System evaluation and improvement: To check the effectiveness of the implemented system, its performance should be monitored in order to improve health and safety in organization.

These six health and safety management system assessment tools can be effectively and efficiently implemented to fulfill the needs of organisations.

4 Solutions

- Hospitals being most sensitive to hazards and accidents, fire safety needs to be implemented since the time of construction according to the National Building Code and other safety standards.
- Eliminating storage of hazardous waste and use of hazardous gases or chemicals inside the hospital by controlling from an isolated area.
- Every hospital building must have at least four emergency exits and a maximum of six [17].
- High rise buildings should be easily reachable by road [18].
- Equipment operation and maintenance through regular inspection.
- Hospital buildings must be inspected in every 6 months, and the inspection process should be filmed.
- An air ambulance must be available, helps in rescuing large number of people.
- A map should be present in all the rooms of the hospital guiding the people about their location.
- Automatic sprinkler and alarm systems shall be installed according to NFPA 14.
- Length of the hospital main entrance should be at least 30 m.
- Hospitals buildings must be earthquake proof.
- Provision of underground ground water should be present so that water can be drawn in case emergencies to extinguish or control the fire.
- All the walls in the hospital buildings shall be coated with fire-resistant coatings.
- Canteen buildings should be constructed away from the hospital.

4.1 Evacuation Duties

A competent person shall be assigned the authority for evacuation process and he shall be responsible for executing the evacuation plan and taking onsite decisions [19]. Specific departments must also have evacuation responsibilities. And the responsibilities are discussed below [20];

Employees of the Hospital

If case of a fire accident, employees should not wait for the instructions and shall take away the patients in immediate vulnerability. Patients shall be moved to a nearby harmless zone on the same level if there is a facility for horizontal evacuation. If patients are not within the reach of direct danger and the alarm is active, employees shall wait for evacuation orders. Patients should never be left unattended. For reporting pre-assigned disaster response assignments, the employees must ensure that patients under someone else's care should be taken charge of before they go to report. For an instance, conducting proper hand-off before the patient is left alone for any purpose during emergency [21].

Security Department and Human Resource Department

Workers of the hospital's security unit shall instantly initiate a head count and communicate orders required under emergency situations. Radio or telephone can be used as a means of communication. A variety of duties for the security head includes but not limited to:

- Ensuring the presence of officers in the anterior entry of the hospital building to facilitate the responders and lead them to the hazard.
- Ensuring the presence of officers to look after the entrances/exits and initiate lock-down processes assigned for the building.
- It is to be verified that officials follow the manuals given for security and emergency operations processes.
- Patients and visitors along with the staffs should be addressed with intercoms or other alternate communication facilities.

A representative out of the labor pool shall be appointed to communicate the building's emergency operations unit. During the disaster situation, normal visiting hours should be suspended. It is expected that a number of people including family members, visitors, and nearby residents will volunteer their help during an emergency. The human resource department of the hospital shall undertake the responsibility to cope up with the influx and allocate workers to list the helpers and also allot them to specific areas [22]. A variety of duties of a planning officer shall include:

- Communicating with the representatives assigned at the hospital for head counts under emergency and hand out emergency orders.
- Control centers and staging areas to be managed and established for everyone inside the building (patients, visitors, workers, volunteers, students, etc.).

- Recording fitness levels of the volunteers assigned to assist patients up and down the floors using stairs.
- Ensuring that the assigned workers are providing the emergency unit with details of family members of the patients and volunteers that are in the facility and assigning to stay with relatives of victims in the hospital waiting area.
- Facilitate sign-in of volunteers with their credentials, contact details, and list any special talents (especially fluency in other languages) [23].

5 Summary

There are several concerns in fire safety of a hospital building. Provision of fire prevention and protection plans must be prepared and communicated to prevent losses due to fire. According to the studied data, few hospitals with high-tech protection systems were not competent enough in fire safety management. Faults and difficulties identified in these fire accidents are as follows:

1. Absence of mechanized fire fighting systems.
2. Non-compliance of law enforcement.
3. Myopic planning.
4. Maintenance and management of fire fighting appliances.
5. Incompetency of hospital staff regarding safety issues.
6. Combustible materials used and stored in the building.
7. Poor accident management methodology.
8. Legal and administrative bottlenecks regarding security and safety systems.

It can be concluded that hospital management has a prime responsibility toward safety of patients by implementing control measures to fire accidents. Impact of fire could be reduced with proper and ideal safety management in hospital building. Every single employee of a hospital shall follow her/his responsibility during fire emergencies and try to save as many potential victims as possible without waiting for instructions.

References

1. Gestal, J. 1987. Occupational hazards in hospitals: Accidents, radiation, exposure to noxious chemicals, drug addiction and psychic problems, and assault. *British Journal of Industrial Medicine* 44 (8): 510–520.
2. Nagral, S. 2012. Fire in a hospital. *Indian Journal of Medical Ethics* 9: 76–77.
3. Gregory, K.L., V.F. Malinoski, and C.R. Sharp. 1969. Cleveland clinic fire survivorship study, 1929–1965. *Archives of Environmental Health: An International Journal* 18 (4): 508–515.
4. Muchatuta, N.A., and S.M. Sale. 2007. Fires and explosions. *Anaesthesia & Intensive Care Medicine* 8 (11): 457–460.
5. Childers, A. 2010. *Prioritizing patients for emergency evacuation from a healthcare facility.*

6. Boyd, A., et al. 2014. Emergency planning and management in health care: Priority research topics. *Health Systems* 3 (2): 83–92.
7. Abrahams, R. 1929. Tribute to Dr. Gustav Scholer. *Psychiatric Quarterly* 3 (1): 131–132.
8. Tekulve, S. 2014. Guests on earth: A novel by Lee Smith (review). *Appalachian Heritage* 42 (1): 109–114.
9. Groves, A. 2006. *St. Anthony's Hospital Fire, Effingham-April 4, 1949.*
10. Robert Solomon, P. 2012. To lock or not to lock? *Environment of Care® News* 15 (8).
11. Ong, W.C., and Z. Suleiman. 2015. Fire safety management problems in fire accidents in hospital building. *Advances in Environmental Biology* 43–47.
12. Riccitelli, M., and I.J. Kennedy. 1961. The historical development of proprietary nursing homes in Connecticut. *Journal of the American Geriatrics Society* 9 (8): 705–713.
13. Bowes, P. 1976. Casualties attributed to toxic gas and smoke at fires: A survey of statistics. *Medicine, Science and the Law* 16 (2): 104–110.
14. Ong, W.C., and Z. Suleiman. 2015. Problems in implementation of fire safety management in Malaysia government hospital. *Advances in Environmental Biology* 47–51.
15. Zhu, K., Y. Yang, and Q. Shi. 2016. Study on evacuation of pedestrians from a room with multi-obstacles considering the effect of aisles. *Simulation Modelling Practice and Theory* 69: 31–42.
16. Wabo, N.C., P. Örtenwall, and A. Khorram-Manesh. 2012. Hospital evacuation; planning, assessment, performance and evaluation. *Journal of Acute Disease* 1 (1): 58–64.
17. Bailey, Norman T.J. 1956. Statistics in hospital planning and design. *Applied Statistics* 5 (3): 146.
18. Hardy, O.B., and L.P. Lammers. 1977. *Hospitals, the planning and design process.* Aspen Publishers.
19. Albert, D.E., Innovalarm Corporation, 2007. Enhanced fire, safety, security and health monitoring and alarm response method, system and device. U.S. Patent 7,173,525.
20. Malhotra, H.L. 1987. *Fire safety in buildings.* Borehamwood: Department of the Environment, Building Research Establishment, Fire Research Station.
21. Robson, L.S., Judith A. Clarke, Kimberley Cullen, Amber Bielecky, Colette Severin, Philip L. Bigelow, Emma Irvin, Anthony Culyer, and Quenby Mahood. 2007. The effectiveness of occupational health and safety management system interventions: a systematic review. *Safety Science* 45 (3): 329–353.
22. Cagliuso, N.V., N.E. Leahy, and M. Sandoval. 2010. Developing the hospital emergency management plan. *Health Care Emergency Management: Principles and Practice', Jones and Bartlett, Sudbury*, MA, 89–110.
23. Fernandez, L.S., 2007. *Volunteer management system design and analysis for disaster response and recovery.* The George Washington University.

Need for Better High-Rise Building Evacuation Practices

Suvek Salankar, S. M. Tauseef and R. K. Sharma

Abstract There are various egress components for high-rise building evacuation during an emergency. It is recommended to use these components depending on the type of scenario, but it is important to note that there are certain advantages and disadvantages associated with these egress components. The concept of deciding suitable fire safety system of the building, i.e., available safe evacuation time (ASET) and required safe evacuation time (RSET) is very important while deciding their use. Apart from these egress components, there are few important evacuation strategies which can be used during building evacuation. To study different characteristics of evacuation of the building using various egress components, few computerized fire models are available, but we need further better evacuation strategies, considering the limitations associated with these egress components.

Keywords High-rise building · Evacuation · Egress component ASET · RSET · Evacuation strategies · Fire models

1 Introduction

Due to the commercial and educations development in large cities, urban population and population density are increasing and being the only available solution, high-rise building is becoming popular day by day. Modern day construction

S. Salankar (✉)
College of Engineering Studies, UPES, Dehradun, India
e-mail: suvek.salankar@gmail.com

S. M. Tauseef
Environmental Research Institute, University of Petroleum and Energy Studies (UPES), Dehradun 248007, Uttarakhand, India
e-mail: smtauseef@ddn.upes.ac.in

R. K. Sharma
India Glycol, Kashipur, India
e-mail: rajeevsharma.rk@gmail.com

N. A. Siddiqui et al. (eds.), *Advances in Fire and Process Safety*,
Springer Transactions in Civil and Environmental Engineering,
https://doi.org/10.1007/978-981-10-7281-9_16

technology aided by 3-D modeling using sophisticated software is also supporting this concept to enable maximum occupancy per unit ground area. In India, there will be 251 buildings which are above 60 M by the year 2018 and out of which 176 buildings will be in Mumbai only. Before the year 2000, there were only 26 buildings above 60 m [1]. This shows rapid increase and the demand for high-rise building. It is also essential to understand the meaning of high-rise building. But interestingly, there is no one specific definition of high-rise building. Various standards have defined 'high rise' by different ways [2]:

1. A building having 'many stories' as per the Oxford English Dictionary.
2. 'Higher than 70 feet (21 m)' as per US General Laws.
3. '75 feet or higher' measured from the lowest level of access to fire vehicle to the highest floor as per The International Building Code (IBC 2000), the Building Construction and Safety Code and NFPA of the Life Safety Code®, 2006 edition.
4. In India, there are different definitions as per different development control rules. For example, Thane Municipal Corporation considers any building with a height of 25 m [3] and as per the National Building Code of India, it is a building with a height of '15 m and above' [4].

Apart from 'high-rise building,' there are a few more terminologies of the buildings. As per CTBUH, i.e., Council on Tall Buildings and Urban Habitat [1],

1. Tall buildings are buildings with a height of 50 m or above.
2. Super tall buildings are buildings with a height of 300 m or above.
3. Mega-tall buildings are buildings with a height of 600 m or above.

As high-rise buildings are increasing rapidly all over the world, it is also carrying few concerns along with them. One of the prominent concerns is fire safety, and presently, saving human life and property due to fire is a major threat to such buildings. This is one of the major challenges which need to be resolved on priority. Functional diversification of such buildings is making firefighting and evacuation operations more difficult. There are various vertical components like stairways, elevator shaft, pipes, ducts, electrical shafts through which fire may spread very fast, if proper precautions are not taken. More over, the current cladding system increases the difficulties for the fire fighters to do fire fighting and rescue [5]. This may result in reduction in available evacuation time. It gets further complicated as normal elevator could not be used during the fire and stairs are the only channel for vertical evacuation.

As per available data, fire incidents are increasing year by year, and its impact on high-rise building is also increasing exponentially.

- In China, fire incidents in high-rise buildings show an upward trend because of the rapid development of economic construction. In 2011, residential building fires were a critical concern. Almost 39.7% of total fires occurred in residential buildings [6].
- In India, yearly about 25,000 people lost their lives due to fire and associated causes. According to the survey carried out by Pinkerton and FICCI in 2013,

8.45% of the overall ranking of risks was fire which is fifth highest risk in the industry [7].

- In the USA, out of 1,298,000 fires in 2014, 494,000 fires occurred in structures, which is a 1.3% increase from the year 2013. Of the structure fires, 367,500, i.e., 74% occurred in home structures, which is a 0.5% increase from the year 2013 [8, 9].
- In Brazil, though the numbers of fires in such buildings are less, major concerns are severity and the consequences of such fires. Also, there is no official figure available about existing buildings to compare the exact ratio [10].

2 Required Safe Evacuation Time (RSET)/Available Safe Evacuation Time (ASET)

Evacuation process is nothing but an escape movement that the occupants of the building make under emergency situations. To make this process safer, the basic requirement is to estimate the time when occupants can safely evacuate the building. The concept of ASET and RSET is widely used in such cases [11]. It is obvious that the safe design is established when RSET is less than ASET.

There are a few important factors to understand the concept the RSET (Fig. 1). They are

1. Ignition (IG): Time at which fire ignited.
2. Detection (DET): Time at which fire detected.
3. Alarm (AL): Time at which alarm activated.
4. Recognition (REC): Time at which, it is recognized by the occupants.

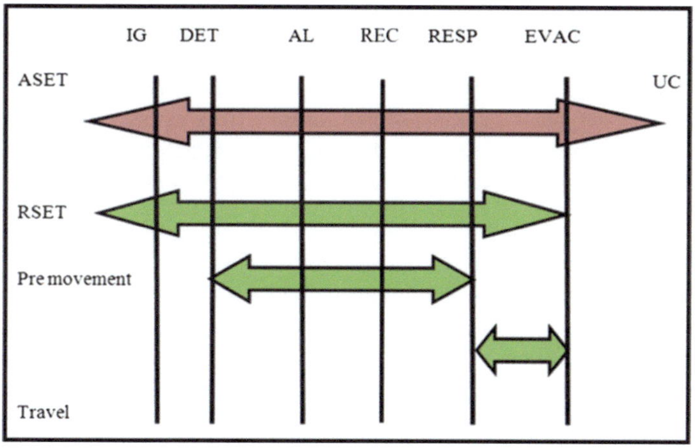

Fig. 1 Important factors of evacuation

5. Response (RESP): Time at which, it is responded by the occupants.
6. Evacuation (EVAC): Time at which, it is completely evacuated.
7. Untenable Conditions (UC): The time when the fire products like narcotic gases, irritant gases, heat, smoke, toxic gases kill the residents.

The delay in intimation and initiation plays a vital role in the evacuation. The margin between ASET and RSET is called 'safety margin', and this can be used to decide the life safety measures in buildings [12].

3 Available Egress Components and Its Challenges

The fundamental purpose of every egress component is safe evacuation of all the occupants during a fire emergency. In case of high-rise building, this problem is more prominent due to its typical characteristics which resulted in difficulty during firefighting, rescue, and evacuation. In the initial stage of fire, the horizontal smoke dispersion is at the rate of 0.3 m/s, and in the fully developed stage, it may be up to 3–4 m/s vertically. It means that if there is a fire to the 100 m high building, smoke will disperse up to the top floor through the vertical shafts in just 30 s [13]. So evacuation time is an important factor in recognizing the time required by the occupants for safe evacuation. Egress components can be of two types, vertical egress components and horizontal egress components. The vertical egress component is a vital factor for the building evacuation procedure but each component has its unique properties which may affect the evacuation process. Few important components are as follows [14].

3.1 Stairs

Stairs are the traditional and safest method to evacuate the building in case of fire emergency. There are two factors which need to be taken care. One is design factor of staircase, and the other is a behavioral factor of the occupants. Few important design factors of the staircases are its numbers, width, length, location, slope, capacity, ventilation, pressurization system, etc. [15, 16]. Few important behavioral related issues include,

- Ergonomics, motivation levels and group behaviors.
- Influence of gender or role from their behavioral perspective.
- Merging stream of evacuees at floor-stair landing.
- Fatigue to the occupants.
- People with disabilities [17].
- Counterflows of firefighters.

All these factors resulted in slow discharge time, which leads to an extensive queuing on the staircase including staircase entrance of each floor. It is also important to note that the average human speed on the stairs is from 0.52 to 0.62 m/s, and it is 0.45 and 0.43 m/s for kids and elderly people, respectively [18]. This low speed might adversely affect the evacuation process in terms of bottleneck, queuing, and stampede. People using wheelchair may also block the main staircase. Ventilation (natural or mechanical) also plays a vital role in the evacuation. If it is not maintained properly, it may increase the difficulties further. Unwanted material kept at staircase is also a common practice which blocks the access of evacuation, which may cause delay and also stampede.

3.2 Evacuation Elevators

The traditional concept of not using elevators in case of fire emergency has been challenged by the existing disadvantages of the staircase and need of faster and effective methods to evacuate high-rise buildings. It is the fastest evacuation method. In various standards, it is permitted and recommended to use elevators, for extreme conditions and with certain precautions [19]. Some of those are National Building Code of India, 2005, the Singapore Fire Safety Code-2013, the British Standard (BS 9999:2008 Code of practice for fire safety in the design, management and use of buildings), Building Construction and Safety Code: the European Standards EN 81, International Building Code (IBC) 2012, NFPA 101-2012: Life Safety Code and American Society of Mechanical Engineering, Safety Code for Elevators and Escalators.

But using elevators for evacuation has a few design and behavioral issues [20]. Some of the issues are

- The limited space in the lift car.
- Fire products like smoke, heat and flame may enter the elevator shaft.
- While elevator moves, smoke may get stuck inside the elevator due to negative pressure.
- Operation of an emergency power supply, water entry into the lift car, emergency communication systems, and spread of contaminants.
- Proper decision regarding pickup locations of elevators and a number of elevator stops.
- The willingness of people to use an elevator during a fire emergency.
- Limited capacity vis-a-vis huge occupant load increases waiting time for using elevators. This is proportional to floor levels and percentage of occupants to use the lift [21]

  To address this issue, the concept of zoning has been introduced. The zoning of the building can be done by dividing it into a certain number of floors where elevators can be allotted. Larger and faster shuttle elevators are generally provided to travel between the refuge floors and the street level without any openings on other floors. This can eliminate a few issues mentioned above.

Considering the increasing number of occupants in high-rise building, use of elevators is a good option for fast evacuation, and with good training, this can be achieved. A lift is normally used for people like disabled or aged occupants who cannot use stairs. It has been evident that roughly 3,000 lives were saved by using both lifts and stairs for WTC-2. There are few other incidents in which elevators were used by occupants for evacuation [22]. It is seen that by using lifts and stairs, 36% (110–70 min) of evacuation time can be reduced. It can be further reduced by up to 58% in the early stage if both lift and stair are used [23]. This strategy is further useful for the scenarios with larger number of occupants. There are a few other important factors like perceived risk, pre-evacuation delays, and actions which depend on gender, age, physical condition, education, knowledge, tenure in the building, location of fire floor, etc., which also need to be taken care [24].

3.3 Sky-Bridges

One of the possible methods is the introduction of a horizontal evacuation means at a certain height, i.e., the use of sky-bridges to link towers. This method has been already executed in several buildings. The immediate benefit of the sky-bridges is minimization of the vertical evacuation travel distance, but it is possible only if adjoining buildings are available. Sky bridges can be combinely used alongwith elevators and stairs but its effectiveness depends on infoamtion provided to evacuees [25].

3.4 Refuge Floor

Refuge floor plays an important role in evacuation, especially for disabled people. Following are few important advantages of refuge floor,

- It is a safe place of rest for the evacuees till further assistance arrives.
- Use to shelter injured and/or disabilities people.
- Use as a command point for emergency teams.
- The use of evacuation lifts from the refuge floor is easier since it acts as a pickup point for accommodating a significant number of evacuees, and the area is safe for the occupants to wait even in emergency situations. There are few factors which may cause failure to this concept, i.e., human behavior issues (overcrowding or nonuse), its use for any other purpose, its maintenance, etc. In commercial buildings, there is a concept of refuge area which is an extension to the normal floor plate.

3.5 Alternative Means of Escape

Use of Helicopters: Though it is one of the available options of evacuation, it is extremely dangerous for landing and the rescue operation due to the air turbulence and updrafts caused by smoke and heat. So it is not generally recommended for evacuation.

There are a few more types of means of escape, i.e., platform rescue devices, escape chute devices, and controlled descent devices [26].

Platform Rescue Devices (Fig. 2): It is an enclosed platform or set of platforms moving along the guides on the exterior of a building. It can be either mobile type brought by the emergency services when needed or permanent type installed on the building. By using this, many occupants can be evacuated at a single time, and special skill is also not required. So it is very useful for aged or physically disabled people. The important factor is, it should be properly maintained.

Controlled Descent Devices (Fig. 3): It is personal equipment which lowers the person at a controlled speed on the outer side of the building. Its operation is very simple to use, and electrical power is also not required. But for more occupancy it may not be very effective.

Escape Chute Devices (Fig. 4): It is a channel-shaped device of fire-resistant fabric. This is vertical or outward sloping. The sloping solution normally serves a specific floor, and the vertical solution normally enables evacuation from number of floors. It is also provided in the separate shaft near the staircase. It is very easy to use if properly maintained.

As per CTBUH global news, there are few more escape means. Mr. Zhou Miaorong, who is a retired operator in Shanghai, has invented a slide that allows high-rise residents to evacuate quickly and safely from their buildings (Fig. 5). When it is not in use, the slide can be folded up against the railing. Also, a

Fig. 2 Platform rescue device

Fig. 3 Controlled descent device

Fig. 4 Escape chute

Fig. 5 Evacuation slide

Fig. 6 Escape parachute

Panamanian inventor has demonstrated a parachute, which can open after just 100 feet to be used as an escape option (Fig. 6).

4 Current High-Rise Building Evacuation Strategy

Overall fire safety strategy involves two important factors: building performance and egress strategy. Building performance is further divided into structural performance, i.e., structural strength during the fire and compartmentation, i.e., fire spreads mitigation. Building evacuation strategy is related to the time required for safe evacuation of all the occupants. The building evacuation strategy can be total evacuation, phased evacuation, stay-in-place evacuation, or delayed evacuation [27]. The use of suitable strategy depends on the behavioral aspects of occupants and their preparedness, which in turn depends on the type of occupancies, for example, occupants in an office building or educational institute or assembly area are always ready to move but it is not always true for occupants in residential or hospital buildings. Occupants are familiar with escape routes in case of residential and an office building, but not the case for hotels, hospital, assembly buildings. Well-trained staff is available in office, hotels, and hospital buildings, but may not be in residential buildings. The building use is very important to predict the possible behavior of the occupants and to provide an adequate strategy [14]. The WTC incident showed the importance of a robust means of egress strategy and the need for further requirement. It is also the need to check if combination of different components is used for better results. Though suitable egress components of any high-rise building are very important, it is only the initial step for achieving an acceptable level of safety. The systematic use of the available egress components plays a fundamental role in building evacuation system. Strategies available are as follows.

4.1 Total Building Evacuation

This is an older, safest, and traditional way of building evacuation. This type of evacuation allows occupants of all the floors to be evacuated simultaneously. This is the simplest strategy to implement, but there are a few challenges, especially for a large number of occupants. It may result in an extensive queuing on the staircases, and it could take long time to descend downwards using stairs [23]. Also, people may cross the affected areas of smoke and may expose to more risk.

4.2 Phased Evacuation

It has been observed that the single-staged total evacuation is not always practicable in high-rise buildings. Therefore, phased evacuation is preferred where the occupants on the most critical floors and its nearby floors are given priority to evacuate first on staircases. The remaining occupants can be evacuated consequently as necessary. In this strategy, occupants at risk are removed quickly [28]. Its effectiveness depends on available fire protection installations, the training level of the people, and communication system of the building.

4.3 Stay-in-Place Approach

Normally, it is suggested that the occupants should stay in their locations, if all the exits from a floor are blocked during fire. They should close the door and seal the cracks and ask for help. This is very useful, especially for disabled people. It was suggested that this strategy is most appropriate if the building has the following main characteristics [27],

- The building is above six floors.
- The building is residential and has enclosed compartments.
- The building has non-combustible construction.
- An alarm and communication system for communication to occupants about the fire and to guide them time to time.

4.4 Delayed Evacuation

A delayed evacuation is suggested when evacuees are temporarily waiting in some identified dedicated area like refuge for help from rescuers. This strategy is very useful to rescue people with temporary or permanent disabilities because of their own limitations.

5 Computer Modeling for Fire

Computer modeling for fire is to predict fire and its related characteristics by the use of a mathematical method which is expressed as a computer program. It is categorized into fire modeling, detector response modeling, fire endurance modeling, egress modeling, and miscellaneous modeling [29].

5.1 Fire Modeling

Fire model predicts the fire process and its characteristics [30], such as temperatures, smoke obscuration, gas flow rates, heat fluxes, toxic gas, sprinklers and detectors activation time, reduction in strength of the building, structural failure of building elements. Fire modeling is of two types, physical fire modeling and mathematical fire modeling [31]. Physical fire modeling is actually burning the objects to evaluate their effects which can be either full scale or small scale. Mathematical models are sets of behavior of a physical system which can be divided into probabilistic model and deterministic models. Probabilistic models are on the basis of a series of sequential events without the use of physical and chemical equations describing the fire process. Deterministic models are on the basis of physical and chemical equations describing the fire process which is further divided into three categories [14, 32],

1. Hand calculations,
2. Zone models,
3. Field models.

Hand calculations are algebraic equations which are developed on experimental correlations. Zone models are the software for assessing fire dynamics. It is divided into two zones, generally referred as the upper and lower zones. Field models, also referred as computational fluid dynamics (CFD) models, separate a compartment into thousands of small cubes depending on user inputs. Fire modeling can be useful for the design and assessment of fire-protection systems.

5.2 *Egress Modeling*

Egress models envisage the time for occupants to evacuate the building. It is able to analyze the occupants' movement through complex structures. Few advanced egress models are able to import the actual building floor plans, and occupants' movement can be seen. The speed of occupants during egress and their interaction can be calculated. An important criterion for assessing the suitability of the models is their ability to represent various egress components. It should be able to simulate both vertical and horizontal components. The decision for use of elevators should also be addressed by the models. Evacuation time can be calculated by the egress models which can be helpful for taking appropriate decision. Egress models is of following types on the basis of its grid/structure [14]

(1) Course network models

In this type, the space is a network of arcs and nodes, representing different sections of the infrastructure (e.g., rooms, stairs, etc.). This is the simplest technique for simulating an evacuation scenario. The major advantage is its representation of complex structure. It can calculate the time very fast, even for the simulation of a complex evacuation scenario. The main limitations are not representing the behaviors that may occur during an evacuation.

(2) Fine network models

In this model, the space is represented as a grid of uniform cells. This can be occupied by one person at a time. The movement of the agents is simulated through a series of steps in the cells of the network. In these models, the occupant's location can be tracked during the evacuation based on a fine network representation.

(3) Continuous models

Continuous models simulate the agents through a system of some references within the environment. Here, occupant behavior can be simulated more flexibly, with respect to their location, orientation, and distance among the agents. Such characteristics are important to simulate high densities. For this model, more computational time is needed to simulate complex scenarios as it requires to re-calculate the referral points of the agents every time.

5.3 Detector Response Models

These models calculate the time to activate the detector or any other initiating device.

5.4 Fire Endurance Models

It simulates the response of structural elements of the building to fire exposure.

5.5 Miscellaneous Models

The models which are not covered under any of the previous categories are termed as miscellaneous.

6 Conclusion

At present, one of the major challenges of high-rise buildings is fire and life safety. Therefore, it is necessary to establish the certain methods to prevent and control human loss during fire emergencies. In high-rise buildings, fire spread very rapidly, normally in few seconds and the required evacuation times is normally much more, in minutes or even hours. Various egress components are available in the high-rise buildings, but every egress component has its own limitations. Vertical escape routes are very important for high-rise building evacuation. But due to its long distance, evacuation time is very long. If it is more than ASET (available safe evacuation time), then occupants could not evacuate safely and human losses will be more. It is also important to note that walking speed of children and old aged people are less as compared to others. Due to their slow movement, the problems such as bottlenecks, queuing, and stampede during emergency evacuation can be encountered which might adversely affect the evacuation of other occupants. Wheel-chaired people may block the way for descending occupants. While using an elevator, there are not only design issues, but also few behavioral issues, such as the willingness of the occupants of using the elevator, should be taken into account. It is observed that it is not safe to use any single component during every emergency. Depending on the height of building and fire floor, the occupants will have to decide the best strategy. It may not necessarily same for all the occupants of every floor.

References

1. www.skyscrapercentre.com.
2. Kavilkar, R., and S. Patil. 2014. Study of high rise residential buildings in indian cities (a case study–Pune city). *International Journal of Engineering and Technology* 6 (1): 86–90.
3. Bombay, G. 1991. Development control regulations. 1991 (5): 1–110.
4. National building code of India, Part IV.
5. Ma, Q., and W. Guo. 2012. Discussion on the fire safety design of a high-rise building. *Procedia Engineering* 45 (3): 685–689
6. Xin, J., and C. Huang. 2013. Fire risk analysis of residential buildings based on scenario clusters and its application in fire risk management. *Fire Safety Journal* 62 (Part A): 72–78.
7. Nair, R.R. 2013. Fire safety in India—An overview. *Industrial Safety Review* 10.
8. Hylton, J.G. Haynes. 2015. Fire loss in the United States during 2014. *National Fire Protection Association* 9 (15): 7–12.
9. Hall, J.R. Jr. ed. 2013. *High—Rise building fires.*
10. Ono, R. 2003. An analysis of high-rise building fires and human fire safety consciousness through fire statistics and field survey. *Fire Safety Science* 1073–1084.
11. Poon, S.L. 2014. A dynamic approach to ASET/RSET assessment in performance based design. *Procedia Engineering* 71: 173–181.
12. Ng, C.M.Y., and W.K. Chow. 2006. A brief review on the time line concept in evacuation. *International Journal on Architectural Science* 7 (1): 1–13.
13. Liu, X., H. Zhang, and Q. Zhu. 2012. Factor analysis of high-rise building fires reasons and fire protection measures. *Procedia Engineering* 45: 643–648.
14. Ronchi, E., and D. Nilsson. 2013. Assessment of total evacuation systems for tall buildings-literature review 1: 45.
15. Wang, P., Q.Y. Ban, J. Liu, and C.S. Zhao. 2013. Impacts of staircase windows on pressurized ventilation system. *Procedia Engineering* 52: 363–370.
16. Bellido, C., A. Quiroz, A. Panizo, and J.L. Torero. 2009. Performance assessment of pressurized stairs in high rise buildings. *Fire Technology* 45 (2): 189–200.
17. Cai, N., and W.K. Chow. 2011. Fire safety requirements on lift system for evacuation in super tall buildings. *International Journal on Engineering Performance-Based Fire Codes* 10 (2): 17–23.
18. Kuligowski, E.D., R.D. Peacock, P.A. Reneke, C.R. Hagwood, K.J. Overholt, R.P. Elkin, J.D. Averill, B.L. Hoskins, and E. Wiess. 2014. Movement on stairs during building evacuations. *NIST Technology Note* 1839: 212.
19. Kinateder, M.T., H. Omori, and E.D. Kuligowski. 2014. The use of elevators for evacuation in fire emergencies in international buildings. *NIST Technical Note* 1825.
20. Kinsey, M.J., E.R. Galea, and P.J. Lawrence. 2009. Investigating the use of elevator for high-rise building evacuation through computer simulation. *Human Behaviour in Fire Symposium* 85–96.
21. Ronchi, E., and D. Nilsson. 2014. Modelling total evacuation strategies for high-rise buildings. *Building Simulation* 7 (1): 73–87.
22. Huang, L., T. Chen, and H. Yuan. 2014. Simulation study of evacuation in high-rise buildings. *Transportation Research Procedia* 2 (2005): 518–523.
23. Luo, M., and K.H.L. Wong. 2006. Evacuation strategy for super highrise buildings. In *5th annual seminar, tall building construction maintenance.* Hong Kong.
24. Sherman, M.F., M. Peyrot, L.A. Magda, and R.R.M. Gershon. 2011. Modeling pre-evacuation delay by evacuees in world trade center towers 1 and 2 on September 11, 2001: A revisit using regression analysis. *Fire Safety Journal* 46 (7): 414–424.
25. Ronchi, E., and D. Nilsson (Lund U). 2013. Assessment of total evacuation systems for tall buildings. *Literature Review* 1: 45.
26. Shimshoni, J.Y. 2005. Tall building emergency evacuation: 'Time to think differently'. *Fire & Safety Fire Safety CTBUH.*

27. Ronchi, E., and D. Nilsson. 2013. Fire evacuation in high-rise buildings: A review of human behaviour and modelling research. *Fire Science Reviews* 2 (7): 1–21.
28. Kuligowski, E.D., R.D. Peacock, and J.D. Averill. 2013. Modeling the evacuation of the world trade center towers on September 11, 2001. *Fire Technology* 49 (1): 65–81.
29. Olenick, S.M., and D.J. Carpenter. 1995. An updated international survey of computer models for fire and smoke. *Journal of Fire Protection Engineering* 3 (July): 10–12.
30. Cranford, D. 2012. *From the publisher: From the publisher* pp. 1–23.
31. Gorbett, G.E. 2008. Computer fire models for fire investigation and reconstruction. In *International symposium on fire investigation science and technology*, 23–34.
32. Walton, W.D., and E.K. Budnick. 1997. Deterministic computer fire models. *Fire Protection Handbook, 18th Edition*. 52–61.

Design and Analysis of Firewater Network for a Typical Onshore Gas Processing Plant

J. Razeen, V. Venkata Krishnakanth and Shagufta Ejaz

Abstract Gas processing plant is having many hazards, which are inherent to the facility. Even though gas processing installations are generally located in remote areas, experience shows that residential/industrial units come up in close proximity with the passage of time. Hence, these installations, which store, process and handle large quantity of flammable materials, pose threat to surroundings as well, in addition to their own safety. Such conditions, therefore, necessitate the introduction of inbuilt fire protection facilities. It is impossible to design the fire protection facility to control catastrophic fires. Normally fire protection systems will prevent the spread of fire and prevent emergencies to the installations. So designing an effective firewater network plays an important role in controlling the spread of fire. This paper deals about the design and analysis of firewater network for a typical onshore gas processing plant based on OISD-116.

Keywords Firewater network · Fire zone · Gas processing plant Hydraulics

1 Introduction

Firewater network is the most common and economical fire suppression system, and water is the best medium for controlling the spread of fire. Firewater network usually contains three major parts. They are Firewater storage tank, Piping-distribution system and Fire Protection Systems like hydrants, monitors and fixed foam systems, deluge systems and fixed monitors and fire pump.

Generally, availability of continuous supply of water is of prime importance for operation, which is either met by provision of a storage system and pumps or by

J. Razeen · S. Ejaz
HSE Design, Technip India Ltd., Noida, India

V. Venkata Krishnakanth (✉)
Department of HSE, University of Petroleum & Energy Studies, Dehradun, India
e-mail: vkrishnakanth@ddn.upes.ac.in

N. A. Siddiqui et al. (eds.), *Advances in Fire and Process Safety*,
Springer Transactions in Civil and Environmental Engineering,
https://doi.org/10.1007/978-981-10-7281-9_17

nearby natural source. Firewater supply can be derived from process systems like water injection system, if the pressure and flow can be maintained under emergency conditions and if possibility of hydrocarbon contamination can be ruled out. As firewater is a very critical aspect of oil and gas industries, sizing and routing of firewater network are of prime importance.

The firewater system should be designed in such a way that the total demand calculated will be maintained at the outlets with a minimum required pressure. The firewater distribution piping shall be designed in such a way that water flows evenly to all parts of the network and shall be able to isolate any section of a network without affecting other loops.

1.1 Gas Processing Plant

Gas processing plant belongs to the upstream activity to describe the production unit performing the first transformation of the crude oil/gas after the production wells. The gas processing plant is situated as close as possible to the production wells or offshore platform. The crude oil/gas just collected from the wellheads can be directed by the shortest way to the central processing facility. Hydrocarbons produced from the wells are transported to onshore gas processing facility through pipeline. Hydrocarbons produced from the well contain natural gas, natural gas liquids and several contaminants like H_2S, CO_2. Hydrocarbon is passed through several unit operations to produce clean, dry natural gas free of contaminants to meet end user requirements. The natural gas liquids which come as a by-product are used as it is or fractionated to different lighter hydrocarbons like ethane, propane.

2 Methodology

The methodology to design the firewater network involves the following steps (Fig. 1).

2.1 Fire Zone Delineation

Fire zone is a geographical area of the plant, where certain minimum requirements shall be respected, so that in case of fire occurring within a fire zone, the potential for fire spread to other fire zone is limited.

Typically, a fire zone shall be an area segregated by road, access way, pipe racks or pipe sleepers, and clear spaces of 30 m width minimum between equipment to equipment.

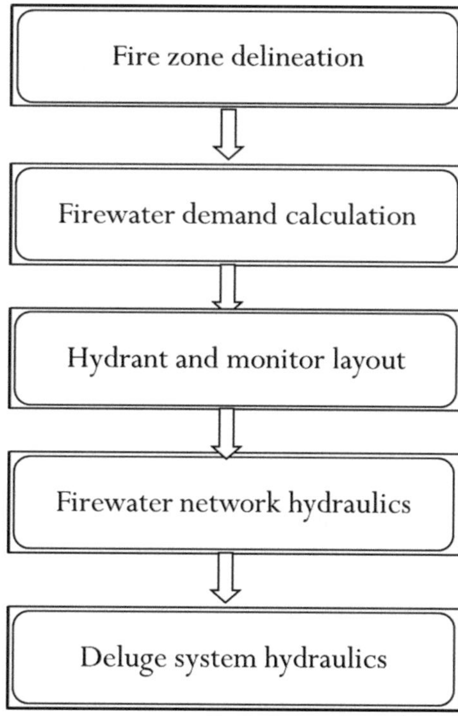

Fig. 1 Flow chart: Design and Analysis of Fire Water Network

2.2 *Firewater Demand Calculation*

Firewater demand for an installation is the total water requirement to fight the major two fire scenarios in the installation (It may be either process area or tank farm area or in transformer area).

Firewater demand for this case study is calculated based on OISD-116.

Three alternative methods are used to calculate the demand.

Alternative-1: Firewater demand for a fire zone.

$$\text{Fire Water Demand for Fire Zone} = \text{Overall Area of Fire Zone} \times \text{Application rate}$$

Application rate: As required by OISD-116, application rate to be considered for each fire zone area is 1 L/min/m^2

Fire zone surface: The surface of each fire zone is the sum of the surfaces of all sub-areas included in this fire zone

Allowance: An allowance of 372 m^3/h shall be considered for supplementary hose stream protection and/or use of mobile fire-fighting equipment to protect adjacent equipment.

Alternative-2: Firewater demand for 10 m × 10 m area.
Firewater demand is calculated by considering 10 m × 10 m area of process unit on fire and provides cover on area of 30 m × 30 m area:
Application rate: As required by OISD-116, application rate to be considered for each fire zone area is 10.2 $L/min/m^2$
Allowance: An allowance of 372 m^3/h shall be considered for supplementary hose stream protection and/or use of mobile fire-fighting equipment to protect adjacent equipment.

Alternative-3: Firewater demand for water spray system (deluge system).
Firewater demand for the deluge protection for each equipment is assessed:
Allowance: An allowance of 372 m^3/h shall be considered for supplementary hose stream protection and/or use of mobile fire-fighting equipment to protect adjacent equipment

2.3 Hydrant and Monitor Layout

2.3.1 Hydrants

(a) Hydrants shall be located in different areas of the facility to provide complete protection for the plant
(b) For hazardous area provide hydrant post for every 30 m around the plant and for building and utility area provide hydrant post for every 45 m
(c) Hydrants should be placed at 15 m from the edge of storage tank and hazardous equipment
(d) For process plants, location of hydrants shall be decided based on coverage of all areas.

2.3.2 Monitors

(a) Monitors shall be located at strategic locations for protection of cluster of columns, heaters, gasifiers, etc., and where it is not possible to approach the higher levels
(b) A minimum of two monitors shall be provided for the protection of each such area
(c) Water monitors for protection of heaters shall be installed so that the heater can be isolated from the remainder of the plant in an emergency

(d) Monitors shall provide protection to firemen in case of fire, and it is also placed in such a direction to direct water on the object
(e) Monitors should not be installed less than 15 m from hazardous equipment
(f) There should be proper planning for the placement of HVLRs so that it delivers its intended purpose
(g) The maximum distance of monitors from equipment protected should be 45 m.

2.4 *Firewater Network Hydraulics*

A detailed analysis of the firewater network has been carried out, and critical parameters like supply pressure, available pressure at the remotest point, velocity and head loss were analysed. Firewater network must be sized for 120% of the total water demand. Proper selection of flow rate allows sufficient water availability in emergencies. Several combinations of flow requirements must be assumed for design of network. The firewater system shall be designed to provide a minimum residual pressure of 7.0 kg/cm^2 g for the most hydraulically remote point of the firewater ring main.

Pipe network problems are usually solved by numerical methods using software since any analytical solution requires the use of many simultaneous equations. Simple methods used to solve pipe network problems are by using Hazen–William equation. PIPENET Standard/Spray module is used for hydraulic analysis of fire-water systems in compliance with NFPA13, NFPA15 and NFPA16 rules. This addresses the hydraulic analysis requirements of virtually all national and international standards.

Hazen–Williams equation:
Pressure drop inside the pipe can be calculated using Hazen–William's formula [1],

$$P = \frac{\left(4.25 \; * \; Q^{1.85}\right)}{\left(C^{1.85}\right) d^{4.87}}$$

where P is the pressure drop or friction loss (per 100 ft) inside the pipe (psi),
Q is the volumetric flow rate (gpm),
C is the Hazen–William's friction loss coefficient,
d is the pipes internal diameter (in.),

(or)
in SI units,

$$P_{\mathrm{m}} = 6.05 \times \frac{Q_{\mathrm{m}}^{1.85}}{C^{1.85} d_{\mathrm{m}}^{4.87}} \times 10^{5}$$

where P_m velocity pressure in psi,
Q is the volumetric flow rate (gpm),
d is the pipes inside diameter (in.).

2.5 Deluge System Hydarulics

The hydraulics of deluge system lines from main header to the equipment connected through deluge valves is done using PIPENET VISION Spray/Sprinkler module.

Pressure drop inside the pipe can be calculated using Hazen–William's formula. The following considerations have been taken into account:

- The firewater pressure in the range of 1.4–3.5 bar (g) is to be achieved for all the water spray nozzles in the system
- The maximum allowable velocity in the header shall be 5.0 m/s for the water spray pipes and ring pipes
- Hazen–William coefficient is considered as 120.

Nozzle Discharge Formula: The pressure drop at the nozzle discharge can be calculated by the following formula [1],

$$Q = K\sqrt{P}$$

where Q is the volumetric flow rate from the nozzle (gpm),
K is the nozzle K-factor,
P is the pressure drop across the nozzle.
Deluge valve modelling equation.

The pressure drop across the deluge valve can be calculated by the following formula [1],

$$P = QX/K$$

where P is the pressure drop across the deluge valve,
Q is the volumetric flow rate through the valve,
K is a constant for the valve,
X is a constant for the valve (with typical values being 1 and 2).

3 Results and Discussion

3.1 Fire Zone Delineation

In this case study, an onshore gas processing plant is considered and is delineated into 15 fire zones (Fig. 2) based on OISD – 116 [2] as follows:

1. Slug catcher area
2. Caustic dosing area
3. Gas compressor area

Fig. 2 Fire zone delineation

4. Scrubber area
5. Gas dehydration area
6. Refrigeration area
7. MEG refrigeration area
8. MEG storage area
9. Produced water storage area
10. Methanol storage area
11. Diesel storage area
12. Hot oil heater area
13. Air and nitrogen area
14. Substation
15. Liquid propane area.

3.2 Firewater Demand Calculation

- In this case study, the firewater demand was calculated for two fire scenarios
- Largest two firewater demands are in *Fire zone-01(Slug catcher area)* and *Fire zone-03(Gas compressor area)*
- Thus, firewater demand calculation for the selected case study is as follows:

Demand for slug catcher area $= 550.8 + 372$*(supplementary stream)* $= 922.8\ m^3/h$

Demand for gas compressor area $= 600 + 372$*(supplementary stream)* $= 972\ m^3/h$,

Total firewater demand $= 922.8 + 972 = 1894.8 = 1895\ m^3/h$,

Therefore, firewater demand is $1895\ m^3/h$

Firewater pump capacity $= 680\ m^3/h$

No of pumps $= 3$

50% of standby pumps shall be provided; so no. of standby pumps $= 2$

Total pumping capacity $= 2040\ m^3/h$

Firewater storage capacity $= 2040 \times 4 = 8160\ m.^3$

3.3 Hydrant and Monitor Layout

- For hazardous area provide hydrant post for every 30 m around the plant and for building and utility area provide hydrant post for every 45 m
- The total no. of hydrants required to protect the onshore gas processing plant is 81

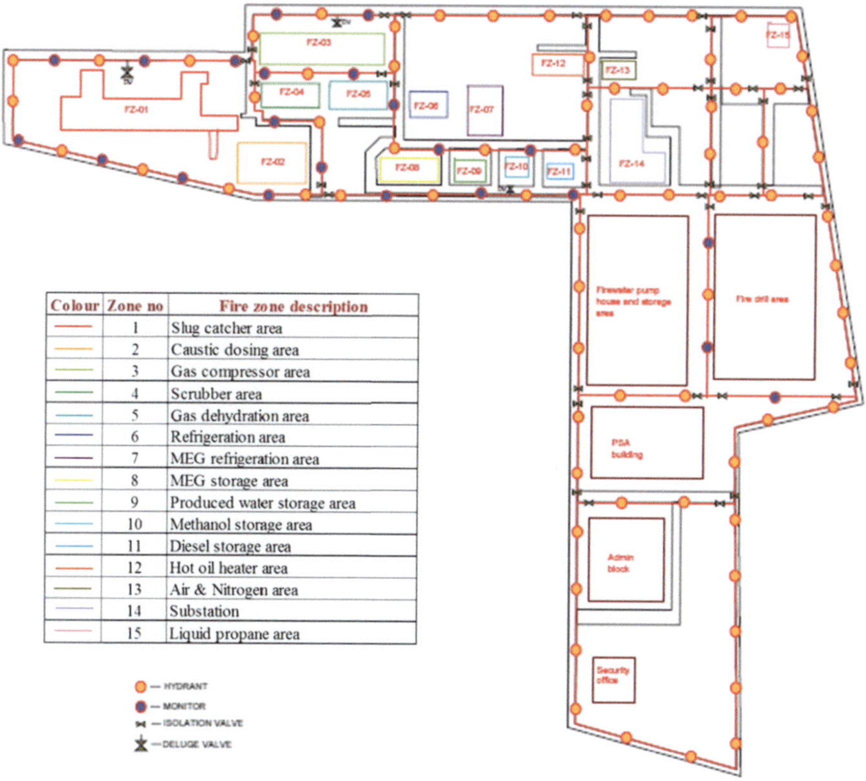

Colour	Zone no	Fire zone description
	1	Slug catcher area
	2	Caustic dosing area
	3	Gas compressor area
	4	Scrubber area
	5	Gas dehydration area
	6	Refrigeration area
	7	MEG refrigeration area
	8	MEG storage area
	9	Produced water storage area
	10	Methanol storage area
	11	Diesel storage area
	12	Hot oil heater area
	13	Air & Nitrogen area
	14	Substation
	15	Liquid propane area

Fig. 3 Hydrant and monitor layout

- Monitors shall be located at strategic locations for protection of cluster of columns, heaters, gasifiers, etc., and where it is not possible to approach the higher levels. A minimum of two monitors shall be provided for the protection of each such area
- The total no. of monitors required to protect the onshore gas processing plant is 22 (Fig. 3).

3.4 Firewater Network Hydraulics

The hydraulics is done by taking into account that [3–5]:

- The length of pipes is considered as 1.5 times the layout dimensions to cater the fitting losses
- Maximum velocity in the firewater network should not exceed 5 m/s

- The minimum available pressure at the remotest point shall not fall below 7 kg/ cm^2(g)
- Hazen–Williams coefficient is considered as 120.

In this case study, the firewater demand calculated was 1895 m^3/h. The firewater network is sized 120% of the required water demand, i.e. 2274 m^3/h. The network hydraulics is done for two major fire scenarios. The two major demands are in gas compressor area and slug catcher area.

3.5 Deluge System Hydraulics

3.5.1 Deluge System Hydraulics for Gas Compressor [3–5]

- The theoretical firewater demand for gas compressor was found out to be 600 m^3/h, and length of pipes is considered as 1.5 times the actual layout dimensions to cater the fitting losses.

Operating pressure range for nozzle: Medium velocity water spray nozzle: 1.4–3.5 kg/cm^2 g (Tables 1, 2 and 3).

3.5.2 Deluge System Hydraulics for Flare KO Drum

The theoretical firewater demand for gas compressor was found to be 137.28 m^3/h.

Assumptions: same as mentioned in Sect. 3.5.1.
Operating pressure range for nozzle: Medium velocity water spray nozzle: 1.4—3.5 kg/cm^2 g (Tables 4, 5 and 6).

Table 1 Nozzle specification [1]

Type	*K*-factor	Flow rate at 1.4 barg (lpm)	Flow rate at 3.5 barg (lpm)	Nozzle angle (degrees)
MVWS	70	82.82	130.95	120

Table 2 Theoretical flow rate [1]

Equipment protected	Length (m)	Diameter (m)	Surface area (m^2)	Application rate (lpm/m^2)	Theoretical flow rate	
					lpm	m^3/h
Gas compressor (7)	14	5	70(7)	20.4	9996	600

Table 3 Installed flow rate [1]

Equipment protected	Average flow rate per nozzle (lpm)	Numbers of nozzles (nos.)	Installed flow rate	
			lpm	m^3/h
Gas compressor	113	112	12,600	756

Table 4 Nozzle specification

Type	*K*-factor	Flow rate at 1.4 barg (lpm)	Flow rate at 3.5 barg (lpm)	Nozzle angle (degrees)
MVWS	42	49.69	78.57	120

Table 5 Theoretical flow rate

Equipment protected	Length (m)	Breadth/dia. (m)	Surface area (m^2)	Application rate (lpm/m^2)	Theoretical flow rate	
					lpm	m^3/h
Flare KO drum	17	4.2	224.32	10.2	2288	137.28

Table 6 Installed flow rate

Equipment protected	Average flow rate per nozzle (lpm)	Numbers of nozzles (nos.)	Installed flow rate	
			lpm	m^3/h
Flare KO drum	68.03	42	2857.66	171.46

4 Conclusion

The firewater network for a typical onshore gas processing plant is designed as per OISD-116 “Fire Protection Facilities for Petroleum Refineries and Oil/Gas Processing Plants”. Effective operation of the ring main is modelled to acquire adequate flow, velocity and pressure for the smooth operation during emergency conditions like fire breakout. The simulation and hydraulic design of the firewater ring main system of an onshore processing terminal to determine the pipe sizes is accomplished using PIPENET software. The assessment of the fire protection system meeting the OISD standards is analysed by using PIPENET Spray/Sprinkler module. The pipe sizing (diameter) is modelled in the software. The typical output illustrates the flow, velocity and direction in each pipe segment and pressures at each node.

References

1. NFPA 15 standard for water spray fixed systems for fire protection 2001 edition.
2. OISD 116. 2012. Fire protection facilities for petroleum refineries and oil/gas processing plants, 2nd ed.
3. Mckinnon, G.P., and Keith Tower. Fire prevention handbook, national fire protection association. In *Hydraulics for Fire Protection, National Fire Protection Association.* Hickey, H.E.
4. Cunha, M.C., and J. Sousa. 1999. Fire water network design optimization: Simulated annealing approach. *Journal of Water Resources Planning and Management* 125 (4): 215–221.
5. Lansey, K.E., and L.W. Mays. 1989. Optimization model for fire water distribution design. *Journal of Hydraulic Engineering* 115 (10): 1401–1418.
6. Sam, M. (eds.). 2004. Lees' Loss prevention in the process industries: Hazard identification, assessment and control. UK: Butterworth-Heinemann.
7. Savic, D.A., and A. Godfrey. Walters. An evolution program for pressure regulation in water distribution networks, centre for systems and control engineering. In *Water Distribution System Handbook.* UK Mays, L.W. Mc Graw Hill Handbooks.
8. NFPA 20 standard for the installation of stationary pumps for fire protection 2003 edition.

Swing Tower of Loader Backhoe Arm for Dynamics and Stress Analysis by Modeling and Simulation

Jitendra Yadav, Praveen Pitta and Akshay Maan

Abstract As the construction work is increasing throughout the world at a rapid speed, the requirement of the automated equipment's is also increasing at the same rate to assist this large amount of construction work. The earth-moving equipment's falls in this category of the automated equipment's and provides a great deal of saving of time in these works. The performance of these earth-moving equipment's relies heavily on their sub-parts if one of them fails during the operation, then it will be drastic and may result in loss of life, money, and time. Thus, the manufacturers carry out variety of tests and analyses in house on these machines to insure their safe working of the machine and security of the others. The loader backhoe is one of the parts of the earth-moving equipment's family. In the present work, the dynamic analysis is carried out on the swing tower which is one of the parts of the loader backhoe machine. For simulating the dynamic response of the swing tower during loading and unloading of the backhoe assembly due to its working, a finite element analysis tool, ADAMS 10.0, is used. After carrying out the dynamic analysis of the swing tower, the stress analysis of the complete backhoe arm is also performed using finite element analysis tool, ANSYS 10.0, to predict the performance of the swing tower mechanism and backhoe assembly more accurately at the design stage itself.

Keywords Loader backhoe arm · Swing tower · Backhoe assembly
Earth-moving equipment · Stress analysis · Finite element analysis tool
Dynamic analysis

J. Yadav (✉) · P. Pitta · A. Maan
University of Petroleum and Energy Studies, Dehradun 248007, India
e-mail: jyadav@ddn.upes.ac.in

P. Pitta
e-mail: praveen8423@yahoo.in

A. Maan
e-mail: akshaymaan93@gmail.com

N. A. Siddiqui et al. (eds.), *Advances in Fire and Process Safety*,
Springer Transactions in Civil and Environmental Engineering,
https://doi.org/10.1007/978-981-10-7281-9_18

1 Introduction

Backhoe loader, also called a loader backhoe, and commonly shortened to backhoe, is the engineering vehicle, which consists of a tractor, fitted with a shovel/bucket on the front and a small backhoe on the back. Due to its (relatively) small size and versatility, backhoe loaders are very common in urban engineering and small construction projects such as building a small house, city roads. It is used for loading, grading, ditch cleaning, trenching, and backfilling. It is an utility machine for heavy-duty works like digging and excavation. It is ideal for small operation where heavy machine is not economical. Earth-moving equipment requires swing tower for backhoe swing mechanism. A backhoe assembly consists of swing tower, boom, deeper, and a bucket, and this subassembly is mounted over the slide frame bolted on the machine; a backhoe bucket is located at the rear of the vehicle.

The swing tower is one of the most critical components of loader backhoes. It bears all the loads of the backhoe assembly. Thus, it has become essential to estimate its performance under the influence of different operating conditions and geometrical considerations to ascertain the effect of different influencing parameters on the dynamics of swing tower. The influencing parameters can be the variable load which is caused by the empty bucket and loaded bucket, the operating hydraulic pressure on the piston that is induced by the pump, distribution of the pressure developed by the pump between the two cylinders, and the different stretched position of backhoe assembly.

So that proper design of swing tower is essential for an engineer to assure the efficient, effective, and failure-free working of loader backhoe. One way of obtaining the dynamic response of swing tower is performing experiment on a prototype. This method is time-consuming and cumbersome as well as uneconomical. An alternate option to have the dynamic response is the computer simulation that can be performed in house with the use of computer software. The modern softwares are very accurate and reveal the actual result within no time and save time and money.

A lot of simulation work regarding the kinematic and dynamic analysis and analysis of systems using different software simulation tools has been performed in the past studies.

Noel–Messier-Dowty [1] presents the method in use at Messier-Dowty SA during design process with ADAMS software to modelize a landing gear retraction and extension, in order to optimize its behavior. The result of this analysis is that ADAMS model allows to get a convenient optimization of landing gear retraction and extension behavior. Arborio, Munaretto, Velardocchia [2] implemented ADAMS Car on a mathematical model of a new car. The most important results obtained adopting the described strategy are presented through animations. Farzin,

Montazersadgh, and Fatemi [3] performed study to investigate weight and cost reduction opportunities for a forged steel crankshaft with the help of FEM analysis. Kaparthi [4], designed a four-bar mechanism using Burmester theory and the four-position synthesis method with the use of MATLAB. According to Dilluvio [5] suggested to analyze the dynamic behavior of mechanisms without leaving the CATIA environment with the use of CAT/ADAMS Singh, Sharma, Thakur [6] performed Kinematic Synthesis and Optimization of Four-bar Linkage with the use of MATLAB 5.3. Hofstra, Hemmen, Miedema, and Hulsteyn [7] studied the influences of the flexibility of the hydraulic fluid and the steel structure on the achievable accuracy using MATLAB and ADAMS. Hong and Tessmann [8] suggested computerized design analysis of hydraulic systems using a computer program HyPneu which is capable of integrating hydraulic, pneumatic, electronic, and mechanical components, thus permitting the design analysis of complete hydraulic systems. Munzer [9] tried to improve the control of mobile hydraulic cranes.

The first objective of the present paper is to perform kinematic analysis of swing tower with the help of simulation by the computer tool ADAMS 10.0 for variation of different influencing parameters such as total load on the swing tower, which differs because of empty bucket and loaded bucket and the different stretched position of backhoe assembly. The second objective of the study is to perform stress analysis on backhoe assembly to compute the factor of safety for the assembly design to ensure the safe working of the backhoe assembly.

2 Modeling Parameters of Swing Tower

Here a model is developed which has one fixed link, one ternary link (swing tower), and two flexible links which are represented as a cylinder. Cylinder is pivoted on the fixed link through revolute joints, the eye end of the cylinder is also connected to the swing tower through the revolute joints and the motion of the cylinder is angular as well as linear. The representation of the swing tower is triangular, and the one corner of the triangle is fixed by revolute joints as shown in Fig. 1.

The hydraulic pressure produced by the pump is of the order 210 bar which imposes a total force of the magnitude 200,000 N on the cylinders. Furthermore, for the analysis, it has been considered that the total force is distributed between the cylinders of the magnitude 175,000 N on left cylinder and 25,000 N the right cylinder. The simulation is performed for the selected model with consideration of above-mentioned conditions of operating pressure. The sample results are shown in Fig. 2.

The parameters selected for simulation are shown in Tables 1 and 2.

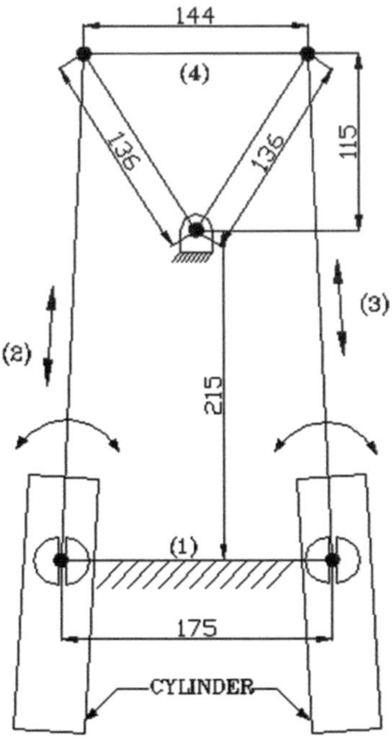

Fig. 1 Detailed view of swing tower mechanism with dimension (mm)

3 Results and Discussion

The simulation on swing tower is performed for to load condition, i.e., when the bucket is empty and when it is filled with soil along with three different stretched positions of backhoe arm, i.e., full stretched backhoe assembly, semi-stretched backhoe assembly, and fully closed backhoe assembly. The simulation is performed to obtain the dynamic response of swing tower in terms of displacement, linear velocity, linear acceleration, time period, frequencies, angular velocity, angular acceleration, and forces variation in cylinders.

From the results of simulation, data are collected and graphs are plotted from Figs. 3, 4, 5, 6, and 7.

From Fig. 3, it can be concluded that the time period of the cycle is maximum in case of fully stretched condition and minimum in case of fully closed condition for both the cases of either empty bucket or filled bucket, though it is higher in case of filled bucket for all the position of loader backhoe arm. It can be observed from Fig. 4 that the angular velocity is minimum in case of fully stretched condition and maximum in case of fully closed condition for both the cases of either empty bucket or filled bucket, though it is higher in case of empty bucket for all the position of

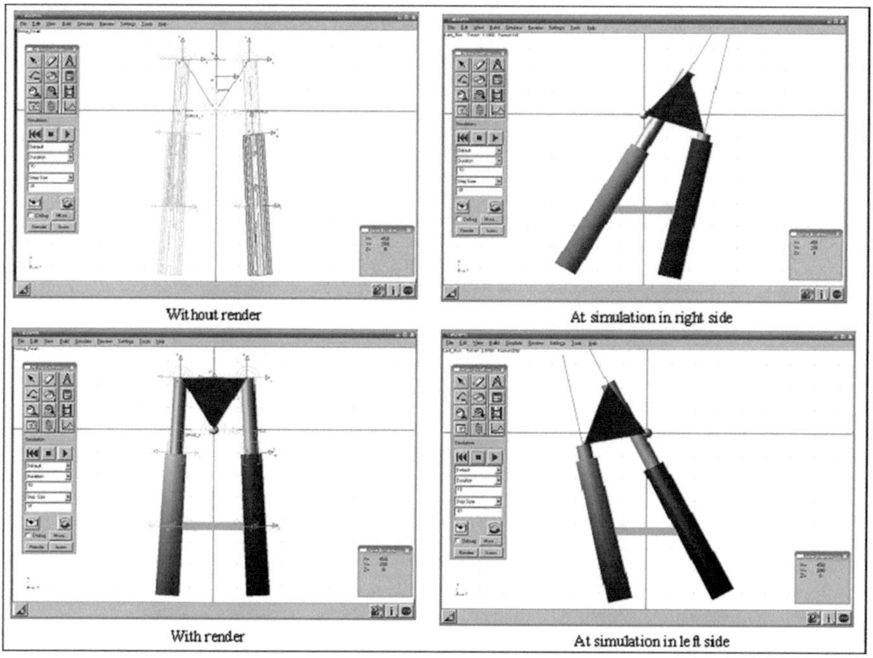

Fig. 2 Schematic model of swing tower mechanism in ADAMS software tool

Table 1 Parameters for simulation for empty bucket

S. No.	Description		Case-I	Case-II	Case-III
			Fully stretch	Half stretch	Fully closed
1	Total Mass of swing tower with empty bucket (kg)		1065	1065	1065
2	Mass moment of inertia (kg m^2)	I_{xx}	11,133	9234	2783
		I_{yy}	0	0	0
		I_{zz}	11,133	9234	2783
3	Length (m)		5.60	5.10	2.80

Table 2 Parameters for simulation when bucket filled with soil

S. No.	Description		Case-I	Case-II	Case-III
			Fully stretch	Half stretch	Fully closed
1	Total mass of swing tower when bucket filled with soil (kg)		1387	1387	1387
2	Mass moment of inertia (kg m^2)	I_{xx}	14,499	12,025	3625
		I_{yy}	0	0	0
		I_{zz}	14,499	12,025	3625
3	Length (m)		5.60	5.10	2.80

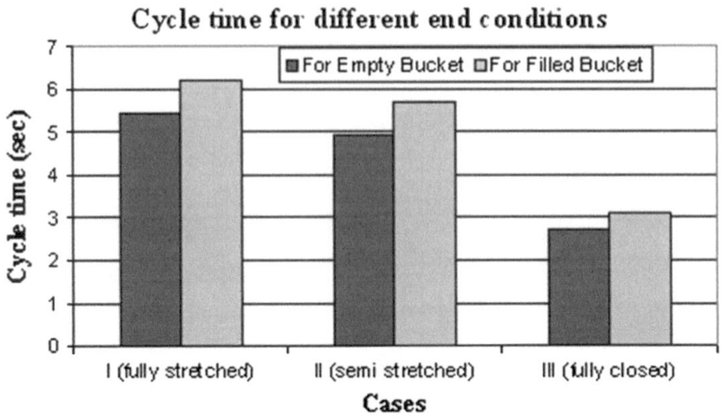

Fig. 3 Cycle time for different end conditions

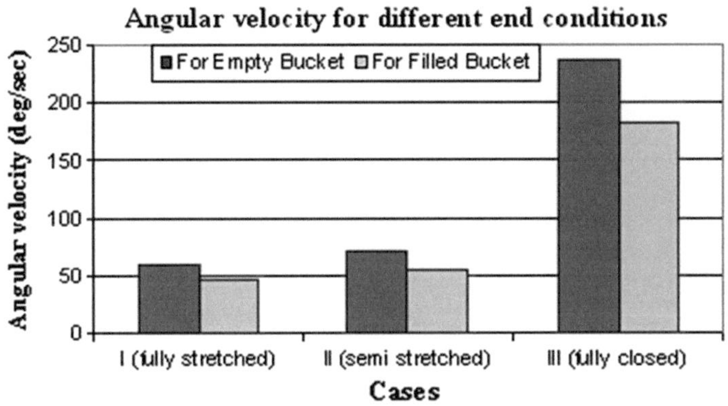

Fig. 4 Angular velocity for different end conditions

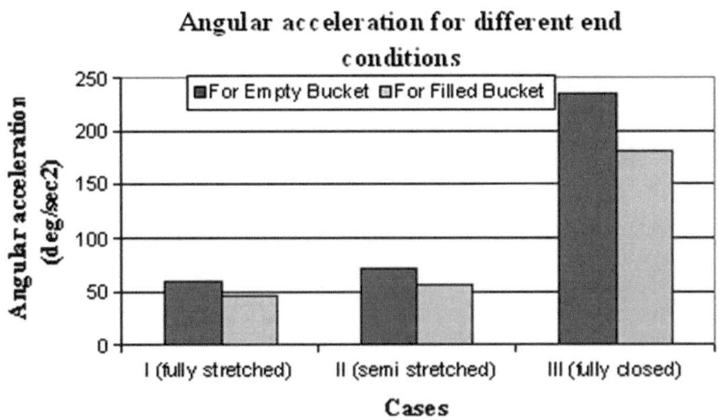

Fig. 5 Angular acceleration for different end conditions

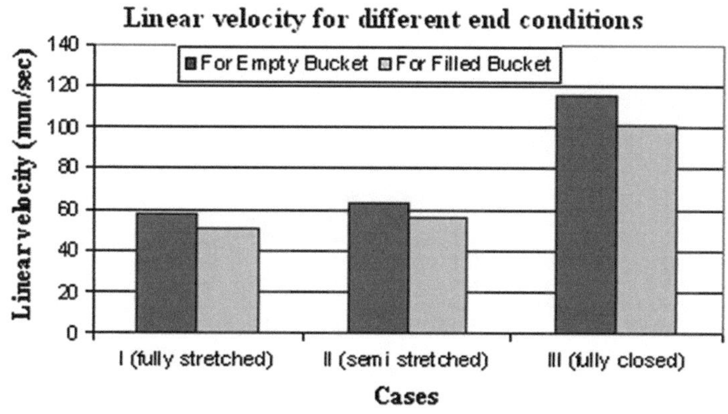

Fig. 6 Linear velocity for different end conditions

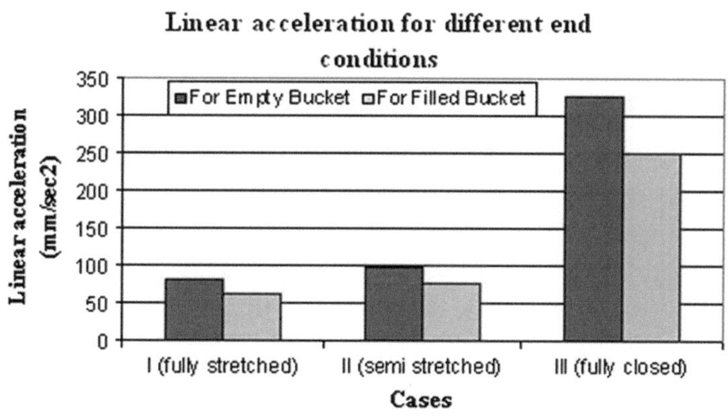

Fig. 7 Linear acceleration for different end conditions

loader backhoe arm. It is clear from Fig. 5 that the angular acceleration is minimum in case of fully stretched condition and maximum in case of fully closed condition for both the cases of either empty bucket or filled bucket, though it is higher in case of empty bucket for all the position of loader backhoe arm. It is palpable from Fig. 6 that the linear velocity is minimum in case of fully stretched condition and maximum in case of fully closed condition for both the cases of either empty bucket or filled bucket, though it is higher in case of empty bucket for all the position of loader backhoe arm. It can be seen from Fig. 7 that the linear acceleration is minimum in case of fully stretched condition and maximum in case of fully closed condition for both the cases of either empty bucket or filled bucket, though it is higher in case of empty bucket for all the position of loader backhoe arm. So for

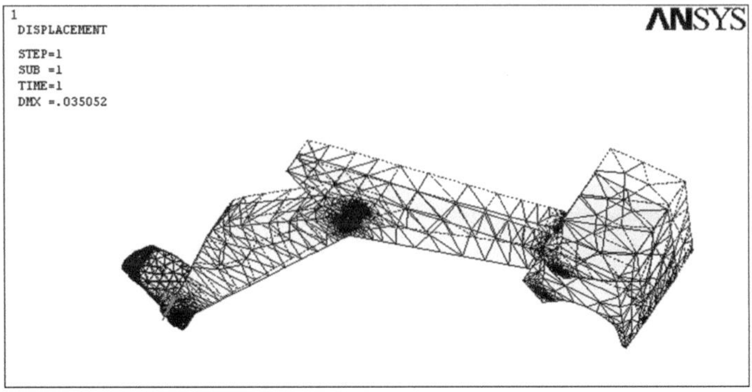

Fig. 8 Deformed shape on application of forces

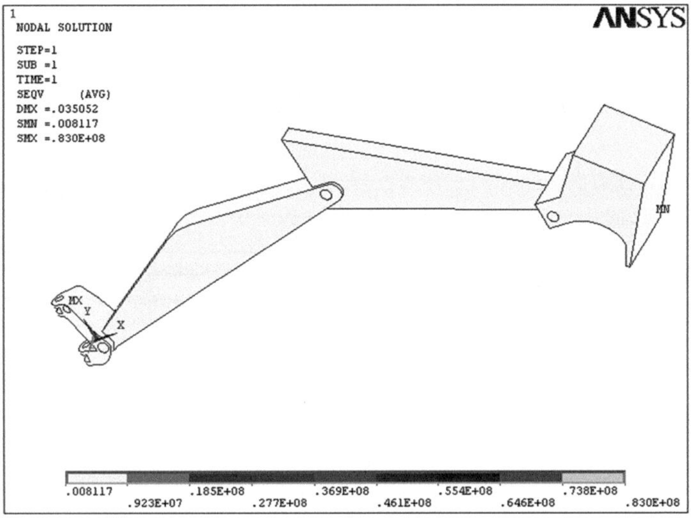

Fig. 9 Nodal solution of the backhoe assembly

from above discussion, it can be concluded that the frequency of oscillation is higher in case of empty bucket and higher for fully closed position of backhoe arm.

The results of dynamic analysis of the backhoe assembly performed on the model by computer simulation tool ANSYS 10.0 are shown in Figs. 8 and 9. The effort is concentrated toward the understanding of the stress analysis i.e. finding the maximum-induced stress in magnitude, direction, and point of application.

From the stress analysis, it has been observed that the maximum stress induced is of the magnitude 83 N/mm^2 at the intersection of boom and swing tower. The material for the application is mild steel whose yield stress is of the order 230 N/mm^2. Further, analyzing the safety aspects of the application the factor of safety is evaluated. The factor of safety evaluated is justifiable, as in the present application, the load is a dynamic variable load and the environmental conditions are dusty; so for the design of this application to be safe, the factor of safety must be greater than 2.5, which is of the order 2.77 in the present case, so the design is safe.

4 Conclusion

It is palpable from the present modeling and simulation that different working conditions for the loader backhoe arm can be selected to get the actual dynamics of the system so to ensure the failure-free operation. Furthermore, from the stress analysis, one can be able to know the critical regimes of operation to reveal the severity which indeed will help in selection of proper factor of safety to make the design safest without actual fabrication. In the present work, a methodology to ascertain the dynamic behavior of a mechanical system is proposed with the help of computer simulation. In such practices, one need not to go for implementation of the actual physical system for the same which indeed is a costlier and time-consuming affair. One can invoke with the methodology to get the idea of the dynamic behavior of the system at the design stage so that selection of proper material for different parts of the systems can be done properly. In this manner, the designer can ensure the optimum performance of the system. The dynamic analysis by introducing more influencing parameters and validation of the results are the future scope of the present work.

References

1. Noel, O. 1998. Use of ADAMS in dynamic simulation of landing gear retraction and extension. In *13th European ADAMS Users's Conference*. France: [s, n].
2. Arborio, N., et al. 2000. Vehicle dynamics and stability analysis with matlab and adams car. In *Titolo volume non avvalorato, 15th ADAMS European Users Conference*, Rome, Italy.
3. Montazersadgh, F.H., and Ali Fatemi. 2007. *Stress analysis and optimization of crankshafts subject to dynamic loading, Project Report.* Forging Industry Educational Research Foundation and American Iron and Steel Institute. University of Toledo.
4. Kaparthi, P. 2001. *Kinematic synthesis of a well service machine. Master's thesis*, Morgantown: West Virginia University.
5. Dilluvio, C.J. 2000. Virtual prototyping of automotive open air systems using CAT/ADAMS. In *Proceedings international ADAMS user conference.*
6. Dr. Singh, V.P., S. Sharma, and B.S. Thakur. 2005. Kinematic synthesis and optimization of four-bar linkage. *Journal-Institution of Engineers India Part MC Mechanical Engineering Division* 85 (1): 199–205.

7. Hofstra, C.F., A.J.M. van Hemmen, S.A. Miedema, and J. van Hulsteyn. 2000. *Describing the position of backhoe dredge buckets, lecture note*. Netherland: Technical University Delft.
8. Dr. Hong, Ing T., and Dr. Richard K. Tessmann. *Computerized design analysis of machine tool hydraulic system dynamics*. FES/BarDyne Technology Transfer Publication.
9. Marc, E.M. 2000. Control of mobile hydraulic cranes. In *Proceedings of 1st FPNI-PhD symposium*, 475–483. Hamburg.

Recent Development in Machine Safeguarding for Protecting Humans from Complicated Machines

Appil Ora, K. Dharani Kumar and Rishi Dewan

Abstract As we are in twenty-first century, industries are growing at rapid rate and thus evolution of advanced and more complicated machinery takes place. Basically humans are the ones who need to operate such machinery although some of them are operated remotely, but many demand direct human interaction. Every machine has its own unique hazard such as crushing, shearing, entanglement, cutting and drawing-in which may result in mechanical injuries such as abrasion, laceration, avulsion, and others. In order to prevent such injuries and protect human from such complicated machinery, we need to discover new and advanced protective equipment's in order to safeguard human life. This paper discusses about such recent trends in protecting humans and advance technique to provide automatic cutoff if anything goes wrong from the design intent. There are various types of guards available such as interlock guard, fixed guard, adjustable guard and use according to the type of machinery. Apart from the guards, there are different types of protective appliances we are having such as presence sensing, pullback, restraint, tripwire cable, two-hand controls and gates. All such protective devices are designed for safeguarding specific types of hazard and can be installed only on particular types of machines. Each type of guards is having its own advantages and limitations, but it shall fulfill the basic requirement of a guard-such as preventing contact between human and hazardous rotating part of machine, firmly secure machine and is difficult to tamper with, it should not create any new hazard, and must not interfere with the operation. The recent sensitive protective equipment such as trip bars, pressure mats and photoelectric devices help improve safety. In hierarchy of control, safeguarding and protective devices come under engineering control. Once the possibility to eliminate or substitution is exhausted, these protective devices are placed to protect human. Machine guarding and other protective equipment's are placed to remove or eliminate only mechanical hazards associated with the machinery, and they can not eliminate nonmechanical hazard associated

---

A. Ora · K. Dharani Kumar (✉) · R. Dewan
University of Petroleum and Energy Studies, Dehradun 248007, India
e-mail: dharni.kumar@ddn.upes.ac.in

N. A. Siddiqui et al. (eds.), *Advances in Fire and Process Safety*,
Springer Transactions in Civil and Environmental Engineering,
https://doi.org/10.1007/978-981-10-7281-9_19

with the machine such as electricity, noise, vibration. Accidents/incidents can be prevented once workers have proper knowledge about the machine and the safeguard that is placed to protect them.

Keywords Guards · Protective devices · Mechanical motions · Mechanical actions · Amputation · Power transmission apparatus

1 Introduction

Safety of the manpower working in any industry is one of the major responsibilities of an employer; safety from complicated machinery is also a part of the same [1]. In order to safeguard the machinery installed in industry, we placed different protecting tools such as guards, sensing devices, gates, robots. Such equipment covers the hazardous part of the machine so that injuries can be prevented [2]. All such measures come under engineering control, and there may be several administrative and other controls too, which is not covered in this paper. Here we mainly deal with the mechanical hazard associated with the machines; machines may also associate with the nonmechanical hazards like electrical, noise, vibration [3, 4]. Basically, machine guard should be used in such a manner that it serves its basic function and protect human intervention in running condition. Once guarding is removed or protecting appliance is deactivated, mainly during maintenance work, other measures shall be implemented such as Lockout/Tagout [5]. Such protecting equipment shall be installed in such a manner so that it should not interrupt between the feeding mechanism and its functioning. Hence, selection of such equipment is very necessary because not all equipment are suitable for very type of machinery [6]. It totally depends upon the design and principle function of any machine; moreover it also depends upon the functioning of that particular protective equipment [7]. There are various types of protective equipment available in which some provides a kind of physical barrier between machines and human, some are based on sensing mechanism, many are of interlock types like that will not allow the machine to actuate until prior demand is fulfilled, basically it makes out the human body part from the dangerous section [3, 4].

The primary function of guards and other protective equipment is to prevent the human contact with complicated machines mainly with those parts which has a potential to cause harm to the human, also serve the purpose to secure the machines properly [8]. On the other hand, it should also take care of aspects like it should not create any new hazard nor it should interfere between the smooth functioning of the machines. In case if lubrication required, it shall be from safe end without any risk [9]. Guarding and similar protective equipment can prevent or minimize the risk associated with the mechanical hazards, but for overpressure, electrical shock, fatigue due to high noise, etc., alternative measures are required [10]. Such safety equipment should be comply with standard; quality is to be assured before installation, and training is to be provided to all employees and workers regarding its functioning and use of such equipment [11]. Proper care and maintenance is required so that at that time of any undesired event, they can actuate quickly [12]. Many accidents take place just due to by passing the safety guards and equipment, since sometimes it becomes tedious to engage safety system again and again [13]. Procedure to remove the guard or deactivate the equipment shall include prior permission, and all safety measures shall be implemented before any such activity [14].

Types of mechanical hazard:

- Crushing
- Cutting or severing
- Shearing
- Drawing in or trapping
- Entanglement
- Friction and abrasion
- Impact

Basic types of hazardous mechanical motions are:

- Rotating (including in-running nip points)—It is basically a circular type of motion in which chances of cloth gets caught is higher and even person may exposed to the dangerous part of the machinery [15].
- Reciprocating—It is basically back and forth or up and down motion; chances of trapping between the machine part and fixed object are very higher [16].
- Transversing—It is a continuous and straight motion; pinch and shear point may cause injury to persons [17].

Basic types of hazardous actions are:

- Cutting action
- Punching action
- Shearing action
- Bending action

Types of hazardous mechanical components are:

- Point of operation: It is the actual point where machine performs its principle functioning like, shaping, cutting [18].
- Power transmission apparatus: It is a part of mechanical system assembly which actually transmits the power/energy. E.g., pulleys, belts, couplings, flywheels, shafts [19].
- Other moving parts: It includes the parts which perform hazardous motion involved in the machines like, rotating, transversing, reciprocating, etc. They play dominant role in mechanism of any machinery [20].

There are several activities relating to any machine which has a potential to cause some like clearing jams, cleaning of machines, machine adjustments, and schedule and unscheduled maintenance [21].

2 Methodology

Types of machine safeguarding:

(a) **Guards**

i. **Fixed guard**—It provides a physical barrier between the human and machine in such a manner that human body part cannot contact with hazardous area of the machine (generally a moving part of machine) [22]. It acts as fencing around the machine and design according to the size of the machine. It may be having some gaps in between but in such manner so that it cannot allow the human approach to moving part of machine. Material for construction includes sheet metals, bars, wire cloth, plastic [23].
Applicable to machines: power press, enclosed pulleys and belt, band saw, cartoon folding machine, etc.

ii. **Adjustable guard**—It gets adjusted according to the type and size of feedstock. It provides a good range of flexibility in accordance with the thickness of the feedstock [24].
Applicable to machines: router, shaper, table saw, band saw, etc.

iii. **Self-adjustable guard**—Other than adjustable guard, self-adjusting guard is also available which allows the user with a range of selection to adjust the guard in accordance with the size of workpiece and nature of the task to be carried out [25, 26].
Applicable to machines: radial arm saw, circular saw, jointer, etc.

iv. **Interlock guard**—It follows the principle of ON/OFF like when the guard is in open condition means unsafe then it will not allow the machine to operate, whereas when the guard is in closed position then only power supply gets on, and thus machine comes in operating condition [27]. It means interlock guards prevent access of human body part in dangerous section of machine and prevent the power supply required to run a machine; once it is closed, condition gets safer then only it feeds power to machine [28].
Applicable to machines: automatic bagging machine, picker machine, role makeup machine, etc.

Every type of guards or any machine safeguarding possess advantages and disadvantages both; since every type of machine safeguarding cannot serve to any type of machine, there should be some criteria which requires to be fulfilled before its implementation [29]. Here Table 1 describes about the positives and negatives of different types of guards:

(b) **Protective equipment**

1. **Two-hand controls**: The basic idea behind such type of design is like before actuation of machine, hand should be out of danger area and machine should be only actuated when both the hands of a person simultaneously press the switch [30]. Minimum distance shall be maintained between the trip and the point of

Table 1 Advantages and disadvantages of guards

Type of guard	Advantages	Disadvantages
Fixed guard	It offers maximum protection and requires very less maintenance and very suitable to specific application	Not good visibility it offers, and in order to carry out any maintenance activity, it should be removed and hence other safety factors needs to be included
Adjustable guard	It is very suitable for varying sizes of stock	It requires regular maintenance, and possibility of by passing also exists
Self-adjustable guard	It offers a wide range of positions for the operator to adjust the guard according to the nature of work	It may obstruct the visibility and requires frequent maintenance
Interlock guard	It provides good protection between the user and machine, and maintenance work is easy to carry out in such type of guard	It is very easy to disengage, and it should be adjusted very carefully so that it serves it function smoothly

operation so that it becomes safer. It may not save the other part of human body, it is intended to safe the hands only [31].
Applicable to machines: power actuated machinery.

2. **Pullback device**: It is basically provided with the machines with stroking action, and cable is attached to the wrists, hands, or arms. When the slide/ram is up, it allows the hands to reach point of operation [25, 26].
Applicable to machines: machines with stroking action.
3. **Restraint device**: Here arrangement includes the cable or straps attached to the operators hand and a fixed point, in such a manner so that operators move hand in a predetermined safe area [25, 26].
Applicable to machines: Activity requires hand feeding.
4. **Safety tripwire cable**: A cable is arranged around the periphery of the machine, accessible to the operator, and it can be pulled in order to stop the machine. It is easy to reset the cable in order to regaining its normal operation [32].
Applicable to machines: conveyors.
5. **Gate (movable barrier device)**: It is basically a type of barrier which is synchronized with the operation of machine like when gate is open machine will not actuate, and similarly when it is off, proper functioning of machine will be there. Good visibility is around the machine, and it enclosed the machine in such a way so that no dangerous part of machine can be accessible [33].
Applicable to machines: It is a movable barrier device and applicable to those machines which has a potential to human access.
6. **Photoelectric cell**: It comprises of an emitter and receiver, uninterrupted light beam between the emitter and the receiver is to be maintained at all times, and in case any human interruption the device is deactivate. In order to enhance its sensitivity, it may be supplied along with the amplifier and relay system also [34].
Applicable to machines: electrically operated machines.

7. **Pressure sensitivity body bar**: As the name suggests, basically when a person working on any machinery losses balance or trips he or she applies pressure (depressed) on this particular device, which deactivates the machine and the person is saved from protentiathe accident. Here position of the device plays critical role since device shall be activated before the operator is harmed by the machine [35].
   Applicable to machines: rubber mill, etc.
8. **Voice actuated control system**: It works on the principle of radio frequency; actually it comprises of two radio frequency unit, transmitter and receiver. Operator spoke the command (audio signal) which get transmitted by the transmitter to receiver and converted into the digital signal by the voice interpreter; finally it gets to the machine tool and acts accordingly [36].
   Applicable to machines: computer numerically controlled (CNC) machine tool.
9. **Switching device (foot pedal)**: It is a device used for two main purpose, one is to reactivate the system that requires a command after power failure and second for emergency shutdown [37]. It consists of a switching device provided with a restart arrestor to prevent the unintended startup of the machine. It could also be used to disconnect the electric circuit in case of an emergency [38].
   Applicable to machines: portable apparatus, for handling tubular/rod-shaped workpiece.
10. **Translation stopper**: Actually it is applicable for machinery which is working as to cut the workpiece by the rotating blades and involves human interaction also [39] Whenever there is any chance of dangerous position like contact between human and blade, sensor detects it and immediately stops the blades for any further rotation [40].
    Applicable to machines: cutting saw.

Here below given the merits and demerits of above-mentioned protective devices (Table 2).

As per the U.S. Bureau of Labor statistics of 2005 survey data shows total 8,450 number of amputation cases (nonfatal) took place for all private industry. 44% of all amputation cases occurred in the manufacturing sector and rest in other sectors. 60% (5080) of the amputation cases are found while use and care of machines such as conveyors, saws, presses, bending, rolling, powered, and non-powered tools.

Another survey 2003 through 2010 shows data like, 94% of the machine-related fatalities are involving mobile machines for agriculture/forestry/fishing industries, and 78% machine-related fatalities involve mobile machine for construction, whereas 64% of machine-related fatalities for manufacturing industries are due to stationary machines. There is decline in number of deaths by 22% for mobile machinery whereas 56% decline for stationary machine from 1992 to 2010.

Amputation injury mainly occurred in manufacturing industries; more than half of the amputation injury is due to use or care of machine, and below given chart is comparing the nonfatal amputation injury occurred in different types of private industries in which manufacturing industries are at the top (Fig. 1).

Table 2 Advantages and disadvantages of protective devices

Protective device/method	Advantage	Disadvantage
Photoelectric/ radio frequency (presence sensing)	It is not very complicated and allows multiple user safety at a time and provides free movement	Limited to the machines on which it can be applicable and not meant for safety from mechanical damage
Pullback	No requirement of auxiliary barrier for dangerous part and allows the hand to enter point of operation as per the feeding requirement	Not offer free movement and may result in obstruction to work area. It will work only when adjusted properly
Restraint	Person exposed cannot reach the dangerous area, and it requires less maintenance	Requires regular supervision and inspected before every new task. May obstruct workplace and becomes ineffective when not adjusted properly
Safety tripwire	Simple to use and found near/ periphery of machine to be protected	Only intended to protect operator, reset manually
Two-hand control	It is safer to protect the operators hand and allows free movements of hand	Safety system can be bypassed and save operator only
Gate	Prevent access to danger area	Requires maintenance and may interfere with the visibility

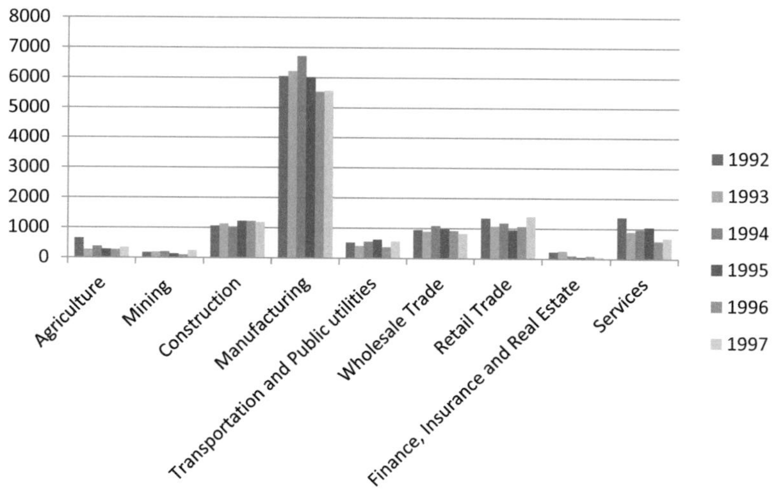

Fig. 1 Number of nonfatal amputations by major industry division, private industry, 1992–97. *Source* U.S. Department of Labor, Bureau of Labor Statistics, Survey of Occupational Injuries and Illnesses and Census of Fatal Occupational Injuries (1992–1997)

3 Results and Discussion

Following are the machines which possess high potential to cause the amputation injury when worker is exposed to the work in manufacturing industries. Here primary function of all those machinery is also defined, basically the intent for which they are designed (Table 3).

All the above-mentioned machines carry out hazardous mechanical motions and actions which may result in mainly amputation injury, and such cases can be avoided by taking prior preventive actions. Guarding and other protective equipment can be installed in place so that machine can be safer for the workers who are supposed to work on that [41] Here below various protective devices (except guarding) are mentioned based on the study which are most preferable or suitable for all such machinery (Table 4).

Table 3 Functions of different machinery

Name of machine	Primary function
Mechanical power presses	To reform the work piece in desired way by reciprocating motion
Power press brakes	It is similar to mechanical power press, but here press brake operation is mechanical/hydraulic
Conveyors	Conveyors are used to transport materials horizontally, vertically, or inclined in many industries
Printing presses	Printing presses are used for printing newspapers, magazines, and books. It may vary from simple to complex
Roll-forming and roll-bending machines	To perform metal bending, rolling, or shaping functions
Shearing machines	To perform numerous functions such as squaring, cropping, and cutting to length
Food slicers	It is electrically operated machine, with rotatory blade which applies force to the food against the slicer blade
Cutting band saws	To cut wood, plastic, metal, or meat
Drill presses	To cut holes in wood/metal by rotating bit
Milling machines	To cut metal by rotating cutting device
Grinding machines	To alter the size, shape, and surface finish of the metal by rotating wheel
Slitters	To perform functions such as cropping, squaring by rotatory knives

Table 4 Suitable protective devices for machinery

Name of machine	Pullback	Restraint	Gates	Two-hand control	Presence sensing	Mats	Tripwire
Mechanical power presses	Suitable	Suitable	Suitable	Suitable	Suitable		
Power press brakes	Suitable	Suitable	Suitable	Suitable	Suitable		
Conveyors							Suitable
Printing presses				Suitable	Suitable	Suitable	
Roll-forming and roll-bending machines				Suitable	Suitable	Suitable	
Shearing machines	Suitable	Suitable		Suitable	Suitable		
Food slicers							
Cutting band saws							
Drill presses							
Milling machines				Suitable	Suitable		
Grinding machines							
Slitters					Suitable	Suitable	

As above study showing that every type of protective equipment cannot be applicable to any type of machines, and in many machines such protective equipment are even not applicable so we need to look forward for some alternative in which guarding is one of the best options in all such cases since it provides a kind of physical barrier between the machine and men [42]. Here below description of different types of guards is given which can be suitable to use in order to cover the dangerous parts of the machines and also defines the portion which needs to be covered with all those types of guards (Table 5).

Another research shows the statistics of 562 severe accidents (i.e., fatal or permanent disability) take place in mining industry of USA during 2000–2007 in which 41% accidents are accounted for machine-related accidents in mining industries. It mainly includes machines like conveyors, milling machines, rock bolting machines. To be more precise, 14%, i.e., 80 out of 562 severe accidents is due to conveyors (mostly at surface operation), 25% of injuries recorded while carrying out maintenance and repair of all types of machinery. Particularly looking forward to maintenance activity, it includes four fatal accidents and 21 amputation accident, machine getting restarted unexpectedly while someone exposed to perform such activity.

Table 5 Applicable guards for machinery

Name of machine	Type of guard	Portion required to cover
Mechanical power presses	Not specific	Point of operation
Power press brakes	Not specific	Point of operation
Conveyors	Fixed	Nip point, shear point, power transmission apparatus, and other moving parts
Printing presses	Fixed	All mechanical hazards point including nip points, chains, sprockets
Roll-forming and roll-bending machines	Fixed or adjustable and interlock	Point of operation (feed in and out). Interlock guard is for other moving parts including power transmission apparatus
Shearing machines	Fixed or adjustable	In feed, pinch point, and shear point
Food slicers	Not specific	To cover portion of blade that is not in use, i.e., top and bottom
Cutting band saws	Self-adjustable	Over entire blade except working portion of blade
Drill presses	Fixed and adjustable	Motors, pulleys, and belts. Adjustable guard used to cover the portion which is not in use of bit and chuck
Milling machines	Fixed, movable, or interlock	Point of operation and around power transmission components
Grinding machines	Fixed	Power belts, spindle end, nut etc
Slitters	Fixed or adjustable and interlock	Point of operation. Interlock or fixed guard can be used to cover power transmission apparatus

Source Accident database maintained by United States Department of Labor, Mine Safety and Health Administration

Above discussed different type of machine safeguarding including protective equipment cannot assure the full proof safety when a person exposed to any machinery to work; several other factors need to be considered while taking into safety consideration, here below some other safeguarding are given which should be taken care along with the machine safeguarding which already been discussed [43]:

- Ensure proper isolation procedure and LOTO system has been followed at the time of maintenance and servicing work of the machinery.
- Follow safe work procedure.
- Ensure compliance of the safety system installed by conducting regular inspection [44].
- Ensure all activity related to machine including maintenance work should be carried out by authorized person only.
- Restriction on wearing jewelry, loose clothing, etc.
- Emergency stop button should be clearly marked and unobstructed [45].
- Ensure proper training and other safety meeting are conducted in advance and all the hazardous activities are carry out under supervision till compeletion of assigned job.
- Regular maintenance of machinery should be arranged as per safe operating procedure.
- Arrangement of warning sign and audible sound is required to inform all people in case of emergency [46].
- Special type of shielding arrangement is required for specific activity to control chips, etc.
- On completion of work, all other auxiliary equipment should be kept back to their original place.
- Special training of the operators is also required including refresher training.
- Distance/location and automatic feed/ejection technique can also be applied in order to enhance safety [47].

4 Conclusion

As world is moving toward generating world class science and technology, machines are playing dominant role since from past when most of the work were done by humans only, which is not the case in present scenario; in order to make the task perfect and fast, we are adopting new machine technology to do it in best possible way but humans are exposed to different kinds of such machinery not only while use of machinery but also for maintenance of it. Today machines are involving very hazardous processes, and from study also it is quite clear that huge number of accidents occur while doing maintenance and repair work of machine, whatever the reason is, but injury that resulting from such accidents are too dangerous mainly it results in fatal, amputation, or permanent nature injury.

It is very much necessary to provide all possible machine safeguarding in order to protect the worker expose to any such type of hazardous machines and all activity related to machine carried out by authorized, trained, and competent person only. Training is also one of the important parts of secondary safeguarding since if person is not having the proper knowledge of machine safeguarding, or other safety aspects then it becomes too difficult to avoid accidents.

This paper can serve as a guide for proper safeguarding of the machines mainly which has a potential to cause amputation or severe injuries. Recorded machine-related injuries are mainly occurred in manufacturing industries; hence machines discussed here are of same kind.

Management should be strict toward the installation of all applicable machine guarding and also look forward to the proper functioning and maintenance of the machine on regular interval. Each person shall undergo prior training and must have enough knowledge about the functioning and safeguarding installed on any machinery.

References

1. Courtney, A., and M. Ng. 1984. Hong Kong female hand dimensions and machine guarding. *Ergonomics* 27 (2): 187–193.
2. Vaillancourt, D.R., and S.H. Snook. 1995. A review of machine-guarding recommendations. *Applied Ergonomics* 26 (2): 141–145.
3. Davies, B., A. Abada, et al. 1980. A comparison of hand anthropometry of females in three ethnic groups. *Ergonomics* 23 (2): 179–182.
4. Davies, B., A. Abada, et al. 1980. Female hand dimensions and guarding of machines. *Ergonomics* 23 (1): 79–84.
5. Parker, D.L., L.M. Brosseau, et al. 2009. A randomized, controlled intervention of machine guarding and related safety programs in small metal-fabrication businesses. *Public Health Reports* 90–100.
6. Grzan, J., and R. Keats. 2006. Machine guarding system having a sensing mat with status indicator lights. Google Patents.
7. Roozbazar, A. 1977. Ergonomics of machine guarding. *National Safety News* 116 (1): 53–59.
8. Chew, D.C. 1988. Effective occupational safety activities: Findings in three Asian developing countries. *International Labour Review* 127: 111.
9. Heinrich, H.W., D.C. Petersen, et al. 1980. *Industrial accident prevention: A safety management approach*. McGraw-Hill Companies.
10. Munshi, K., D. Parker, et al. 2005. Machine safety evaluation in small metal working facilities: An evaluation of inter-rater reliability in the quantification of machine-related hazards. *American Journal of Industrial Medicine* 48 (5): 381–388.
11. Brubaker, S.A. 1997. Optimum machine guarding. *Occupational Health & Safety (Waco, Tex.)* 66 (12): 37.
12. Mendeloff, J., and W.B. Gray. 2005. Inside the black box: How do OSHA inspections lead to reductions in workplace injuries? *Law & Policy* 27 (2): 219–237.
13. Yablonsky, J.E. 1993. Machine guard safety gauge/Guardchek™. Google Patents.
14. Bernard III, T.W., and A.J. Keown. 1978. A goal programming application of capital project selection in the production area. *AIIE Transactions* 10 (1): 52–57.
15. Etherton, J.R. 2007. Industrial machine systems risk assessment: A critical review of concepts and methods. *Risk Analysis* 27 (1): 71–82.

16. Richards, E.G. 1981. Machine guarding apparatus. Google Patents.
17. Lincoln, J.M., D.L. Lucas, et al. 2008. Reducing commercial fishing deck hazards with engineering solutions for winch design. *Journal of Safety Research* 39 (2): 231–235.
18. Jiang, B.C., A.Y. Lio, et al. 1991. An evaluation of machine guarding techniques for robot guarding. *Robotics and Autonomous Systems* 7 (4): 299–308.
19. Parker, D., L. Brosseau, et al. 2007. A comparison of the perceptions and beliefs of workers and owners with regard to workplace safety in small metal fabrication businesses. *American Journal of Industrial Medicine* 50 (12): 999.
20. Haviland, A., R. Burns, et al. 2010. What kinds of injuries do OSHA inspections prevent? *Journal of Safety Research* 41 (4): 339–345.
21. Bello Jr, R. (2002). Lighted automotive floor mats, Google Patents.
22. Davies, B.B., and B. Mebarki. 1981. Hand movement times and machine guarding. *Ergonomics* 24 (5): 387–391.
23. Barnett, R.L., and D.B. Brickman. 1986. Safety hierarchy. *Journal of Safety Research* 17 (2): 49–55.
24. Yamin, S., D. Parker, et al. 2014. National machine guarding program: Design of a machine safety intervention. Safety Science Monitor (Sweden) 18 (1).
25. Parker, D.L., S.C. Yamin, et al. 2015. National machine guarding program: Part 1. Machine safeguarding practices in small metal fabrication businesses. *American Journal of Industrial Medicine* 58 (11): 1174–1183.
26. Parker, D.L., S.C. Yamin, et al. 2015. National machine guarding program: Part 2. Safety management in small metal fabrication enterprises. *American Journal of Industrial Medicine* 58 (11): 1184–1193.
27. Gray, G.C. 2009. The responsibilization strategy of health and safety neo-liberalism and the reconfiguration of individual responsibility for risk. *British Journal of Criminology* 49 (3): 326–342.
28. O'Rourke, D. 2000. *Monitoring the monitors: A critique of PricewaterhouseCoopers (PwC) labor monitoring*. Department of Urban Studies and Planning Massachusetts Institute of Technology.
29. Rasmussen, A. 1929. Barbecue-machine guarding device. Google Patents.
30. Becker, J.E. 2001. Implementing 5S: To promote safety & housekeeping. *Professional Safety* 46 (8): 29.
31. Okunribido, O.O. 2000. A survey of hand anthropometry of female rural farm workers in Ibadan, Western Nigeria. *Ergonomics* 43 (2): 282–292.
32. Rautiainen, R.H., J.L. Lange, et al. 2004. Injuries in the Iowa certified safe farm study. *Journal of Agricultural Safety and Health* 10 (1): 51.
33. Blundell, J. 1987. *Safety engineering—Machine guarding accidents*. Del Mar, CA: Hanrow Press.
34. Montgomery, W. 1956. Machine guarding. *The Journal of the American Society of Safety Engineers* 57–60.
35. Roberts, V.L. 1980. *Machine guarding: A historical perspective*. Institute for Product Safety.
36. Switches, S.S.I. 1998. *Machine guarding safety products catalog & handbook*. NY: Hawthorne.
37. Marshall, R., and E. Bingham. 1977. *Essentials of machine guarding*. USA: Occupational Safety and Health Administration.
38. Safety, M.L.W. 1989. *Guidelines for machine guarding*. Canada: Health and Support Services Division.
39. Imrhan, S.N., M.-T. Nguyen, et al. 1993. Hand anthropometry of Americans of Vietnamese origin. *International Journal of Industrial Ergonomics* 12 (4): 281–287.
40. Courtney, A. 1984. Hand anthropometry of Hong Kong Chinese females compared to other ethnic groups. *Ergonomics* 27 (11): 1169–1180.
41. Brown, G.D., and D. O'rourke. 2007. Lean manufacturing comes to China: A case study of its impact on workplace health and safety. *International Journal of Occupational and Environmental Health* 13 (3): 249–257.

42. Sorock, G.S., E. Smith, et al. 1993. Hospitalized occupational finger amputations, New Jersey, 1985 and 1986. *American Journal of Industrial Medicine* 23 (3): 439–447.
43. Ezenwa, A.O. 2001. A study of fatal injuries in Nigerian factories. *Occupational Medicine* 51 (8): 485–489.
44. Camper, W.J. 1976. Sharpener for commodity slicing machine. Google Patents.
45. Shie III, H.J. 1976. Sharpener for commodity slicing machine. Google Patents.
46. Mendeloff, J. 1984. The role of OSHA violations in serious workplace accidents. *Journal of Occupational and Environmental Medicine* 26 (5): 353–360.
47. Samuel, R.J. 1975. Apparatus for selectively guarding a rotary blade. Google Patents.

Application of Rapid Tooling for Vacuum Forming to Reduce Cycle Time

Beporam Iftekhar Hussain, Mir Safiulla and B. Khaleelu Rehman

Abstract Manufacturing technology has been continuously improving with effective research to reduce the cycle time of production of an object. Vacuum forming is one such process of manufacturing simple parts from thin sheet of material usually made of plastic. This work emphasizes the adoption of rapid tooling concept to vacuum forming process with an objective of reducing the cycle time of production. The cooling phase of the entire cycle time is identified as a means of improvement through its reduction. It contributes to the reduction of manufacturing lead time and time to market the product. The comparative study is conducted by inserting tube inserts of two different materials both of traditional type and rapid tooling type in order to cool the mould. The results conjecture the improvement of cooling time in tools with rapid tooling concept in comparison with those of traditional system of cooling.

Keywords Manufacturing · Vacuum · Tooling · Materials · Cooling

B. I. Hussain (✉)
Department of Mechanical Engineering, Bapatla Engineering College, Bapatla, Guntur 522101, Andhra Pradesh, India
e-mail: iftekharhussainb@gmail.com

M. Safiulla
Department of Mechanical Engineering, Ghousia College of Engineering, Ramanagaram 562159, Karnataka, India
e-mail: safiullamir@gmail.com

B. Khaleelu Rehman
Department of EIC, CoES, University of Petroleum and Energy Studies, Dehradun 248007, Uttarakhand, India
e-mail: krehman@ddn.upes.ac.in

N. A. Siddiqui et al. (eds.), *Advances in Fire and Process Safety*, Springer Transactions in Civil and Environmental Engineering, https://doi.org/10.1007/978-981-10-7281-9_20

1 Introduction

Manufacturing technology is as old as human life itself. The attempt to improve the available manufacturing technologies has been the need of the hour of every time. Several technologies are available today to manufacture the parts made of plastic material as it is gaining popularity and usage because of its less density and low cost of material and manufacturing. Amongst them, a technique called vacuum forming has been used to fabricate simple parts of convex or concave shapes or even hollow shapes of predetermined specifications. Vacuum forming comes under the umbrella of thermoforming process of heat application and creating negative air pressure towards the cavity from the plastic sheet material heated up to forming temperature. The principle of vacuum forming process is visualized in Fig. 1.

The stages consisted in a manufacturing cycle of vacuum forming process are moving (plastic) sheeted roll, heating with electric filaments, suction through vacuum pump to get desired shape and cooling the mould for continuous production. Amongst these stages, cooling of mould is one where there is scope for improvement that is sought through considerable study. The present investigation aims at thoroughly studying the cooling behaviour of the mould in mass production of sample parts made through vacuum forming process.

2 Literature Survey

Several works are reported in the literature pertaining to enhancement of cycle time for various manufacturing process through mould cooling. Peng et al. [1] had studied the steady-state temperature distributions in two kinds of continuous casting moulds, taking into account the effects of the properties of the work and tool materials through the rearrangement of the cooling water slots. Hassan et al. [2] had studied the effect of the cooling system on the shrinkage rate of a polystyrene product produced through injection moulding process. Hu et al. [3] had conducted numerical investigation on the cooling performance of hot stamping tool with various cooling channel designs, i.e. straight hole, longitudinal CCC (conformal cooling channel), transversal CCC, parallel CCC, serpentine CCC. On the other

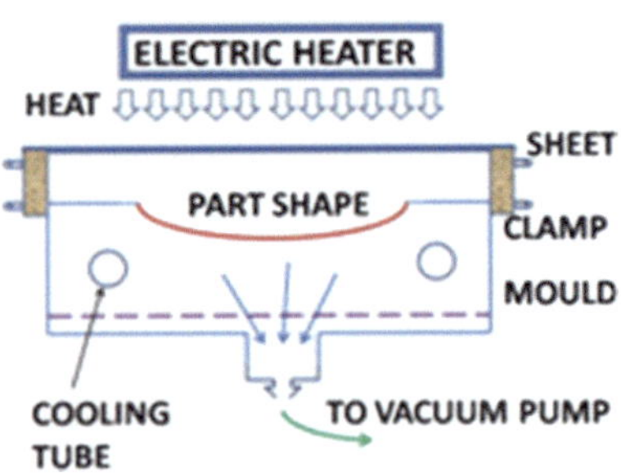

Fig. 1 Principle of vacuum forming process

side, Wang et al. [4] had conducted the numerical simulation of ABS material's vacuum forming process. Dimitrov et al. [5] had investigated to find the most suitable layout like conventional cooling, conformal, or surface cooling of a moulded part with an objective of optimizing and predicting the mould cooling cycle time. Garcia et al. [6] had presented a case study on the usage of additive manufacturing philosophy to improve the quality and properties of an injected part with a motto to reduce the injection cycle time. Wu et al. [7] had suggested a framework for optimizing design of additive manufacturing-based injection mould with conformal cooling to produce plastic parts. Saifullah et al. [8] had investigated the bimetallic conformal cooling channel design with high thermal conductive copper tube insert for injection moulds. The present investigation is aimed at comparing the conventional and conformal cooling channels inside the mould with case studies of two different materials, viz. copper alloy and stavax supreme in order to enhance the productivity by reducing the cooling time and hence the total cycle time.

3 Methodology

Study has been conducted by considering a container made of an amorphous, glassy polymer called polystyrene (PS) which is a form of plastic material. The container to carry some sort of low-density foodstuff is having the specifications of 85 mm length, 75 mm width and 30 mm height with an extra projection or canopy of 5 mm to hold it. It is shown in Fig. 2. The fillet radius is of 15 mm at the vertical

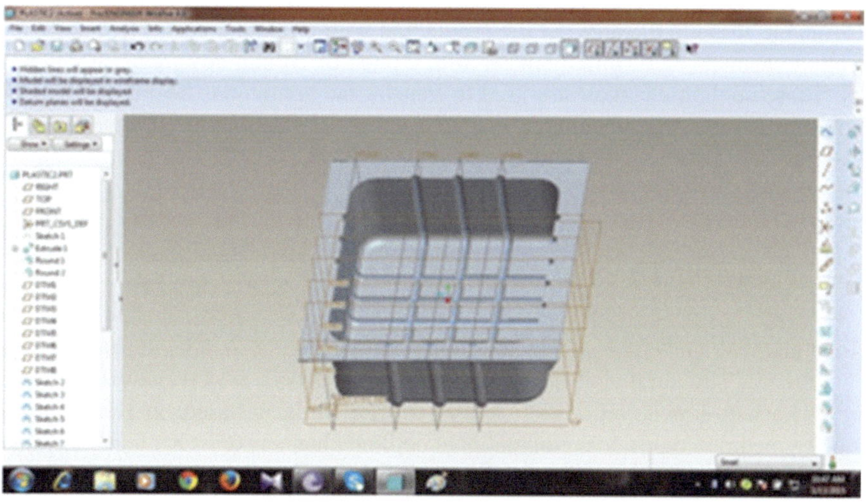

Fig. 2 CAD model of container modelled in Pro/ENGINEER 4.0 PLM software

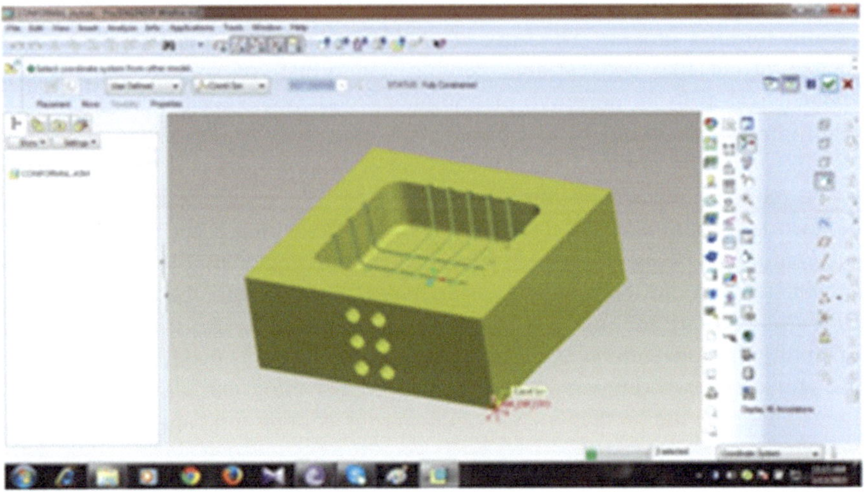

Fig. 3 CAD model of the vacuum forming mould

sides and 3 mm at the four corners of the object. Alternatively, grooves are designed (as shown) for getting the sufficient structural strength of the component. Coming to the properties of the material, at 20 °C temperature, it has a density of 1050 kg/m^3 and thermal conductivity of 157 K(W/m K) with a specific heat of 896 J/kg K.

The mould for fabricating the container using vacuum forming process is also modelled in Pro/ENGINEER 4.0 as shown in Fig. 3. The mould dimensions are designed based on [9] with specifications of a square of 150 mm and height of 50 mm. The mould is made of Aluminium alloy as it offers low density and reasonable strength to carry out the production with minimum cost. The mould has cooling channels of 10 mm diameter with a pitch of 20 mm designed according to [9]. These cooling channels are used to transfer the continuous heat applied to the mould to the sink via conduction and convection. The coolant used for analysis is water at 20 °C with an ambient temperature of 25 °C.

3.1 Transient Thermal Structural Analysis

Finite element analysis is conducted to extract the results. A time-dependent transient thermal structural analysis is conducted using ANSYS 14.0 analysis software in workbench simulation module for 30 s. The study is based on two types of cooling methodologies, viz. conventional cooling channels (CNCC) and conformal cooling channels (CFCC). The difference between the two is understood that CNCC

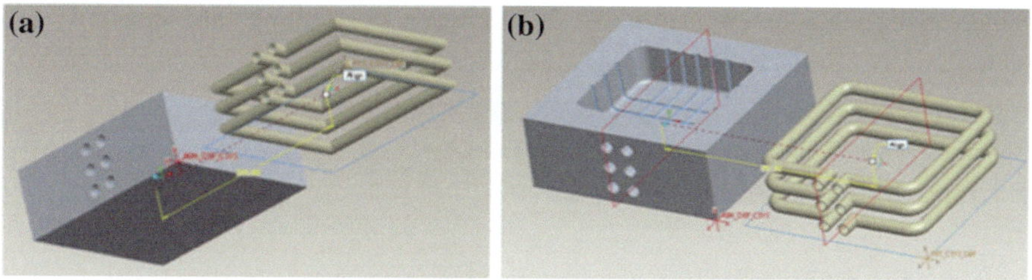

Fig. 4 **a** CNCC within the mould (exploded), **b** CFCC within the mould (exploded)

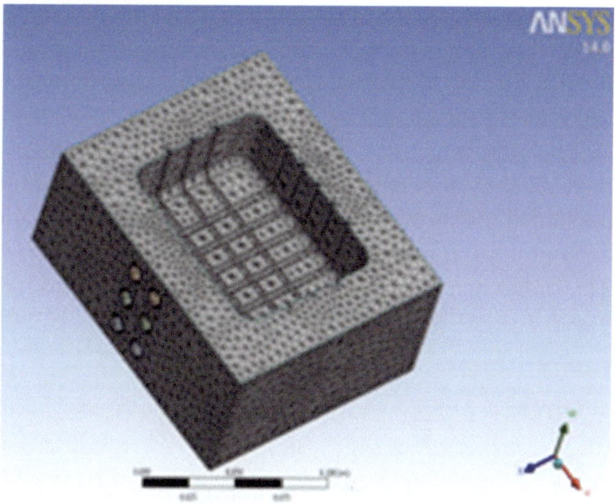

Fig. 5 Free meshed model of mould in ANSYS 14.0

corresponds to the cooling channels that are drilled through drilling machine to get the holes of required diameter (in this case 10 mm), whereas CFCC corresponds to the cooling channels that are obtained from rapid tooling concept of additive manufacturing system without any tools/jigs or fixtures directly from the 3-D CAD model. Further, the study is conducted by inserting tube inserts of two different materials, viz. copper and stavax supreme within the cooling channels. So, altogether four classes of results are expected from the study. The exploded views (for better visualization) of the three-dimensional CAD model of CNCC and CFCC are shown in Fig. 4a, b, respectively. The properties of the mould material used for analysis are taken from [10]. Free mesh is carried out to conduct the analysis. The meshed model is displayed in Fig. 5. The element type used is Solid 187 which is available in ANSYS library of elements. Coming to the boundary conditions, all the side surfaces and bottom surface of the mould are constrained and a forming temperature of 150 °C is applied in the mould-part interface while applying a chilled coolant water temperature of 20 °C inside the cooling tube inserts to dissipate the heat from the mould.

4 Results and Discussion

From the present investigation, the thermal flux for each of the case is determined, which is the energy transfer per unit time from the given surface. The thermal flux results using copper tube insert for conventional cooling channels (CNCC) and conformal cooling channels (CFCC) are portrayed in Figs. 6 and 7, respectively. In the same way, the thermal flux results using stavax supreme insert for conventional cooling channels (CNCC) and conformal cooling channels (CFCC) are portrayed in Figs. 8 and 9, respectively. A comparison of thermal flux results of all the four cases is tabulated in Table 1.

From the results, it is apparent that CFCC has a greater value of thermal flux than CNCC indicating a higher energy being transmitted to the sink through coolant. This shows an inference that conformal cooling channels are better than the conventional form in terms of heat transfer between the mould and coolant (water). This inference is depicted by both the type of insert materials conducted.

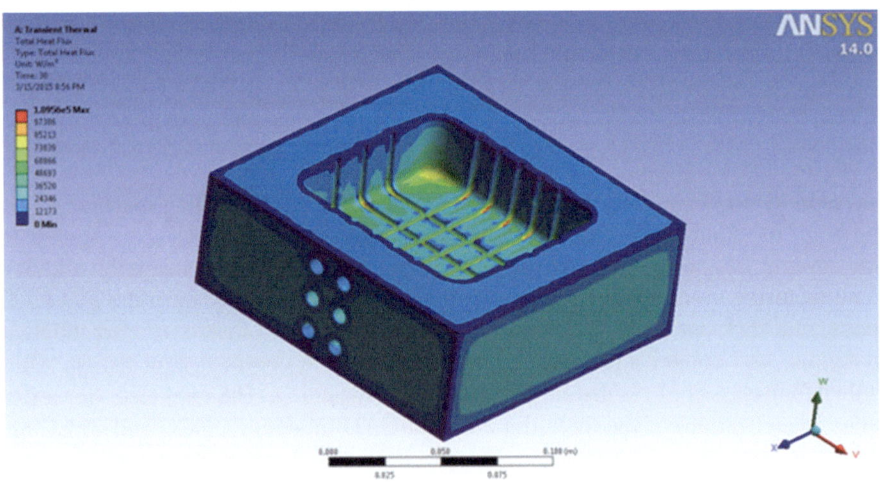

Fig. 6 Total thermal flux for CNCC with copper tube insert

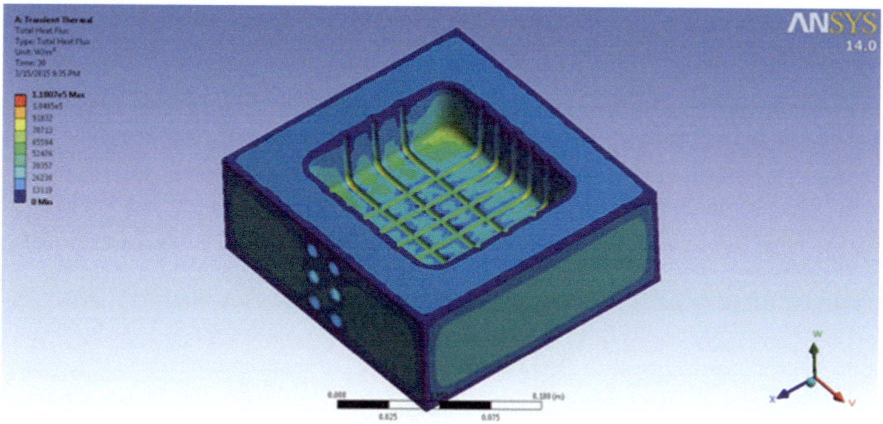

Fig. 7 Total thermal flux for CFCC with copper tube insert

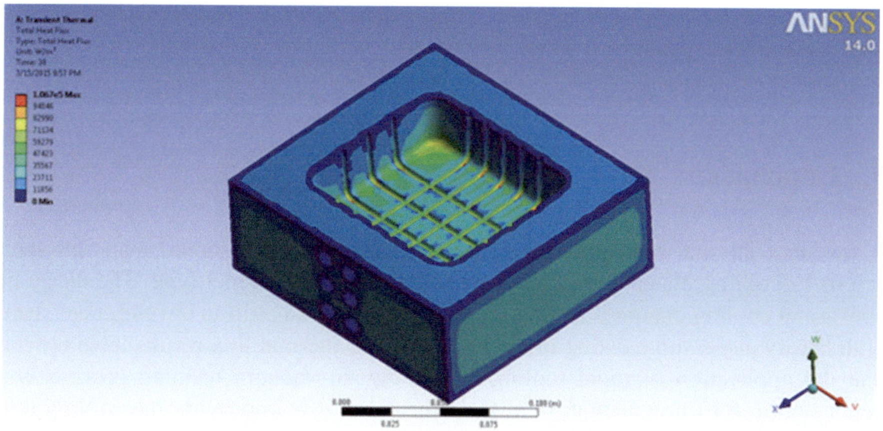

Fig. 8 Total thermal flux for CNCC with stavax supreme insert

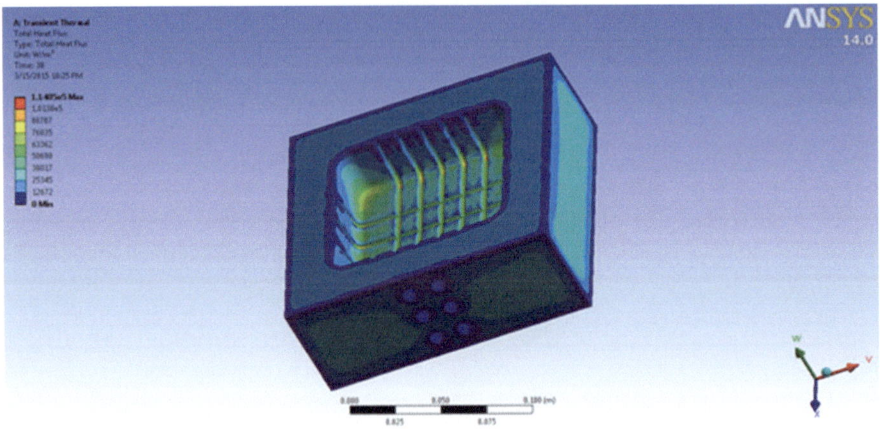

Fig. 9 Total thermal flux for CFCC with stavax supreme insert

Table 1 Total thermal flux values for each case in W/m^2

	Copper tube insert	Stavax supreme insert
Conventional cooling channel (CNCC)	109,560	106,700
Conformal cooling channel (CFCC)	118,070	114,050

5 Conclusion

A transient thermal structural analysis is conducted on two mould materials with inserts in cooling channels of conventional type and conformal type. The usage of conformal cooling channels in vacuum forming process is still in development stage with hardly any setup existing in the industry. The thermal flux results have proven that the application of rapid tooling technology in vacuum forming process is a better option for implementation in the industry as it is improving the cooling rate of the mould by 7.7% for high thermal conductive copper material and 6.9% for stavax supreme mould material. The vacuum forming industry should give the ready answer for enhancement of total cycle time of the process.

References

1. Peng, X., J. Zhou, Y. Qin. 2005. Improvement of the temperature distribution in continuous casting moulds through the rearrangement of the cooling water slots. *Journal of Materials Processing Technology* 167: 508–514.
2. Hassan, H., N. Regnier, C. Pujos, E. Arquis, and G. Defaye. 2010. Modeling the effect of cooling system on the shrinkage and temperature of the polymer by injection molding. *Applied Thermal Engineering* 30: 1547–1557.

3. Hu, P., B. He, and L. Ying. 2016. Numerical investigation on cooling performance of hot stamping tool with various channel designs. *Applied Thermal Engineering* 96: 338–351.
4. Wang, S., A. Makinouchi, T. Tosa, K. Kidokoro, M. Okamoto, T. Kotaka, and T. Nakagawa. 1999. Numerical simulation of acrylonitrile-butadiene-styrene material's vacuum forming process. *Journal of Materials Processing Technology* 91: 219–225.
5. Dimitrov, D., A. Moammer, T. Harms. 2010. Cooling channel configuration in injection moulds. In *Innovative developments in design and manufacturing*, ed. Bartolo et al., 355–360. London: Taylor & Francis Group.
6. Garcia, M.A., C. Garcia-Pando, C. Marto. 2012. Conformal cooling in moulds with special geometry. In *Innovative developments in virtual and physical prototyping*, ed. by Bartolo et al., 409–412. London: Taylor & Francis Group.
7. Wu, T., S.A. Jahan, P. Kumar, A. Tovar, H. El-Mounayri, Y. Zhang, J. Zhang, D. Acheson, K. Brand, and R. Nalim. 2015. A framework for optimizing the design of injection molds with conformal cooling for additive manufacturing. *Procedia Manufacturing* 1: 404–415.
8. Saifullah, A.B.M., S.H. Masood, and I. Sbarski. 2012. Thermal–structural analysis of bi-metallic conformal cooling for injection moulds. *International Journal of Advanced Manufacturing Technology* 62: 123–133.
9. Lin, Z., M. Chou. 2002. Design of the cooling channels in nonrectangular plastic flat injection mold. *Journal of Manufacturing Systems*, 21/3, 167–186.
10. Mat Web. *The online material information resource homepage*. http://www.matweb.com.

Design and Implementation of On-Site Emergency Planning for Liquefied Petroleum Gas Utilizing Automobile Industries

S. Abishek and R. K. Elangovan

Abstract Emergency in an industry is an unexpected scenario created by the major accident whose damages and consequences are very high in magnitude, and the existing facilities are inadequate to cope up with the situation. The On-site emergency is the one whose consequences are confined within the industry premises, and they does not cross boundary of the factory. On-Site emergency planning is an important activity to prevent, control and mitigate the on-site emergencies in the industry. LPG is widely used as a fuel, which is a reliable, clean, environment-friendly and less polluting. LPG is widely used in the automobile industries as a fuel for providing energy required in the processes such as heat treatment, paint dryer. This paper describes the design and implementation of on-site emergency planning for LPG utilizing automobile industries. On-site emergency planning is an integral and essential part of loss prevention strategy in industries and conforms to the applicable statutory requirements. It assesses risks in these industries, and accordingly appropriate response procedures are described. Consequence analysis is used for calculating damage distances of scenarios like major fire, explosion, toxic release. DNV software is widely used for consequence analysis studies, which give damage distances due to fire, explosion and toxic release. The results of the consequence analysis are incorporated in the on-site plan and appropriate response procedures are laid down in the on-site emergency preparedness plan. The plan also identifies team leaders along with members and provides for their assigned duties and responsibilities. Conducting periodic mock

S. Abishek (✉)
Coimbatore Institute of Technology, Coimbatore 600014, India
e-mail: abishek@senthilsafety.com

R. K. Elangovan
DGFASLI, Government of India, Mumbai 400022, India
e-mail: elanrk@gmail.com

N. A. Siddiqui et al. (eds.), *Advances in Fire and Process Safety*,
Springer Transactions in Civil and Environmental Engineering,
https://doi.org/10.1007/978-981-10-7281-9_21

drills and evaluating the outcomes with the criteria assess the effectiveness of the designed On-site emergency plan. The benefits and constraints in designing and implementing On-site emergency plan are also discussed in the paper.

Keywords Emergency · On site emergency · LPG · Automobile industry Mock drill

1 Introduction

Modern industry characterized by complex process and technology is open to an ever-increasing danger form disasters, which can seriously affect the safety, security and stability of the organization [1, 2]. Since LPG is a hazardous material, subsequently damage effect is also very extensive. LPG utilizing automobile industry is taken for the designing, preparation and implementation of on-site emergency plan along with risk assessment report as part of the mandate requirement [3, 4]. Based on the damage effect arrived from consequence analysis, the citation of building location and other hazardous locations or processes is designed. Usually disasters are of two types,

1. Natural disasters such as earthquakes, floods, tsunamis, cyclones, lightning.
2. Man-made disasters included dangerous spills and leak of chemicals, fires and explosions, hit by external objects, contamination and poisoning of food, terrorist attacks.

The result has been extensive damage to men and material on either way. Primarily emergencies are grouped into two types one is major accidents/disasters, which include potential cause of series injury or loss of life and leads to extensive damage to on-site as well as in off-site and the other one is minor which leads to less impact or may be repairable.

An emergency plan is an informative document containing all critical operative emergencies and control measures and which acquaints the occupants of a factory or occupancy with procedures to be implemented, during an emergency.

2 Types of Emergency

Emergency planning is an integral and essential part of loss prevention strategy. Emergency is a general term implying hazardous, both inside and outside the installations. There are two types of emergency. They are on-site emergency and the other one is off-site emergency [2]. On-site emergency is defined as the damage effect confined within the premises, and off-site emergency is the emergency extends beyond the factory premises. The systematic planning for controlling the on-site and off-site with reference to the legal requirements is mandate.

3 Objectives of the On-site Plan

Emergency planning or preparedness is a comprehensive response plan to react to a number of foreseeable emergencies anticipated in the works and to contain the loss of human life, property and provide speedy and effective remedial measures [5, 6]. Identification of scenarios and their consequences is important content of the emergency planning. The main objectives of the plans are

1. Identifying the disaster potential scenarios and advance planning to combat and minimize the damage.
2. Disaster phase, i.e. warning, protective action like evacuation of personnel.
3. Containment of disaster by isolating, firefighting etc.
4. Rescue, relief assistance to the people affected in the works/community effectively and efficiently based on the actual needs and on the information collected locally both in advance of the disaster and as soon as possible after the disaster occurred.
5. Finally when the situation is contained, efforts are to be returned back to near normal conditions.

Of the above points, the first four are most relevant to the immediate attention to works management. The areas affected by each accident scenario can be identified by their consequences like pool fire, flash fire and toxic gas release. It would be appropriate to classify the hazards around the plant and to provide emergency measures in the area both inside and outside.

4 Legal Provisions Applicable for Preparing On-site Plan

The On-Site emergency plan is a mandatory document under various statutes of India. By virtue of the provision under

1. Section 41-B (4) of the Factories Act, 1948 and its amendments of 1987 [3],
2. Rule 13 and 14 of the Manufacture, Storage and Import of Hazardous Chemicals (MSIHC) Rules, 1989, framed under the Environment (Protection) Act, 1986.

Under the above section, the occupier is required to draw up the On-Site emergency plan along with detailed control measures for his factory. The occupier should also make the plan known to the workers and general public in the vicinity of the factory, with safety measures required for them in case of an emergency. The same plan shall be submitted to Chief Inspectorate of factories and to obtain necessary approvals.

5 Main Elements of On-Site Emergency Plans

The main elements of on-site emergency plans are as follows,

1. **Leadership and Administration**: occupier of the company has the power entire responsibility of the factory in case of any emergency.
2. **Emergency Organization**: defining responsible persons for emergency organization are key element and this force shall tackle and control any emergencies arising due to industrial operational activities.
3. **Role and Responsibilities of Key Personnel**: assigned key personnel in the organization and their roles and responsibilities shall be clearly defined how to act during emergency without a second delay. This includes top management to bottom-level employees.
4. **Communication Officer**: Human Resources head may act as communication officer. Communicating to various authorities and information spread out to the various departments, district administration and nearest factories for taking control measures.
5. **Emergency Control Centre (ECC)**: main emergency control centre shall be located in the site main controller room, and the alternate ECC room shall provide in the appropriate location within the facility.
6. **Mutual Aid**: Obtaining mutual aid agreement with nearest factories is mandate requirement. Agreement may be raised based on the types of support extended by the neighbouring industries such as firefighting equipments, manpower, ambulance, medical officer and water supply.
7. **Emergencies and Control Measures**: hazard identification and risk assessment (HIRA) may be studied for the hazardous installation and operations in the industry. Appropriate study may be conducted with the approved expertise, and document may be prepared with necessary recommendations and mitigation control measures.
8. **Emergency Evacuation Plan**: Preparation of emergency evacuation plan is a major task in on-site plan. This shall contain details of wind sacks, entry and exits, escape routs, emergency assembly points, OHC, eyewash and showers, fire extinguishers, first-aid box, sand bucket, fire water storage tank and pumping room, fire hydrants and water monitors, sprinklers, call points and emergency control centre.
9. **Source of Energy Control**: during emergency, all the sources of energy such as light and power may be shut down.
10. **Personnel Protective Equipment**: personnel protective equipment shall be procured and stored in the facility at the easily accessible places. Applicable PPE are safety shoes, facemask, helmet, self-contained breathing apparatus and fire suit.
11. **Occupational Health Centre**: full-fledged occupational health centre may be established as per the legal requirement with full-time medical officer, para-medical staff and dedicated ambulance with drivers.

12. **Rescue and Rehabilitation**: safety officer and medical officer are always taking roles and responsibilities of rescuing. Management, non-governmental organization and government authorities shall take care for rehabilitation.
13. **Training**: on-site plan shall contain applicable training for various emergencies due to incidents. Normally fire mock drill and evacuation mock drills shall be conducted with prescribed periodicity. Records shall be maintained.
14. **Revision of On-Site Plan**: whenever a change in the process, construction and additional installation occurs, revision shall be carried out. Time-to-time plan shall be updated and training shall be given.

6 Consequence Analysis for LPG Bullet

Generally, risk assessment is divided into two types, qualitative and quantitative risk assessments [5–10].

1. **Qualitative Risk Assessment**: 'Qualitative Risk Analysis' is an approach of identifying a hazard through a qualitative review of possible accidents that may occur based on previous accident experience and judgment. Other techniques are hazard and operability study, safety audit, what if analysis etc.
2. **'Quantitative Risk Analysis' (QRA)**: QRA means the systematic development of numerical estimates of the expected frequency and/or consequence of potential accidents associated with a facility or operation based on engineering evaluation and mathematical techniques. One of the quantitative techniques is consequence analysis. LPG-related catastrophic failures and damages in terms of distances are calculated by globally using software tool DNV Phast.

Catastrophic failure of 5 MT bullet in an automobile industry is taken for calculation purpose.

Storage parametres are as below:

1. Storage capacity	5 MT LPG Bullet
2. Storage pressure	2 kg/cm^2
3. Pipeline pressure	1 kg/cm^2
4. Temperature	35 °C
5. Wind speed	5 D—Neutral

With the above-said parametres, catastrophic failure of LPG bullets and applicable maximum credible scenarios is as below,

1. Thermal radiation due to fire for unloading LPG bullet
2. Shock wave pressure due to explosion for unloading LPG bullet
3. LPG unloading line rupture
4. LPG pool fire

5. Thermal radiation due to fire in the LPG bullet
6. Shock wave pressure due to explosion in the LPG bullet
7. Pressure relief valve/safety relief valve failure
8. Pool fire due to LPG bullet

We have selected two critical scenarios taken into account out of the above for calculation purpose. Catastrophic failures of LPG bullet and the scenarios are fire and explosion.

DNV phast software is used for calculation, and the results are tabulated in the below chapters.

7 Results of Thermal Radiation Distance Due to Fire

Thermal radiation due to fire is calculated, and its damage distance with relevant damage contour is shown below.

The damage effects are classified as follows (Fig. 1).

Radiation levels	Damage effects	Results obtained
37.5 kW/m^2	Major damage and 100% fatality	Not reached
12.5 kW/m^2	Minimum energy required for melting plastic and 1% fatality	118.71 m
4.0 kW/m^2	Cause pain for prolonged exposure	246.156 m

The results show damage effects in the plant due to fire (Fig. 2). Based on the results arrived, emergency evacuation plan may be prepared.

Results of shock wave pressure distance due to explosion are plotted below:

Shock wave pressure due to explosion is calculated, and its damage distance with relevant damage contour is shown below.

Radiation Effects: Fireball Ellipse

Path: Study\LPG 5MT Bullet\Catastrophic rupture

			Distance (m)		
			Category 1.5/F	Category 1.5/D	Category 5/D
Radiation Level	4	kW/m2	246.156	246.156	246.156
Radiation Level	12.5	kW/m2	118.171	118.171	118.171
Radiation Level	37.5	kW/m2	Not Reached	Not Reached	Not Reached

Fig. 1 DNV results for fireball due to catastrophic rupture

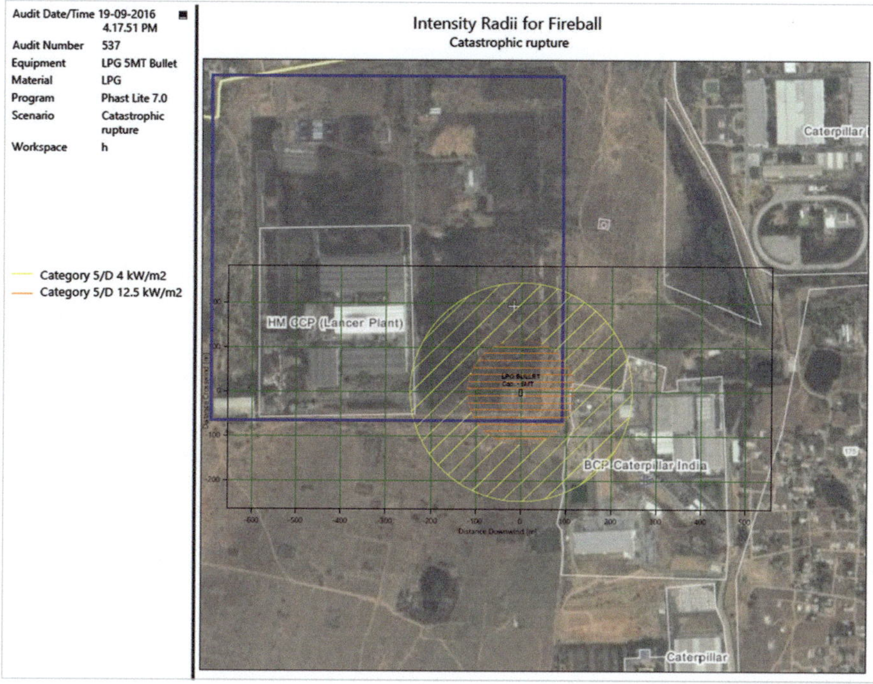

Fig. 2 Intensity radii for fireball

The damage effects are classified as follows (Fig. 3).

Shock wave pressure	Damage effects	Results obtained
0.2 bar	Heavy damage	141.522 m
0.13 bar	Moderate damage	168.319 m
0.02 bar	Significant or minor damage	531.482 m

Explosion Effects: Late Ignition

Path: Study\LPG 5MT Bullet\Catastrophic rupture

Explosion Model Used : TNT
Explosion Location Criterion: Cloud front (LFL fraction)
All distances are measured from the Source
All flammable results are reported at the cloud centreline height

			Maximum Distance (m) at Overpressure Level		
			Category 1.5/F	Category 1.5/D	Category 5/D
Overpressure	0.02068	bar	489.174	491.972	531.482
Overpressure	0.1379	bar	154.372	151.242	168.319
Overpressure	0.2068	bar	128.498	126.077	141.552

Fig. 3 Explosion effect: Late ignition

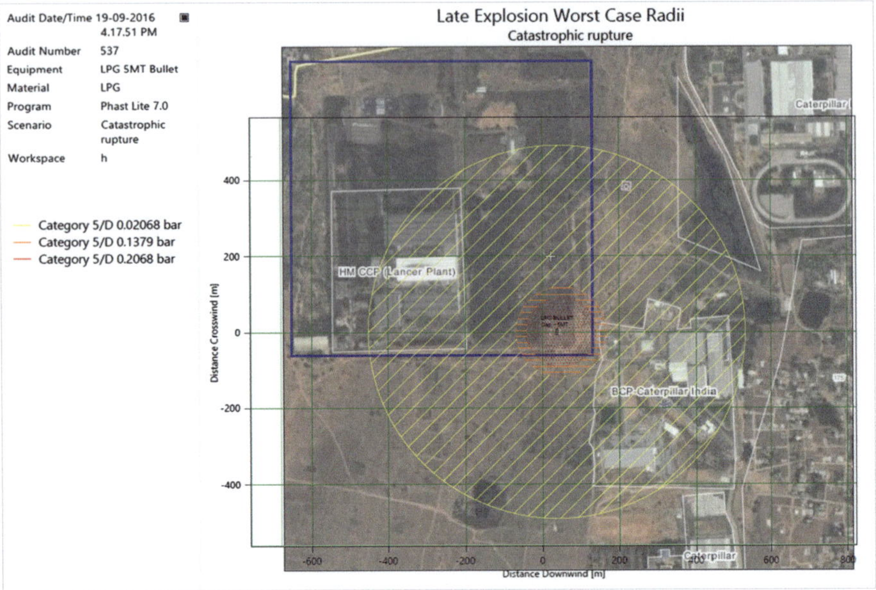

Fig. 4 Late explosion worst case radii

The contour map shows damage effects in the plant due to explosion (Fig. 4). Based on the results, emergency evacuation plan may be prepared.

8 Emergency Evacuation Diagram

LPG bullet-installed automobile industry emergency evacuation diagram is shown in Fig. 5.

Study covers pool fire, jet fire, catastrophic failures of unloading LPG bullet, rupture of pipelines, LPG bullet storage catastrophic failures etc. Study results in damage distances due to fire and shock wave pressure due to explosion under various credible scenarios in metres. This result shall be incorporated in the plot plan, and then subsequently citation of buildings, location of certain hazardous storages and process lines may be finalized.

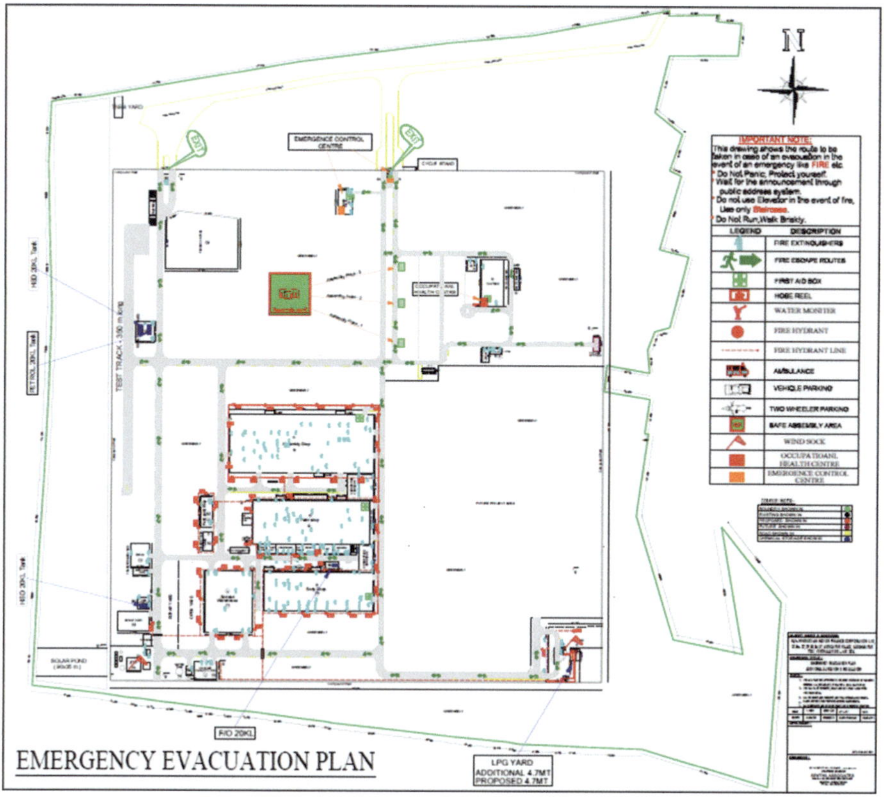

Fig. 5 Emergency evacuation plan

9 Conclusion

On the whole, this paper summarized the elements, method and design of on-site emergency plan of LPG handling automobile industries. In case of catastrophic failures of LPG storage loss of human life on-site as well as off-site, damage to property and financial loss is a glaring reality and therefore stakeholders are duty bound to ensure their installation, operation and functioning fall within the legal and societal framework. At the same time, incidents in the industries are inevitable, and the controlling and preventing of emergencies are possible by systematic preparation of on-site emergency plan in line with legal requirements.

References

1. National Disaster Management Guidelines Chemical Disasters (Industrial). 2007. *National Disaster Management Authority Government of India 2007.*
2. Details to be furnished in the Onsite emergency plan. 1989. *Schedule 11 (Rule 13(1)) of the manufacture, storage and import of hazardous rules, 1989, Government of India.*
3. Compulsory disclosure of information by the Occupier. 1987. *Section 41-B (4) of the Factories Act, 1948 and its amendments of 1987.*
4. Petroleum and Natural Gas Regulatory Board. 2010. (*Codes of Practices for Emergency Response and Disaster Management Plan (ERDMP*)) *Regulations, 2010.*
5. Center for Process Safety of the American Institute of Chemical Engineers (AIChE). 1996. *Guidelines for use of vapor cloud dispersion models.* 2nd ed. New York: AIChE.
6. Center for Process Safety of the American Institute of Chemical Engineers (AIChE). 1995. *International conference and workshop on modeling and mitigating the consequences of accidental releases of hazardous materials, September 26–29, 1995.* New York: AIChE.
7. Gautam, S.S., and P.K. Saxena. 2001. Survey of criticality of risk from LPG storage tanks at user-sites in North India. *Indosh News*, 6 (1).
8. Han, Z.Y., and W.G. Weng. 2011. Comparison study on qualitative and quantitative risk assessment methods for urban natural gas pipeline network. *Journal of Hazardous Materials* 189: 509–518.
9. Alencar, J.R.B., R.A.P. Barbosa, and M.B. de Souza Jr. 2005. Research paper on evaluation of accidents with domino effect in LPG storage areas. *Engenharia Termica (Thermal Engineering)*, 4 (1): 8–12.
10. Antonioni, Giacomo, Gigliola Spadoni, and Valero Cozzani. 2009. Application of domino effect quantitative risk assessment to an extended industrial area. *Journal of Loss Prevention in the Process Industries* 22: 614–624.

Use of QRA to Manage SIMOPS Operations

R. Kannan and Nihal A. Siddiqui

Abstract The simultaneous operations (SIMOPS) are major construction activities or major shutdown activities carried out in oil and gas plants while a part of the plant is in operation. The SIMOPS possess various risks due to the hazardous operations carried within proximity of the running plant. For example, heavy lifting, huge welding and cutting near flammable storage areas, construction/shutdown team performing SIMOPS inside the operational field unaware of the operational hazards, and operational team unaware on the construction hazards and vice versa. Hence, the SIMOPS activities pose greater risk to the plant and people due to hazardous materials and hazardous activities. This paper will provide an overview of how SIMOPS operations are managed safely.

Keywords SIMOPS · Risk · Hazard · ALARP · QRA

1 Introduction

Driven by energy requirements, most of the oil and gas industries are constructing new plants. Most of the new oil and gas plants start commissioning and operations while a part of the plant is still under construction. Some plants introduce new units within the operating plant in order to enhance the quality or quantity of the hydrocarbon. Some plants conduct major shutdown activities while a part of the plant is still producing.

All of the above discussed are forming SIMOPS operations that involve simultaneously construction and commissioning/operation of the plant facilities.

R. Kannan (✉)
Risk 2 Safe Consultants, Chennai, Tamil Nadu, India
e-mail: consultant@risk2safe.org

N. A. Siddiqui
University of Petroleum and Energy Studies, Dehradun, Uttarakhand, India
e-mail: nihal@ddn.upes.ac.in

N. A. Siddiqui et al. (eds.), *Advances in Fire and Process Safety*,
Springer Transactions in Civil and Environmental Engineering,
https://doi.org/10.1007/978-981-10-7281-9_22

The most common construction work activities are involving the following:

- Large number of workers present inside the process plant premises—increased plant overall risk due to escape and evacuation
- Heavy load lifting—potential high-risk activity of dropped object on live hydrocarbon carrying lines
- Welding and cutting activity—source of ignition
- Vehicle movement in and around the operational areas—source of ignition.

The commissioning and start-up activities are involves:

- Introduction of energy sources, e.g., steam, power, utility chemicals.
- Introduction of flammable gas for boiler/furnace/heater start-ups.
- Introduction of raw feed which contains highly toxic materials like hydrogen sulfide gases.

With this mixture of hazardous materials and hazardous activities, safety management during SIMOPS activities is a paramount for oil and gas plant safety. Many devastating incidents took place around the world that shows the importance of safety management during SIMOPS. The following paragraphs will address the QRA approach to achieve the safety management during SIMOPS stage.

2 Detailed Approaches

2.1 Plant Sitting and Layout Decision—Design Stage

It will be always good to design the plant with considering the one train under production and another train under construction or considering a major shutdown of one train with another train under production. But most of the time it never happens, management decides to construct/commission/operate all the trains at one time and they forget to consider the expansion/shutdown/SIMOPS. Quantitative risk assessments (QRAs) [1] are used to analyze the risk due to the SIMOPS operation within the production plant.

2.2 Quantitative Risk Assessment—QRA

Most of the gas plant uses risk-based approach to persuade the management and shareholders to proceed SIMOPS operations. Companies use proficient software, e.g., PHAST RISK, SHEPHERD, ITEMSOFT, etc. To perform QRA studies, the inputs are plant process conditions (flow, pressure, and volume), leak sources and number of people present during the SIMOPS and operators present during normal operations. Additional details like number of ignition sources, dropped objects due to lifting operations are considered too. The results of a QRA are the individual risk and the societal risk.

The individual risk is nothing but a death of an unprotected individual due to loss of containment (LOC) event and the individual exposed at the time of event. The individual is assumed to be unprotected and to be present throughout the event.

The societal risk characterizes the frequency of an accident resulting in *N* number of fatalities instantaneously. It is assumed that the personal exposed to the event are having certain degree of protection. This is presented in an *F-N* curve, in which *N* represents number of deaths and *F* is aggregate frequency of accidents.

(a) Societal Risk and Manning

It is obvious that construction requires more number of people than to operate the plant; it is because the plant construction involves huge manual handling, scaffoldings, structural fitting work activities, but the plant operations require limited number of skilled operators as the plant operations have been modernized and are operated in automode. Therefore, number of personnel present during the SIMOPS stage is important to perform QRA. As an outcome of QRA, there will be a *F-N* curve produced to compare with the benchmarks based on the region and industry. Various *F-N* curves are given in Fig. 1. Societal risk requires information regarding the population around the plant, specially their location and density. If the worker population is more in any particular area, the likelihood of more people getting harmed in case of an event (e.g., toxic gas release) is high.

(b) Ignition Controls

Ignition controls are used especially in gas plant, where a flammable gas leak may lead to explosion or flash fire or jet fire situation. The equipment used within the hazardous areas are zone classified, but the construction activities are carried out

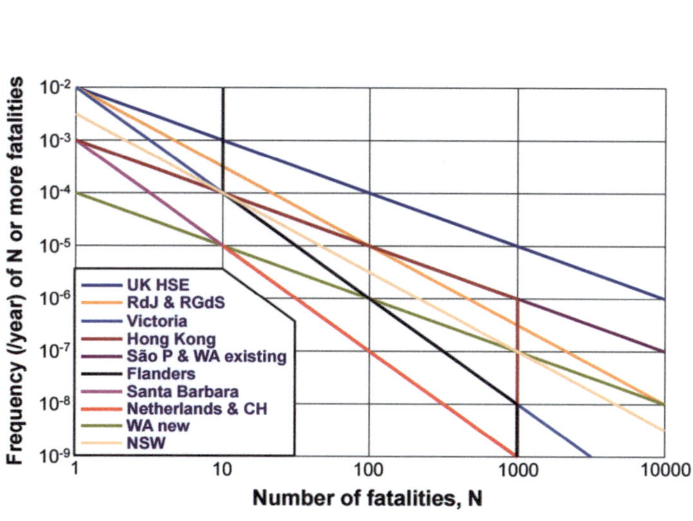

Fig. 1 *F-N* curves

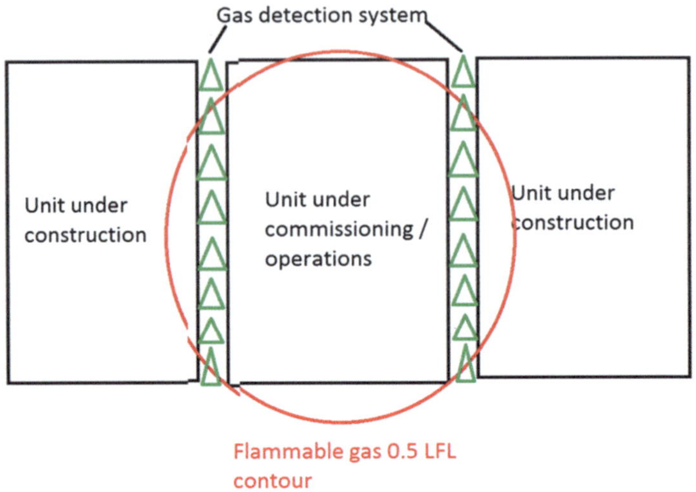

Fig. 2 Flammable gas detection arrangement

using non-explosion protected equipment, which could be the source of ignition in the event of flammable gas leaks. To control the ignition sources, the construction works should be linked with limited power sources and the power sources are linked with automatic shutdown system coupled with fire and gas detection system; this way, it will ensure cutoff of the ignition sources whenever a flammable gas leak is detected (do a test before SIMOPS starts). Based on the gas dispersion contour, a distance from the release source to gas concentration of lower flammability limit (LFL) and 0.5 LFL are identified. The distance is used as the minimum spacing to prevent ignition of released gas. In some cases, the gas detection shall be established along the boundary of the unit where construction is ongoing Fig. 2.

(c) Thermal Radiation

Based on the heat radiation contour, several distances are identified according to levels of the heat flux, each with different potentials to harm people or damage equipment. The heat flux levels are based on relevant standards and industry guidelines. Table 1 provides thermal radiation exposure effects [2].

(d) Knowledge, Awareness, and Skills of SIMOPS Workers

The presence of construction workers during SIMOPS pose the biggest challenge because of their limited knowledge and understanding of the running plant. Therefore even though these workers are present during plant construction, it is highly recommended to conduct plant induction training before the SIMOPS work starts. And the contents include, not limited to, the following:

Table 1 Thermal exposure

Thermal radiation (kW/m^2)	Effect
2	Exposure to 1 min will cause pain
<5	Exposure to 15–20 s will cause pain
>6	Exposure to 10 s will cause pain required rapid escape
12.5	Medium exposure can cause fatality
25	Likely fatality for extended exposure
37.5	Minimum distance to avoid damage to equipment in a short-term period

1. Worker understanding on hazardous areas,
2. Hot work activities in hazardous areas,
3. Use of safety equipment (e.g., fire extinguisher, safety shower),
4. Action to be taken in the event of fire and gas alarms,
5. Waste management and housekeeping,
6. Lifting equipment and their safe use,
7. Escape and evacuation arrangements and escape way routes,
8. Use of respiratory escapes equipment.

2.3 Emergency Escape, Rescue Arrangements

During SIMOPS stage, the biggest challenge is to train the construction workers against operational emergency response and it is paramount for any gas processing company to implement the same. If the emergency response requirement [3] is not adequate, then there will be probability of a huge number of fatalities or workers with serious injuries due to the confusion or unawareness of emergency response arrangements during the SIMOPS stage. Therefore, to implement an effective SIMOPS ERP, the following requirements need to be fulfilled.

- Consequence distances due to major accident scenarios to be measured (i.e., for toxic gas leak, thermal radiation, and explosion over pressure ranges for credible and large leak sizes).
- Once the distances are established, if the SIMOPS areas are within the impacted zone, then escape, evacuation, and rescue strategy needs to be analyzed.
- SIMOPS emergency response plan to be developed, which shall consist of; o Location of construction workers' assembly point for various scenarios (i.e., for toxic, it should be a safe location outside the plant or if inside it should be airtight shelters, for blast overpressure and thermal radiation, the room should be sufficiently designed to withstand the temperature and pressure).

- Escape equipment, e.g., respiratory protective emergency escape masks which can provide sufficient breathable air for the distances up to assembly points to be reached by worker.
- The respiratory protection equipment should be determined based on the concentration of toxic gas present in the air during toxic gas release scenario, and using the available protection factor (APF), the RPE could be selected [4–6].

As per OSHA, assigned protection factor (APF) means the workplace level of respiratory protection that a respirator or class of respirators is expected to provide to employees when the employer implements a continuing, effective respiratory protection program as specified by this section [5].

Maximum use concentration (MUC) means maximum atmospheric concentration of a hazardous substance from which an employee can be expected to be protected when wearing a respirator, and is determined by the assigned protection factor of the respirator or class of respirators and the exposure limit of the hazardous substance. The MUC usually can be determined mathematically by multiplying assigned protection factor specified for a respirator by the NIOSH recommended exposure limit (REL), permissible exposure limit, short-term exposure limit, ceiling limit, peak limit, any other exposure limit used for the hazardous substance [6].

- In the event of plant evacuation, a sufficient number of transportation arrangements to shift the worker from plant to safe location (it may be their accommodation camps).

3 SIMOPS MOPO Matrix

Before SIMOPS stage starts, the operations and construction team will prepare a SIMOPS matrix that will list all those activities within SIMOPS period with “go, no -go and go with caution” condition. An example is given in Fig. 3. The SIMOPS activities shall be controlled through the manual of permitted operations (MOPO) matrix, and the PTW issue points are keeping the MOPO to verify that the personnel strictly follow the rules.

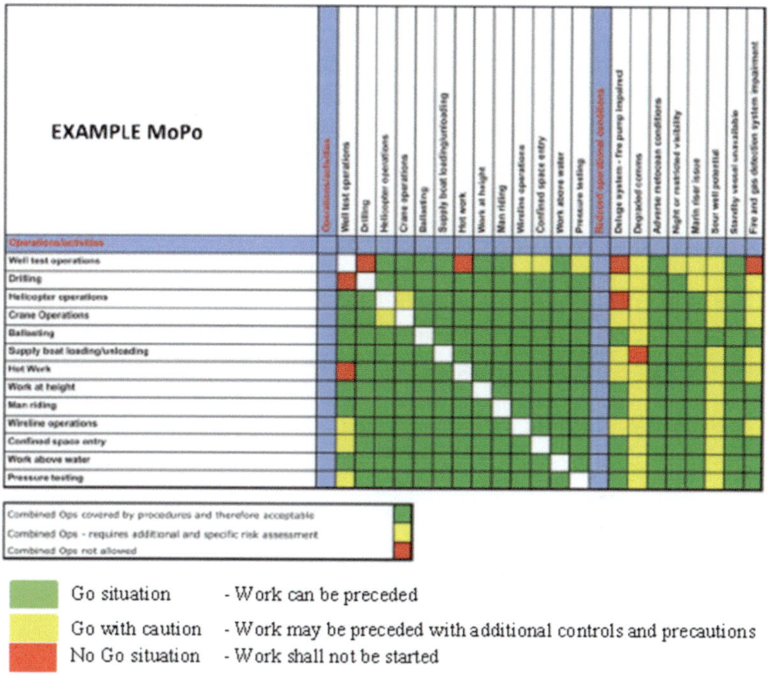

Fig. 3 Manual of permitted operation matrix

4 Conclusions

The SIMOPS activities always have an inherent risk that needs to be clearly understood by the SIMOPS personnel and the management team. The SIMOPS situation is unavoidable, but the accidents during SIMOPS are avoidable by following the precautions listed in this paper. The above-discussed SIMOPS controls can vary industry to industry; our intention is to have a general understanding of SIMOPS and safety management during SIMOPS.

References

1. Uijt de Haag, P.A.M., and B.J.M. Ale. 2005. *Guidelines for quantitative risk assessment*. Nederland, Dec 2005.
2. Methods of approximation and determination of human vulnerability for offshore major accident hazard assessment, HSE, UK. 2010.
3. The International Association of Oil and Gas Producers, Guidance for Evacuation, Escape and Rescue, Risk Assessment Data Directory. March 2010.

4. Respiratory protective equipment at work, HSE, UK. 2013.
5. OSHA Assigned Protection Factors for the Respiratory Protection standard, OSHA. 2009.
6. Bollinger, Nancy. 2004. *NIOSH respiratory selection logic*. NIOSH, CDC.

Correlating the Factors of Human Error and Behavior-Based Safety Using Pareto Analysis and BBS Observation Application

R. Shristi Vidusha, Y. Sivaji Raghav, Sagarkumar Vaghasia and Bikarama Prasad Yadav

Abstract Incidents cost financially and socially to both the company and the victim. The studies and analysis carried out so far on those industrial incidents reveal many contributing factors. This paper aims at correlating such factors leading to the untoward incidents and unsafe work behavior being its primary contributor at any workplace. The study analyzes the trend and causes of the incidents for a period of 4 years (2013–2016) through Pareto analysis to find the role of unsafe work behavior in industrial incidents. The study also intends to find the most risky and the safest behavior by observing a sample of 100 workers at an MCB manufacturing plant with an aid of a BBS observation application. These observations are usually recorded in written form and are documented for future references which consume a lot of space, time, and materials. To overcome these shortcomings, a computer-based application is developed to record the observations systematically in a mobile or laptop or tablet. The observations can be viewed through BBS dashboard or through the database file. Apart from that, this paper provides possible solutions to add to the existing control measures to improve safe work culture at the workplace .

Keywords Incidents analysis · Human error · Pareto analysis · Behavior-based safety observation · Computer application

1 Introduction

In any company, the concept of behavior-based safety gains greater significance to reduce the human error incidents which cause fatally to men and materials. The sole reason for the evolvement of this concept is due to the nonexistent of a zero incident

---

R. Shristi Vidusha (✉)
Weatherford Drilling Services, Dubai, UAE
e-mail: shriztee@gmail.com

Y. Sivaji Raghav · S. Vaghasia · B. P. Yadav
Department of Health, Safety and Environment Engineering, University of Petroleum and Energy Studies, Bidholi, Dehradun 248007, India

N. A. Siddiqui et al. (eds.), *Advances in Fire and Process Safety*,
Springer Transactions in Civil and Environmental Engineering,
https://doi.org/10.1007/978-981-10-7281-9_23

at a workplace of any nature. In this world of neck-to-neck competition, the companies are running after humungous production numbers. They are indirectly forcing their own employees to bypass the SOPs to achieve the targets quickly. This paves the way for the uncanny attitude toward the job leading to many unwanted incidents. Years before, when industrial incidents were brought to light, lots of reasons have been traced out for the same. It started with the then experts saying that it was the unsafe conditions that led to the accidents. Later, it was said that the unsafe acts caused the unsafe conditions. Again when further measures were carried out and failed, another concept evolved which said that a good work practice is the ultimate key to prevent the unsafe practices and conditions to escape such happenings. Now in recent times, after numerous researches and incident analyses it has been concluded that the good behavior of an employee toward his workplace safety and his personal safety alone can prevent any such incidents. This is when the concept of behavior-based safety came into existence. Behavior-based safety can be defined in many ways but a better way to explain this would be "safe attitude toward personal and workplace safety." This project was influenced by the fact that any kind of incident, be it a first aid injury or a lost time injury, contributes to the loss of production time, damage to the machineries, and other social issues such as disturbed mentality of their colleagues. Out of many kinds of approaches toward reduction of such incidents, behavior-based safety approach yields positive and promising results in developing the safe work culture at workplace [1]. This project tries to trace out the relation between the unending incidents in an industry and the safe work behavior of its employees. Lind [2] says that the reason behind the fatal accidents is dangerous method of working and that of the nonfatal accidents is working at a running process. Human error is now a greater threat to industrial accidents rather than technical error, and further managing the human risks is never fully efficient since the moderation of such human risks is only possible and not their elimination [3].

The project intends to study and provide follow-up actions for the safe and at-risk behavior and provide safety training assessment of all permanent, temporary, and contractor employees of production and warehouse department of a MCB and RCCB production plant. The study has taken 100 sample employees from the firm, and personal interview has been preferred by the author to exactly observe the level of commitment toward safety at the workplace. The study tries to proffer the maximum possible and feasible corrective actions for the identified unsafe behavior observations which have been the cause of injuries. This will benefit the employees and employer if followed in a long run.

2 Literature Review

The study was aimed at reasoning out the industrial incidents through human error management by modifying the at-risk work behavior into safe work behavior. In an attempt to correlate the same, Fam et al. [4] prove that the frequency of accidents and the unsafe behavior are highly correlated. The study interviewed the previous accident victims and has found that certain employees are naturally prone to create accidents. Skalle et al. [5] describe the two types of human errors as active type that is done directly by operators and the latent type which is done indirectly by the operators due to the underlying errors like designing error. While defining an unsafe work place, Choudhry et. al. [6] is of the opinion that, any workplace is unsafe unless the worker along with the management commit to safety and work to achieve zero incidents. Management commitment does not stop in just giving training and spreading knowledge about safety among the workforce. It goes on to keep track of the implementation of the safety management system and effectiveness of the training and awareness. In other words, as concluded in [7], it can be said that management's less commitment toward safety is like giving chances to the workers to take risks. Petersen [8] inferred two important key factors for any kind of accidents from the Heinrich's triangle of accident causation, viz., the basic ground for accidents is man factor and the management is the body with the obligation and ability to prevent such occurrences. According to [9], the managements' responsibility to support and develop the safe work behavior is proven to increase the standard of the safety culture at the workplace. The managements' participation in safety gets benefitted by the workers' cooperation to accomplish the targeted goal. Geller [10] suggests that work behavior not only gets influenced by internal factors such as stress, anxiety but also by external factors such as work environment, management involvement, workplace ambience, peer group. Ramsey et al. [11] have done a study on the effects of thermal conditions on safe work behavior and found that within the preferred zone of temperature at the workplace the occurrences of unsafe work behavior are very minimal. Adding to this, Seo [12] has identified the factors of unsafe behavior as the perception of workers toward safety climate, work pressure, hazard level, risks, and barriers. Hoyos [13] has mentioned that modern technologies have increased the complexity of the process demanding higher cost, complexity, and capacity of safety system. He has also indicated the increasing severity of the unsafe work behavior of the employee at workplace. In his work regarding accidents in aviations, Weigmann and Shappell [14] have stated that major accidents are caused by judgment errors, while the minor accidents occur due to the procedural errors, and these errors are increasing in recent times. There have been other sorts of reasoning out for the incidents which are found in contrary to the concept of human error. Reason [15] has given in his paper two types of approaches toward an accident. One is person approach which deals with the person working, and the other is the system approach which deals with the working condition and environment. These two approaches will give rise to different viewpoints on managing the errors. Brown et al. [16] state that there are many authors who suggest some perspectives that actually spot the light on the operational and social systems for reasoning out the accidents apart from human errors and unsafe acts. Laitinen et al. [17] have adopted the observation method to identify the safety index of the employees so as to relate it with the accident rate at that site. The study also inferred that the observation method paves way for new opportunities for promoting safety in any industry. On assessing the safe work behavior of the employees in the construction sites at Pakistan, Mohamed et al. [18] identified that the risk of the unsafe work behavior of the employees lies in the fact that they unknowingly put themselves and their colleagues at superfluous risks. The culture that has "collectiveness and femininity" is likely to influence safer work behavior than "individualism and masculinity." In an attempt of correlating the behavior change and culture change, DeJoy [19] explains that BBS focuses on identifying the at-risk behavior of the employees and modifying it to enhance the safety culture which can be done through close monitoring and conditioning by reinforcement.

Geller [20] in their publication has mentioned that an unsafe behavior is continuously carried on because of the fact that positive consequence of the same seems certain and immediate but its negative consequence seems a rare probability. As an effective way to motivate positive safety culture, Vredenburgh [21] points out that a good reward system that includes attractive incentives motivates good work behavior and that the culture can be learned by connecting behaviors with its consequences. Yeow and Goomas [22] also proposed a program which links the outcome and incentives in order to improve the work behavior of the employees. This proved to be efficient since the teams were concentrating indirectly on safety in order to win incentives. Human beings get encouraged when someone comments positively on their deeds. The same works in safety improvement also. Williams and Geller [23] in their work point out that the behavior-based feedback method gives a significant rise in the %safe behaviors at workplace and has paved an efficient pathway to take follow-up actions thus reducing the incidents. In support of this, Hermann et al. [24] compared the existing safety methods and BBS method and found great results in two years that reduced the severity rate up to 96 and 99% reduction in lost time injury rates. Similar to such BBS review studies, there have been many other studies that tested the effectiveness of BBS method. When Lingard and Rowlinson [25] tried to implement BBS method in seven housing sites in Hong Kong, he got mixed results in which housekeeping yielded a great positive result but work at height and scaffolding works yielded satisfactory results. DePasquale and Geller [26] studied the effectiveness of BBS program by one-on-one and focused group interviews and observed that the mandatory approach in implementing the program proves to be efficient rather than the voluntary approach. Zhang and Fang [27] have implemented BBS supervisory-based intervention cycle along with BBS tracking and analysis system in construction industry and found that the system proved efficient and provided promising results in achieving sustainable improvement in the workers' safe behavior. Zohar and Luria [28] also prove the same that the supervisory-based intervention proves to be more effective when they receive weekly feedback on their subordinates' improved safe work behavior. Komaki et al. [29] explain in his paper that training alone is not sufficient to improve the safety climate. The training along with the feedback program proves to be effective since it involves supervisory and managerial role in improving the safety climate at any workplace.

3 Methodology

3.1 Pareto Analysis

Pareto analysis is one of the best analyses to find the significant causes to any kind of happening. The principle of this analysis is also called as 80/20 rule since it helps us to select the contributing factors that have a prominent control effect on the incidents. This project uses this analysis to identify the main contributory causes of the incidents on which 20% effort is put to reduce 80% of the incidents.

3.2 BBS Observation Checklists

The BBS observation is usually carried out with the help of a checklist which covers all the personal aspects of a worker and the work environment which helps in assessing the behavior of any worker. The checklists are the simple and the quicker way to review a work environment. Similarly, the method is adopted for the behavior observation of the employees. A new checklist has been framed after referring standard checklists from various firms such as DUPONT. The checklist method at times gives a blinkered vision. To avoid this, the checklist adopted in this project provides option to input any other unsafe conditions or workers' feedback. This helps in improving the observation process to identify the exact antecedents to the behavior.

3.3 PHP Language and BBSO App

The BBSO (BBS observation) application was developed through PHP language. PHP is the acronym for PHP: Hypertext Preprocessor, which is a source code programming language that can be embedded in HTML. This open-source programming language helps in managing dynamic contents, databases, session tracking, and also in developing Web sites. This programming language is used in this project to computerize the BBS observation checklist so that the database can be managed systematically and can be viewed and analyzed anytime in the future. The copies of the developed application and the databases are indicated in Fig. 1. The input for the development of application was just a BBS observation checklist and the format for calculating the safety index. The application asks for the details of the observed employees except for their name since BBS is strongly based on NO NAME NO BLAME culture. The details include operating line, stage, shift, age, and gender. The age and gender were included because of the fact that behavior is influenced by personal and physiological factors also. The application also gives a provision to retrieve the safety index of any observation through "open

BBS" option. The BBS dashboard sums up the total safe and unsafe behavior and safety index for that data on a particular date. The application provides an option to enter any comments to indicate any other extra details of the working condition apart from the checklist.

4 Results and Discussion

4.1 Incident Analysis

Incident data of a period of 4 years have been collected from 2013 to 2016. These data include the majority of first aid incidents and a few lost time accidents. The trend (Fig. 2) shows that the incidents found a decreasing trend in 2015 when the plant was redesigned for efficiency enhancement. At this phase, a lot of equipments were automized and many types of equipment were installed with the safety guards such as machine guard, two hand sensors, push buttons, door interlocks, and light screen. The incidents were found to decrease more in 2016. The records show that the company's introduction of Safety Assurance Program that includes all the aspects of improvisation of safety culture during 2015 has a vital role in the decreasing trend in 2016. The analysis also takes the man-hours lost due to the recorded incidents. The time lost due to first aid incidents has also been taken into consideration to get a wider perception of the total time lost due to the incidents.

4.2 Critical Stages and Factors—Pareto Analysis

Pareto analysis has been done on the incidents with respect to both the working stages on which the incidents occur and the causes due to which the incidents occur. From the analysis made on the working stages, it has been found that more number of incidents occur on conveyors, welding machines (core, connector, contact insertion), and test benches (magnetic, mechanical, omnipolar, thermal benches). According to the analysis (Fig. 3), if the incidents on the conveyors are reduced, then 80% of the total incidents can be eliminated with just 20% effort. The analysis made on the injury causes (Fig. 4) reveals that the major cause to the incidents is workers' negligence toward safe working procedure.

From the analysis made, all the causes can be summarized into the following categories and of them the first four prove to be the aspects of unsafe work behavior.

- Fatigue,
- Rushing,
- Frustration,
- Complacency,

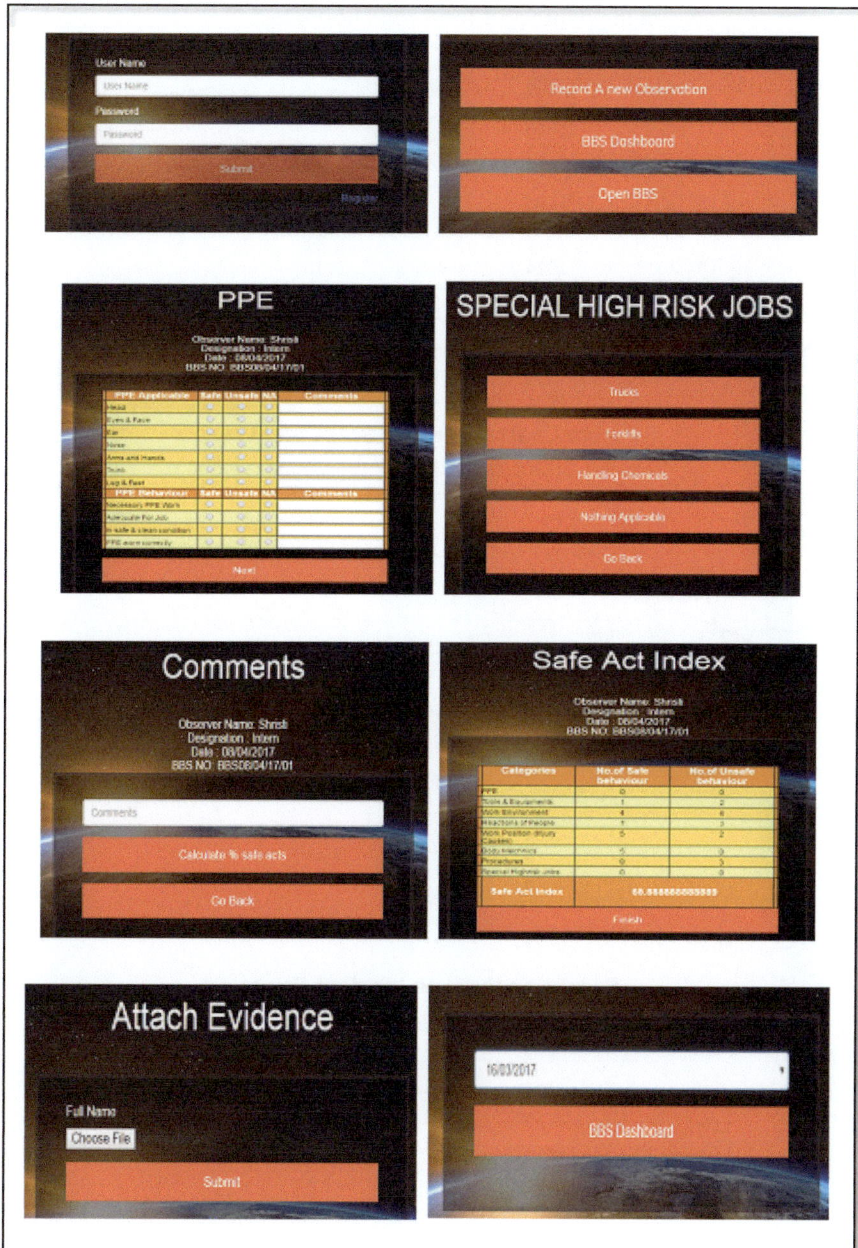

Fig. 1 Snapshots of the BBSO app

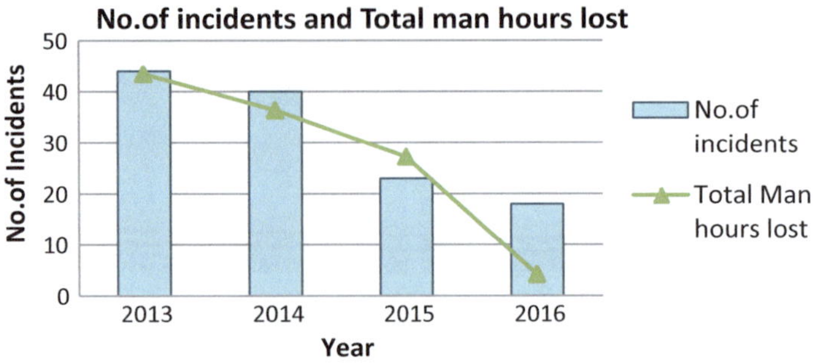

Fig. 2 No. of incidents and total man-hours lost

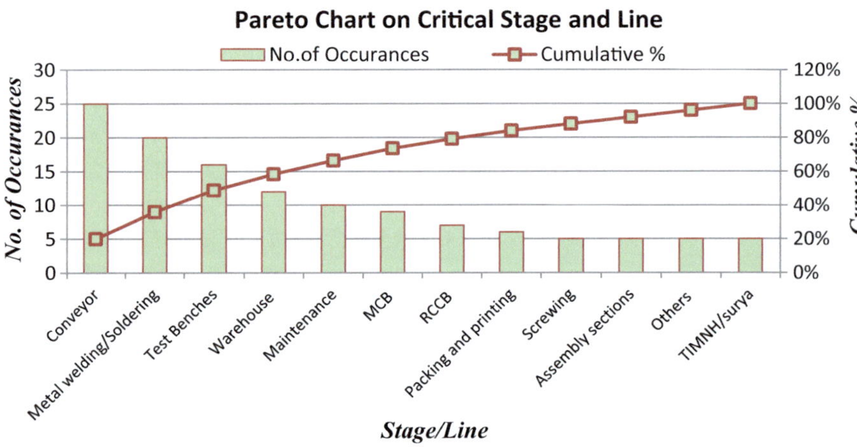

Fig. 3 Pareto analysis on working stages

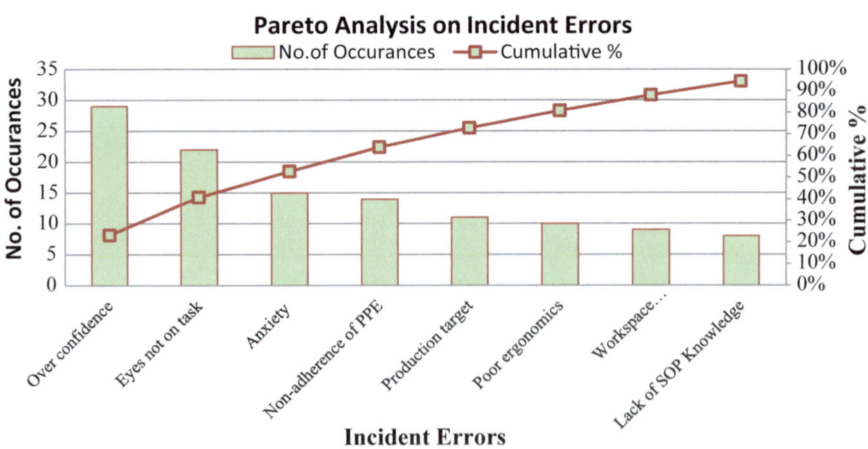

Fig. 4 Pareto analysis on incident errors

- Unsafe condition, and
- Lack of competency.

Hence, it is inferred from the analysis that the reason behind most of the accidents is because of the human error and the unsafe work behavior.

4.3 Behavior-Based Safety Observation

To evaluate the work behavior of the workforce, observations were carried out with the newly developed computer application. The application was developed with PHP programming language. The samples were interviewed, and individual safety indices were identified and analyzed to summarize the safety culture.

Safety index calculation: To quantify the behavior, safe act index has been used in this project which indicates the percentage of safe behavior out of the observed behavior [17]. This is to encourage the employees to work toward good safe act index so as to get appreciation.

$$\text{Safe Act Index} = \frac{100 \times \text{No. of Safe Behavior}}{\text{No. of Safe} + \text{Unsafe Behavior}}$$

Positive reinforcement is more significant than the negative reinforcement since employees always like it when someone praises about them and is discouraged when someone points out their mistakes. The observation was carried out for 100 workers personally by the author. The distribution of the numbers was kept equal between the sectors so that the analysis can result out the exact percentage of safe and at-risk behaviors. The sectors were MCB/M9, RCCB/ADDITIM, TIMNH/SURYA, and warehouse. The inferences made from the analysis are shown below in graphs. In 100 observations, only 26 observations had high-risk jobs (forklifts—4; trucks—6; handling chemicals—16). From Fig. 5, it can be seen that 60% of the observed employees show the safety index ranging from 41 to 70. A workplace with such distribution of safety index indicates a high level of at-risk behavior of the employees (Fig. 6).

The safe act index ranges from 11.11 (thinning station, RCCB) to 88.24 (warehouse). The above graph mentions the average index with maximum of 65.77 indicating the low level of safe work behavior at the workplace. To identify the most unsafe category and the most unsafe behavior further analysis has been done. The age distribution of the samples as shown in Fig. 7 indicates that 73% of the workforce is in the age between 20 and 29. Some of these employees when interviewed showed a higher level of job dissatisfaction and are forced to stay due to family reasons. To identify the relationship between the safety index and the age factor, calculations have been done for the cumulative percentage of observations in each age group. From Fig. 8, it has been found that 38.36% of the samples aged 20–29, 65.22% of the samples aged 30–39, and 50% of the samples aged 40–49 have identified to show the safety index below average.

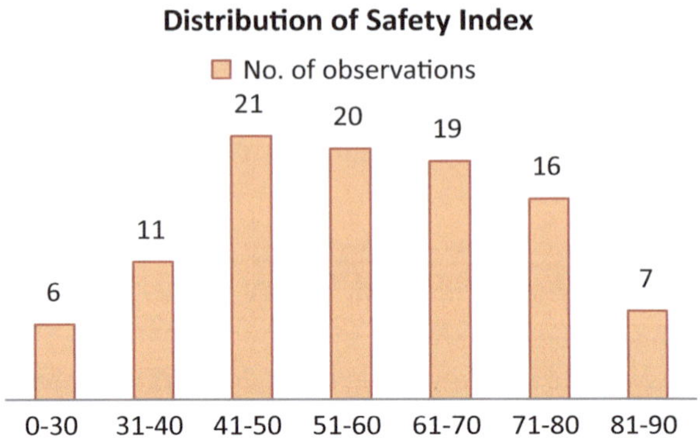

Fig. 5 Distribution of the safety index

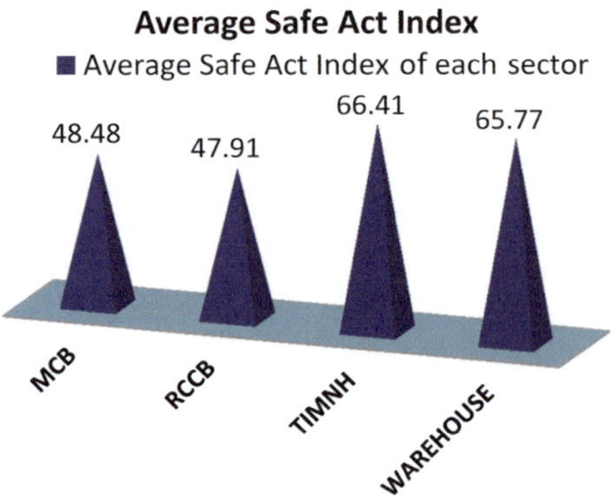

Fig. 6 Average safety index of each sector

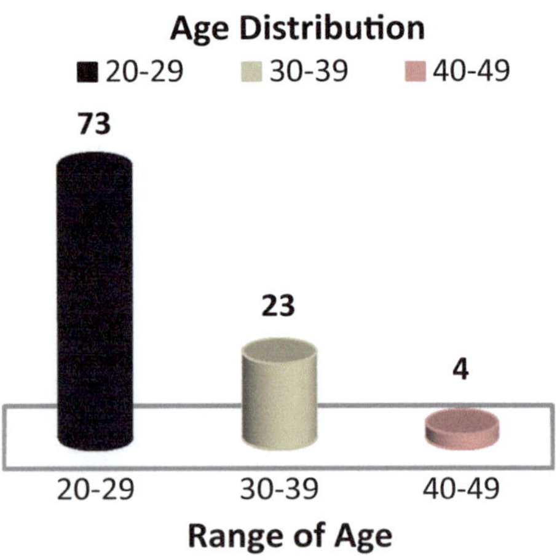

Fig. 7 Age distribution

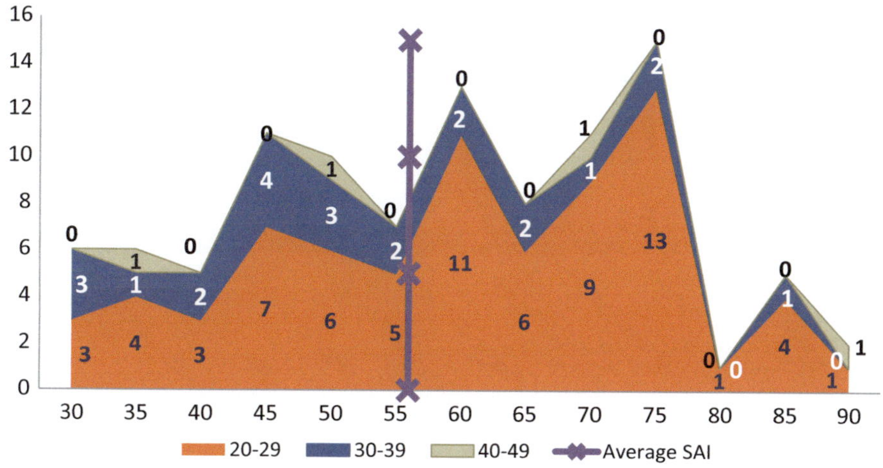

Fig. 8 Correlation of employee's age and safety index

Percentage of safe and unsafe observations: The total number of safe and unsafe behavior from all the 100 observations was categorized as each category requires to acquire the percentage of safe and unsafe behavior out of the total number of observation. This percentage of safe and unsafe behavior has been calculated to identify the category that is most safe and that is most unsafe. Figure 9 shows that the category which has the highest safe % is procedure (66%) and the most unsafe category is the high-risk jobs (78.63%). From all the observations, the percentage of each safe and unsafe behavior has been calculated to identify the top 10 most safe and unsafe observations that are indicated in Figs. 10 and 11, respectively.

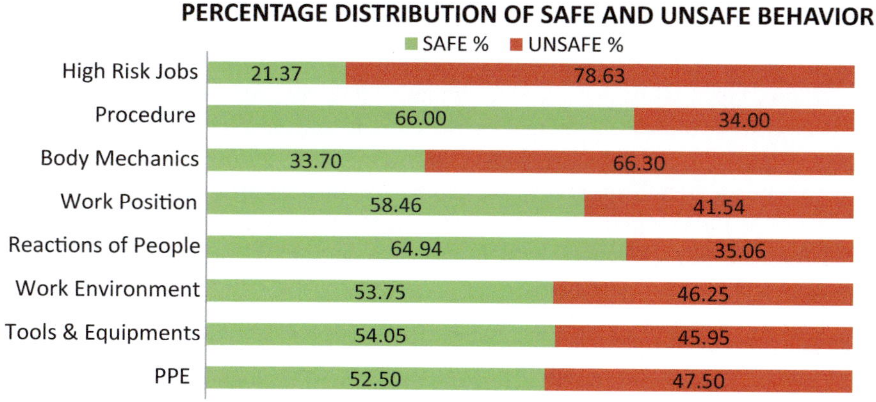

Fig. 9 Percentage of safe and unsafe behavior under each category

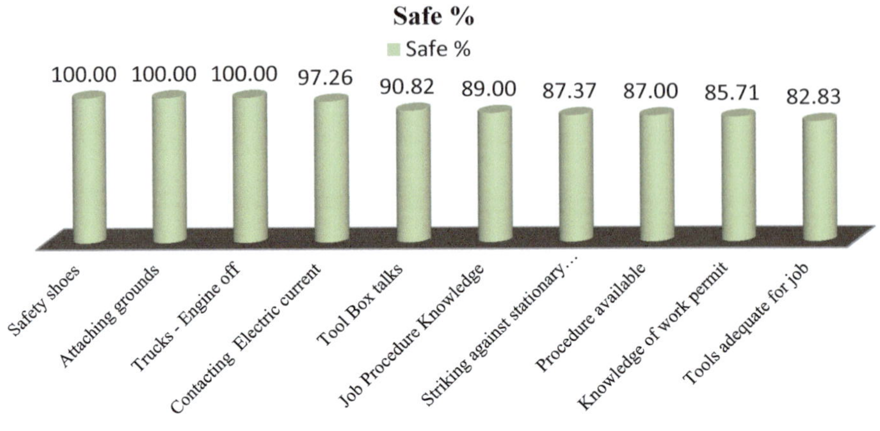

Fig. 10 Top 10 safe observations

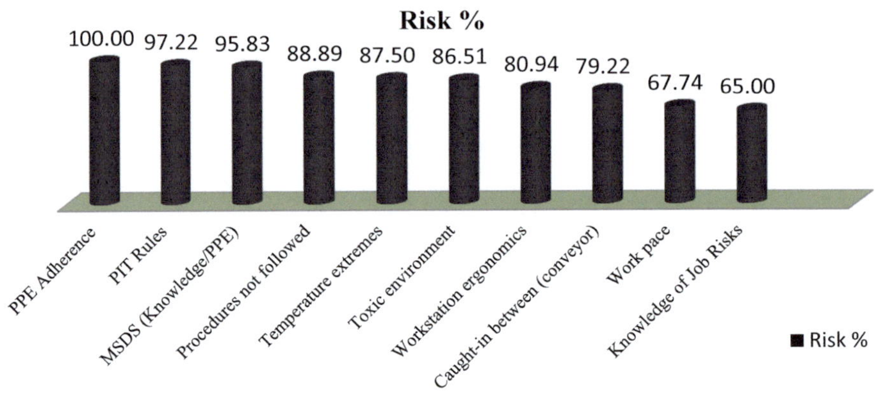

Fig. 11 Top 10 at-risk observations

From Fig. 10, it can be deducted that the workers are aware of the procedures of job, toolbox talks, work permit, etc. Similarly, the working area is safe from the injury causes such as contacting electric current and striking against stationary objects. The trucks are found to be in off position when loading or unloading takes place which is a significant safe observation. Wearing safety shoes at any point of time indicates the discipline at workplace. All the workstations are observed with the display of all applicable procedures such as LOTO, job procedure, PPE applicable, JH (Jishu-Hozen) sheets.

From Fig. 11, it can be observed that other than safety shoes and normal uniform, the PPE adherence to eyes, nose, and hands is very low. In many working stages where the workers ought to wear the PPE, the PPE adherence is very low and the management action seems to be absent against it. The power industrial trucks

(PITs) are driven with at most carelessness and overconfidence thus not following the PIT rules. When it comes to the critical stages where the chemicals such as paints and thinner are handled, there is no proper display of MSDS and sufficient training and knowledge awareness of the same are lacking, thus exposing the workers to slow chronic exposure to those chemicals. When the safe observations were identified, it was inferred that the workers are aware of the procedures at workplace. But the irony is that the workers were found not following the same. When enquired about the situation, the answer from almost everyone was that it takes little more time to follow that procedure and it acts as a deterrent factor in reaching the given production target for the shift. The same reason explains the other risky observations such as unsafe working pace and caught in between the conveyors to move the job manually.

The knowledge about the specific job risks is not the same as that of the job procedure and the shortcuts to get the work done easily. The shortcuts are easily learnt from the peer group since the workers are always rotated from one working stage to another for relaxing their static posture. But the job risks when following those shortcuts are neither learnt nor taught by the peer group. This was one of the main reasons for most of the injuries. The work area under study mostly assembles the components manually rather than through machines and equipments. This assembling involves very small components, and there are lots of stages which are prone to finger injuries because of the sharpness of those components. These stages are not provided with any kind of gloves apart from the bandages which can be worn on the fingers thus not disturbing the production but protecting the fingers which is simply inadequate for the job. While sorting out the top 10 observations, the author was unable to highlight some other important observations which were equally important. Hence, top two observations under each category were listed out as in Table 1.

5 Findings and Recommendations

The study has found the significant at-risk behavior and has provided certain recommendations not only to improve the identified unsafe behavior but also to improve the entire safety climate at the workplace. Though the observations and analysis end up in a wide range of findings, the main inference would be that the workplace is safe and the share of management commitment toward safety is meager. During the interview, many employees were asked for the cause of bypassing the work procedure. Invariably, the answer from all the employees was that the normal work procedure takes more time and the company is all about the production numbers and not about the incident numbers. There were "*n*" number of stages which were hazardous such as thermal adjust benches with a lot of heat emission, thinning stations with open access to hot molten tin, braid, and other welding machines with open spark, soldering machines being operated by already wounded operators with burns. The company does have a system to record

Table 1 Top two safe and unsafe observations under each category

Category	Safe observation	Category-wise percentage	Unsafe observation	Category-wise percentage
PPE	Legs and feet	100.00	PPE worn	93.75
	Trunk	83.00	Adequate for the job	62.00
Tools and equipments	Adequate	82.83	Pre-job inspection	62.24
	Usage	41.41	Used correctly	58.59
Work environment	Signage and warnings	80.00	Toxic atmosphere	91.43
	Proximity	65.98	PPE adherence	61.46
Reactions of people	Attaching grounds	100.00	Workplace	67.74
	Adjusting PPE	89.36	Eyes on task	60.00
Work position	Electric current	97.26	Hazardous substance	93.10
	Fall from height	90.00	Temperature extremes	87.50
Body mechanics	Lifting/bending	68.24	Repetitive motion	82.42
	Reaching/extending	45.24	Standing/sitting	80.85
Procedure	SIM	90.82	Followed	88.89
	Job procedure	89.00	Job risk and hazards known	65.00
Trucks	Engine off	100.00	Loading/unloading	100.00
	Position of load	66.67	PPE	83.33
Forklifts	Training/load	75.00	PIT rules following	100.00
	Backup beeper	50.00	Backup beeper	50.00
Handling chemicals	Exposure	25.00	MSDS/PPE	100.00
	Training	12.50	Training	87.50

incidents but in recent times not all the injuries have been recorded. This proved to be the reason for the drastically reduced number of incidents in 2016 (Fig. 2). Due to such a lack of commitment from the top management, the employees also tend to have a very callous attitude toward work.

5.1 Generic Recommendations

The work culture of the organization has to be changed. There are limitless safety procedures which aid to make the workplace safe with a conducive environment. But the effectiveness of the implemented procedures should be reviewed to continuously improve the safety management system. The study tries to propose certain programs to improve the same in a better way to benefit both the parties.

Rolling trophy: A safety steering committee has to be developed with the members from all sorts of employees from various departments such as maintenance, quality, on-roll, and off-roll. This committee should be trained in BBS observation program. The committee members should be given a fixed line and a fixed target for the week. A weekly feedback program that includes immediate incentives would be a motivating factor for the members to complete the given target within speculated time. Meanwhile, a BBS score board can be displayed at each line indicating the number of safe and unsafe behaviors observed in that particular week. The line with more number of safe behaviors can be given a rolling trophy. This program can be done at least once in a month thus involving the top management. This kind of reward system will motivate the employees to change their at-risk behavior to the desired safe behavior while achieving the production target.

Reporting system: Apart from the incident and near miss reporting, there must be a reporting system for reporting any kind of workplace hazards and unsafe condition. The organization has a "Safety Bubak" system—safety bubak is used as a highlighted representation of a hazard—where the employees are supposed to place a Bubak in order to mark/identify a hazard. But this Bubak system is not followed in the real-time scenario. The supervisors are to be involved in a drill where the one who gets to place more number of Bubak gets a reward. Thus, almost all the unsafe conditions get reported.

Safety walks and talks: The safety walk and talk should be undergone by the safety officer. The walk should not just get limited by identifying unsafe acts and conditions, and it should also involve interaction with the employees to identify their distress. The walk may also include involving the workforce in framing the corrective actions. Thus, the workforce plays a major role in both informing about the unsafe workplace and rectifying the same.

Check and act: The JH sheets are provided at every stage and are to be filled in the beginning of every shift. The study reveals that this is either not done or not properly done by most of the employees. The JH sheets are a platform to inform about the poor condition of the machine/equipment. These sheets can be checked at every toolbox talks, and the sheets that are not done properly can be attended instantly. This helps the employees in learning the importance of the sheets and also the proper filling procedure.

Attraction for attention: To gain the attention of the employees, attractive posters can be placed in the particular work line which has been observed with more number of unsafe behaviors. The posters may contain description of any positive safe behavior which is desired from the particular line. This way, the worker will work to get less number of posters such that of increasing the safety index.

5.2 Specific Recommendations

Apart from the generic recommendations to improve safe work culture, some control measures too are recommended to minimize the incident rates.

1. The conveyors should be guarded and segmented according to the size of the poles. This would prevent the operator from pushing/removing the poles from the conveyor.
2. The welding machines should be placed with vision guard to minimize the effect of the sparks. The fume extractors at each stage should be periodically checked and maintained properly.
3. The PIT operators and the operators handling paints and tin should be trained properly and refreshed periodically. Evaluation of these training should be done appropriately.
4. MSDS sheets are to be displayed wherever necessary, and training regarding the same should be materialized, recorded, and reviewed.
5. The employees can be divided into teams and trained to do HIRAC of their own line to know any unsafe conditions and their control measures.
6. Continuous standing and sitting can be minimized by job rotation where the worker can be given some other job to relax the muscles.
7. Stretching exercises should be taught to the employees during tea breaks so as to relax their tensed muscles.

Apart from the above-mentioned recommendations, control measures for the significant areas of concern have been given in Table 2.

Table 2 Control measures for most unsafe observations under each category

Category	Sub-categories	Unsafe (%)	Control measures
PPE	PPE worn	93.75	• All the stages should be reviewed for the PPE necessity • Proper PPEs are to be given to the employees • Training and motivational programs should be developed to improve the PPE adherence
	Adequate for the job	62.00	
Tools and equipments	Pre-job inspection	62.24	• The pre-job checklists (JH sheets) should be checked before any shift • The toolbox talks should involve the topics of safe usage of tools and equipments • The employees can be divided into teams to do HIRAC of their own line to know any unsafe conditions
	Used correctly	58.59	

(continued)

Table 2 (continued)

Category	Sub-categories	Unsafe (%)	Control measures
Work environment	Toxic atmosphere	91.43	• The MSDS sheets should be displayed prominently • The warning signs should be displayed appropriately at those hazardous stages • Training regarding the proper disposal of hazardous waste should be given • Continuous supervision should be done to ensure the proper handling of hazardous/toxic substance
	PPE adherence	61.46	
Reactions of people	Workplace	67.74	• The production target per day can be optimized such that the supply equals demand • The workers should be motivated to work safe and not fast • Continuous and periodical training and supervision on the working procedures
	Eyes on task	60.00	
Work position	Hazardous substance	93.10	• Proper machine guards to be fixed to avoid any kind of physical contact with the hazardous/heat region • Heat resistant aprons should be given to those at heat-emitting workstations • Nitrile gloves can be given to withstand any burns or cuts
	Temperature extremes	87.50	
Body mechanics	Repetitive motion	82.42	• Short breaks at shorter intervals may be relaxing than longer breaks at longer intervals • Job rotation where the worker can be shifted to a less stressed job for sometime during the shift • Chairs should be provided to maintain balance in the workstation height
	Standing/ sitting	80.85	
Procedure	Procedure followed	88.89	• The stages can be modified to suit the workers' mentality to meet the production demand Light screen in the conveyors to identify the interruption • Continuous training and supervision on the job procedures and risks
	Job risk and hazards known	65.00	
Trucks	Loading/ unloading	**100.00**	• If possible, manual loading/unloading can be modified into automized one • Proper training in manual lifting procedures and importance of PPE • Continuous supervision and reporting of near misses
	PPE	**83.33**	
Forklifts	PIT rules following	**100.00**	• The vehicle conditions should be pre-checked before use • Any faulty condition (no seat belt, no horn) should be immediately informed and rectified • A general supervision on hourly basis to ensure the PIT rules adherence
	Backup beeper	**50.00**	
Han Chem	MSDS/PPE	**100.00**	• Training on MSDS and importance of being safe with chemicals • Training proper and safer usage of the chemicals • Inspection on daily basis to ensure the correct procedures
	Training	**87.50**	

6 Summary and Conclusion

The paper intended to study the relation between the accidents and employee behavior through Pareto analysis of the incidents. The analysis reiterates that the incidents are caused due to the unsafe work behavior of the workers. Hence, the author tried taking the BBS observation of 100 workers to identify the most safe and unsafe observations. To quantify the observations, safety index has been calculated so that the percent of safety of the worker is obtained. The author developed a mobile application using PHP programming language to observe the candidates thus reducing the time and materials for retrieving the observed data through manual recording. The BBSO (BBS observation) application proved to be efficient in recording as well as retrieving the data. The analysis was made easy since the data could be retrieved in the desired open spreadsheet format. According to the literature survey, a combination of dedicated management commitment, supervisory-based intervention, and reward system would improve the safe work behavior and reduce the incidents at any workplace. But unfortunately, the analysis showed the lack of commitment from the top management in implementing the safety procedures.

References

1. Abdullah, M.S., et al. 2016. Safety culture behaviour in electronics manufacturing sector (EMS) in Malaysia: The case of flextronics. *Procedia Economics and Finance* 35: 454–461.
2. Lind, S. 2008. Types and sources of fatal and severe non-fatal accidents in industrial maintenance. *International Journal of Industrial Ergonomics* 38 (11): 927–933.
3. Reason, J. 1995. Understanding adverse events: Human factors. *Quality in Health Care* 4 (2): 80–89.
4. Fam, I.M., et al. 2008. Safety behaviors assessment in process industry: A case study in gas refinery. *Journal of the Chinese Institute of Industrial Engineers* 25 (4): 298–305.
5. Skalle, P., A. Aamodt, and K. Laumann. 2014. Integrating human related errors with technical errors to determine causes behind offshore accidents. *Safety Science* 63: 179–190.
6. Choudhry, R.M., D. Fang, and S. Mohamed. 2007. The nature of safety culture: A survey of the state-of-the-art. *Safety Science* 45 (10): 993–1012.
7. Petersen, D. 2000. Safety management 2000: Our strengths & weaknesses. *Professional Safety* 45 (1): 16.
8. Petersen, D. 1984. *Human-error reduction and safety management.*
9. Jitwasinkul, B., B.H. Hadikusumo, and A.Q. Memon. 2016. A Bayesian belief network model of organizational factors for improving safe work behaviors in Thai construction industry. *Safety Science* 82: 264–273.
10. Geller, E.S. 2001. Behavior-based safety in industry: Realizing the large-scale potential of psychology to promote human welfare. *Applied and Preventive Psychology* 10 (2): 87–105.
11. Ramsey, J.D., et al. 1983. Effects of workplace thermal conditions on safe work behavior. *Journal of Safety Research* 14 (3): 105–114.
12. Seo, D.-C. 2005. An explicative model of unsafe work behavior. *Safety Science* 43 (3): 187–211.
13. Hoyos, C.G. 1995. Occupational safety: Progress in understanding the basic aspects of safe and unsafe behaviour. *Applied Psychology* 44 (3): 233–250.

14. Weigmann, D.A., and S.A. Shappell. 1997. Human factors analysis of postaccident data: Applying theoretical taxonomies of human error. *The International Journal of Aviation Psychology* 7 (1): 67–81.
15. Reason, J. 2000. Human error: Models and management. *Western Journal of Medicine* 172 (6): 393.
16. Brown, K.A., P.G. Willis, and G.E. Prussia. 2000. Predicting safe employee behavior in the steel industry: Development and test of a sociotechnical model. *Journal of Operations Management* 18 (4): 445–465.
17. Laitinen, H., M. Marjamäki, and K. Päivärinta. 1999. The validity of the TR safety observation method on building construction. *Accident Analysis and Prevention* 31 (5): 463–472.
18. Mohamed, S., T.H. Ali, and W. Tam. 2009. National culture and safe work behaviour of construction workers in Pakistan. *Safety Science* 47 (1): 29–35.
19. DeJoy, D.M. 2005. Behavior change versus culture change: Divergent approaches to managing workplace safety. *Safety Science* 43 (2): 105–129.
20. Geller, E.S. 2005. Behavior-based safety and occupational risk management. *Behavior Modification* 29 (3): 539–561.
21. Vredenburgh, A.G. 2002. Organizational safety: Which management practices are most effective in reducing employee injury rates? *Journal of safety Research* 33 (2): 259–276.
22. Yeow, P.H., and D.T. Goomas. 2014. Outcome-and-behavior-based safety incentive program to reduce accidents: A case study of a fluid manufacturing plant. *Safety Science* 70: 429–437.
23. Williams, J.H., and E.S. Geller. 2000. Behavior-based intervention for occupational safety: Critical impact of social comparison feedback. *Journal of Safety Research* 31 (3): 135–142.
24. Hermann, J.A., G.V. Ibarra, and B. Hopkins. 2010. A safety program that integrated behavior-based safety and traditional safety methods and its effects on injury rates of manufacturing workers. *Journal of Organizational Behavior Management* 30 (1): 6–25.
25. Lingard, H., and S. Rowlinson. 1998. Behavior-based safety management in Hong Kong's construction industry. *Journal of Safety Research* 28 (4): 243–256.
26. DePasquale, J.P., and E.S. Geller. 2000. Critical success factors for behavior-based safety: A study of twenty industry-wide applications. *Journal of Safety Research* 30 (4): 237–249.
27. Zhang, M., and D. Fang. 2013. A continuous behavior-based safety strategy for persistent safety improvement in construction industry. *Automation in Construction* 34: 101–107.
28. Zohar, D., and G. Luria. 2003. The use of supervisory practices as leverage to improve safety behavior: A cross-level intervention model. *Journal of Safety Research* 34 (5): 567–577.
29. Komaki, J., A.T. Heinzmann, and L. Lawson. 1980. Effect of training and feedback: Component analysis of a behavioral safety program. *Journal of Applied Psychology* 65 (3): 261.

An Accident Vulnerability Index Based on Fuzzy Logic

Tabassum-Abbasi, Tasneem Abbasi, S. M. Tauseef and S. A. Abbasi

Abstract The vulnerability to an accident in a chemical process industry is not merely a function of the severity of a loss of confinement or the size of the area that can be affected by it. What constitutes the impact area is equally important. Vulnerability is a function of several variables—population density, quality of roads (to help disaster management), proximity to, and quality of, healthcare facilities, etc—besides an overall awareness of the risk and preparedness for the emergency. Additionally, factors such as time and climate which are not directly related to the composition of the impact area also determine it. In order to estimate the risk of an accident more precisely—so that strategies to prevent accidents or to cushion their impacts (when accidents occur) can be put in place accordingly—an index has been attempted which has the provision to factor in the vulnerability aspect. Given the fuzziness associated with the occurrence as well as the impact of any accident, this index is based on fuzzy logic.

Keywords Vulnerability · Accident forecasting · Land use-land cover
Fuzzy logic · Index

1 Introduction

Chemical process industries which handle flammable and/or toxic materials carry the risk of major accidents [1–5]. Such accidents rarely end with a single jeopardy and often cause secondary and higher order accidents, thereby demonstrating 'domino effect' [6–10]. This causes catastrophic escalation of the initial accident [11–14]. This aspects of potential risk makes chemical process industries highly hazardous for the surroudning population and proterty [15–20].

Tabassum-Abbasi (✉) · T. Abbasi · S. A. Abbasi
Center for Pollution Control and Environmental Engineering, Pondicherry University, Puducherry 605014, India
e-mail: tab.abbasi@gmail.com

S. M. Tauseef
Department of Health, Safety and Environment, University of Petroleum and Energy Studies (UPES), Dehradun 248007, Uttarakhand, India

N. A. Siddiqui et al. (eds.), *Advances in Fire and Process Safety*,
Springer Transactions in Civil and Environmental Engineering,
https://doi.org/10.1007/978-981-10-7281-9_24

However, more often than not, the risk assessment studies that are carried out to assess the severity, individual risk, and societal risk due to an accident occurring in a chemical process industry, rarely account for the surrounding land use and/or land cover around the industry, which can be affected by the accident [21–24]. Vulnerability has been defined as "the degree of loss to a given element at risk, or set of such elements, resulting from the occurrence of a phenomenon of a given magnitude" [22]. Tixer et al. [21] developed a methodology to assess the vulnerability of an industrial site by means of an integrated risk index implemented with a geographical information system (GIS). This methodology assessed the vulnerability of an accident to humans, environment, and materials due to four physical effects: overpressure, thermal flux, gas toxicology, and liquid pollution. It assessed the contribution of each on the basis of Saaty's multicriteria decision method to give a final vulnerability index value. Kontic and Kontic [23] also evaluated the environmental vulnerability of the surroundings of an accident as a part of a three-step process to determine a threat index in the accident impact zone. They claim to go a step ahead of Tixer et al. [21] by determining whether the environmental element of interest will withstand the threat intensity level it is going to be exposed to in case of an accident. They expressed the environmental threat vulnerability as a ratio of the expected environmental damage/loss to the maximum possible damage/loss on a scale of 1–5. The scale is based on preset criteria for exposure and received amount of energy/mass. This probability assessment is derived as expert opinion. The vulnerability is then provided for each grid cell. Planas et al. [24] have introduced a 'risk severity index'.

In this paper, a methodology to assess the vulnerability of human population as well as to the land use/land coverage around a particular industry undergoing an accident is developed and applied in terms of a fuzzy accident vulnerability index. Given that a lot of subjectivity is involved in determining the extent of damage, a fuzzy logic-based inference has been used to tackle it.

2 Methodology

In order to compute the vulnerability of an accident, the identification of the elements that will be impacted is the most crucial step. This methodology factors in the accidental vulnerability of the land use/land cover around the industrial site and the population density of human being to evaluate the zones of maximum vulnerability. The accidents are evaluated based on three physical effects i.e., overpressure, thermal radiation, and toxic release.

2.1 Evaluating the Vulnerability of Land Use/Land Coverage

The land use/land coverage around an industrial zone covey a lot of information about the damage that may be caused by an accident. Thus evaluating the vulnerability of a particular land use/land coverage will help decision makers in the planning and selection of sites for new industrial setups, re-planning the area around a hazardous industry and providing adequate preventive and control measures.

The land use/land cover has been categorized as below:

1	Forest cover
2	Water bodies
3	Industries
4	Residential areas
5	Institutional (research areas, schools, public offices, etc.)
6	Commercial areas
7	Agricultural zones
8	Transportation infrastructure (roads, bridges, ports, bus stands, airports, etc.)
9	Public parks
10	Other public utilities (fire station and hospitals)
11	Heritage monuments
12	Protected sites (forests with endangered species, highly sensitive ecosystems, etc.)

Each land use/land cover was evaluated for its importance based on the following factors:

1	Degree of irreversibility of damage
2	Inconvenience caused due to damage of these facilities
3	Time it will take to restore or the time till impact of the accident will be felt
4	Number of persons directly and indirectly effected by the accident and its consequences
5	Economic losses that result due to the accident
6	Economic investment required to rebuild

The opinion of four experts was taken by asking them to rate each of the land use/land cover categories based on the above criteria on a scale of 1–10.

The results were averaged and then fuzzified using a trapezoidal fuzzy membership function of categories low, medium, and high (Fig. 1). The membership values were multiplied to the weight 0.1, 0.35, and 0.55 assigned to the categories low, medium, and high, respectively, and aggregated to get the final weightage of each category.

The weightage for each land use/land cover as defuzzified and normalized is presented in Table 1.

In order to determine the vulnerability to land use/land cover, the area under consideration is divided into four zones of impact i.e., very high, high, medium, and low which are weighted as 0.5, 0.3, 0.15, and 0.05. The percentage area of each category lying within each zone is evaluated using the map of the area. The following expression is used to calculate the vulnerability of each category.

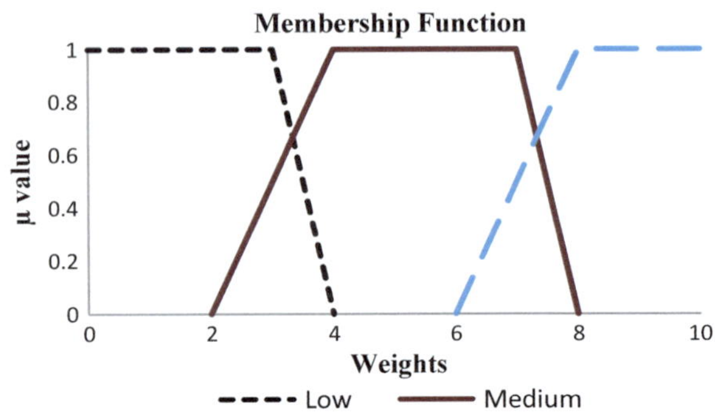

Fig. 1 Trapezoidal membership function for evaluating weights

Table 1 Weightage obtained for each land use/land cover category

Land use/land cover	Normalized weights
Forest cover	0.0755
Water bodies	0.0858
Industries	0.0635
Residential areas	0.0772
Institutional (research areas, schools, public offices, etc.)	0.0840
Commercial areas	0.0909
Agricultural zones	0.0995
Transportation infrastructure (roads, bridges, ports, bus stands, airports, etc.)	0.0978
Public parks	0.0772
Other public utilities (fire station and hospitals)	0.0978
Heritage monuments	0.0669
Protected sites (forests with endangered species, highly sensitive ecosystems, etc.)	0.0840

$$V_c = A_z \times W_z \times W_c \tag{1}$$

where

V_c is the vulnerability of category of land use/land cover in zone
A_z is the percentage area under the zone
W_z is the weight assigned to the zone
W_c is the weight assigned to the land use/land cover.

The total vulnerability for land use is calculated as:

$$V_{lu} = \sum V_c \tag{2}$$

2.2 *Evaluating the Vulnerability of Human Population*

In order to examine the vulnerability of human population in each of the land use/land cover categories, three zones of impact are chosen as very high, medium, and low, and based on the population density, the vulnerability of human population to death is evaluated.

$$V_{hz} = P_z \times W_{hz} \quad (3)$$

where

V_{hz} is the vulnerability of human population to death for the particular land use within the particular zone of impact
P_z is the percentage of total population in the zone
W_{hz} is the weight of the zone of impact.

The total vulnerability of human population to death is:

$$V_p = \sum V_{hz} \quad (4)$$

2.3 *Evaluating the Vulnerability of an Area Around an Industry*

The vulnerability of an area around an industry is calculated as in Table 2.

$$V_{tot} = V_{lu} + V_p \quad (5)$$

The vulnerability index interpretation can be done as in Table 2.

3 Application of the Methodology

The methodology is applied to an area of 25 km^2 having the following land use/land coverage and corresponding population densities. The accident incident is cased due to the physical effect of overpressure (Table 3).

Table 2 Interpretation of vulnerability index

Range of vulnerability score	Interpretation
0.1–0.3	Low vulnerability
0.3–0.6	Moderate vulnerability
0.6–0.8	High vulnerability
0.8–1	Very high vulnerability

Table 3 Area and population density for land use/land cover

Land use/land cover	Area covered (km^2)	Population density per 0.1 km^2
Agricultural zones	5	4
Institutional (research areas, schools, public offices, etc.)	7.5	83
Commercial areas	7.5	50
Residential areas	2.5	20
Transportation infrastructure (roads, bridges, ports, bus stands, airports, etc.)	1.25	10
Water bodies	0.75	0
Industries	0.5	5

3.1 *Evaluating the Vulnerability of Land Use/Land Coverage*

For the given area, the zones of vulnerability of land use are given in Table 4.

Based on the above, applying the weightages for each zone with each category of land use/land coverage, the vulnerability for each is obtained as shown in Table 5.

The total vulnerability score for land use/land cover for the entire area of 25 km^2 V_{lu} is 0.1235.

Table 4 Area percentage of land use/land cover under various zones of vulnerability

Land use/land cover	% area of the land use/land cover within the zone			
	Very high	High	Medium	Low
Agricultural zones	0	10	85	5
Institutional (research areas, schools, public offices, etc.)	0	15	70	15
Commercial areas	0	35	60	5
Residential areas	0	10	90	0
Transportation infrastructure (roads, bridges, ports, bus stands, airports, etc.)	0	10	80	10
Water bodies	0	70	30	0

Table 5 Vulnerability scores for land use/land cover

Land use/land cover	Vulnerability scores
Agricultural zones	0.0249
Institutional (research areas, schools, public offices, etc.)	0.0198
Commercial areas	0.0250
Residential areas	0.0201
Transportation infrastructure (roads, bridges, ports, bus stands, airports, etc.)	0.0064
Water bodies	0.0274

3.2 *Evaluating the Vulnerability of Human Population*

In order to evaluate the vulnerability of human population to death, three zones are considered around the accident point. A very high-risk zone having a radius of 0.5 km around the industry covers an area of 0.78 km^2, the moderate risk zone covers an area of 4.52 km^2, and the low-risk zone covers 19.68 km^2 (Table 6).

The vulnerable population calculated for each zone and category is presented in Table 7.

The vulnerability index score is obtained by calculating the percentage of total population in each zone and multiplying it with the weights assigned to the zone (Table 8).

The total vulnerability score for human fatality for the entire area of 25 km^2 is V_p is 0.1568.

Table 6 Area percentage of land use/land cover under various zones of vulnerability to human population

Land use/land cover	% of the area of the land use/ land cover within the zone of		
	High risk	Moderate risk	Low risk
Agricultural zones	0	20	80
Institutional (research areas, schools, public offices, etc.)	0	25	75
Commercial areas	0	10	90
Residential areas	0	2	98
Transportation infrastructure (roads, bridges, ports, bus stands, airports, etc.)	0	15	85
Water bodies	0	3	97
Industries	100	0	0

Table 7 Vulnerable population under each zone

Land use/land cover	Population affected in very high zone	Population affected in medium zone	Population affected in low zone
Agricultural zones	0	40	160
Institutional (research areas, schools, public offices, etc.)	0	1556	4669
Commercial areas	0	375	3375
Residential areas	0	10	490
Transportation infrastructure (roads, bridges, ports, bus stands, airports, etc.)	0	19	106
Water bodies	0	0	0
Industries	39	0	0
Total	39	2000	8800

Table 8 Calculation of vulnerability score for human population

Human vulnerability zone	Population in the zone	Percentage of total population in the zone	Weight of the zone	Vulnerability score
Very high	39	0.36	0.5	0.0018
Medium	2000	18.45	0.4	0.0738
Low	8800	81.19	0.1	0.0812

3.3 Evaluating the Total Vulnerability of an Area Around an Industry

The total vulnerability score from Eq. (5) is $V_{tot} = 0.2803$, and from Table 2, the score indicated that the chosen industrial area has a low vulnerability.

4 Conclusion

In this study, for the first time, fuzzy logic has been used to develop an accident vulnerability index. Fuzzy logic helps account for the subjectivity involved in making decisions and interpretations. This vulnerability index measures the vulnerability of various land use/land cover area along with that of the vulnerability of human population. The developed index was applied to an example area of 25 km^2 having an industry in which an accident occurs, resulting in damage of the surroundings due to over pressure generated from the accident. Based on the population densities and the land use area, the overall vulnerability index value has been calculated as 0.2803. This value of the index indicates that the vulnerability to the present land use/land cover type due to the industry is low.

This new methodology of calculating the accident vulnerability of an industrial location can help decision makers and land use planners a great deal in modifying existing plants or to select new sites for setting up of industries, etc.

5 Acknowledgement

SAA thanks the Council of Scientific and Industrial Research (CSIR), New Delhi, for the Emeritus Scientist grant (21(1034)/16/EMR-II).

References

1. Khan F.I., and S.A. Abbasi. 1998. *Techniques and methodologies for risk analysis in chemical process industries*, ix + 364 Pages. New Delhi: Discovery Publishing House.
2. Abbasi, T., E.V. Ramasamy, F.I. Khan, and S.A. Abbasi. 2013. *Regional EIA and risk assessment in a fast developing country*, x+433 pages. New York: Nova Science. ISBN 978-1-61942-234-6.
3. Khan, F.I., and S.A. Abbasi. 1997. Accident hazard index: A multi-attribute method for process industry hazard rating. *Process safety and Environmental Protection* 75: 217–224
4. Abbasi, T., and S.A. Abbasi. 2008. The boiling liquid expanding vapour explosion (BLEVE) is fifty… and lives on! *Journal of Loss Prevention in the Process Industries* 21: 485–487
5. Tauseef, S.M., and S.A. Abbasi. 2011. CFD-based simulation of dense gas dispersion in presence of obstacles. *Journal of Loss Prevention in the Process Industries* 24: 371–376.
6. Khan, F.I., and S.A. Abbasi. 1998. Models for domino effect analysis in process industries. *Process Safety Progress (AIChE)* 17: 107–123.
7. Khan, F.I., and S.A. Abbasi. 1998. DOMIEFECT: (DoMIno eFFECT) a user-friendly software for domino effect-analysis. *Environmental Modelling & Software* 13: 163–177.
8. Abdolhamidzadeh, B., T. Abbasi, D. Rashtchian, and S.A. Abbasi. 2010. A new method for assessing domino effect in chemical process industry. *Journal of Hazardous Materials* 182: 416–426.
9. Abdolhamidzadeh, B., D. Rashtchian, T. Abbasi, and S.A. Abbasi. 2011. Domino effect in process-industry accidents—an inventory of past events and identification of some patterns. *Journal of Loss Prevention in Process Industries* 24 (5): 575–593.
10. Khan, F.I., and S.A. Abbasi. 1999. MAXCRED—a new software package for rapid risk assessment in chemical process industries. *Environmental Modeling & Software* 14: 11–25.
11. Khan, F.I., and S.A. Abbasi. 2002. A criteria for developing credible accident scenarios for risk assessment. *Journal of Loss Prevention in Process Industries* 15: 467–475.
12. Khan, F.I., and S.A. Abbasi. 2001. Estimation of probabilities and likely consequences of a chain of accidents (domino effect) in Manali Industrial Complex. *Journal of Cleaner Production* 9: 493–508.
13. Khan, F.I., and S.A. Abbasi. 2001. An assessment of the likelihood of occurrence, and the damage potential of domino effect in a typical cluster of industries. *Journal of Loss Prevention in Process Industries* 14: 283–306.

14. Khan, F.I., and S.A. Abbasi. 1999. The worst chemical industry accident of 1990s: what happened and what might have been—a quantitative study. *Process Safety Progress (AIChE, USA)* 18: 135–145.
15. Abbasi, T., and S.A. Abbasi. 2007. Accidental risk of superheated liquids and a framework for predicting the superheat limit. *Journal of Loss Prevention in the Process Industries* 20: 165–181.
16. Abbasi, T., H. Pasman, and S.A. Abbasi. 2010. A scheme for the classification of explosions in the chemical process industry. *Journal of Hazardous Materials* 174: 270–280.
17. Abdolhamidzadeh, B., T. Abbasi, D. Rashtchian, and S.A. Abbasi. 2010. Corrigendum to a new method for assessing domino effect in chemical process industry. *Journal of Hazardous Materials* 184: 1–3.
18. Khan, F.I., and S.A. Abbasi. 1997. OptHAZOP—an effective and optimal methodology for conducting HAZOP, study. *Journal of Loss Prevention in Process Industries* 10: 191–204.
19. Khan, F.I., and S.A. Abbasi. 1997. TOPHAZOP: A knowledge—based software. *Journal of Loss Prevention in Process Industries* 10: 333–343.
20. Vasanth S., S.M. Tauseef, T. Abbasi, and S.A. Abbasi. (2013). Assessment of four turbulence models in simulation of large-scale pool fires in the presence of wind using computational fluid dynamics. *Journal of Less Prevention in Process Industries* 26: 1071–1084.
21. Tixier, J., A. Dandrieux, G. Dusserre, et al. 2006. Environmental vulnerability assessment in the vicinity of an industrial site in the frame of ARAMIS European project. *Journal of Hazardous Materials* 130 (3): 251–264.
22. Coburn, A.W., R.J.S. Spence, and A. Pomonis. 2000. Vulnerability and risk assessment. 2nd ed. UNDP-Disaster Management Training Programme, The Oast House, UK, 2001. ARPAT. Analisi del rischio per l'area di Livorno e strategie di intervento, ed. Mossa Verre, Firenze.
23. Kontic, Davor, and B. Kontic. 2009. Introduction to threat analysis into the land-use planning process. *Journal of Hazardous Materials* (163): 683–700.
24. Planas, E., J. Arnaldos, B. Silvetti, et al. 2006. A risk severity index for industrial plants and sites. *Journal of Hazardous Materials* 130 (3): 242–250.

IoT- and NDT-Based Bridge Risk Assessment and Identification

Aditya Agarwal, Vishal Sharma, Vishakha Shukla, Bikarama Prasad Yadav and Rajesh Singh

Abstract Bridges are the essential part of the development of infrastructure of a nation. They are the best examples of engineering marvels which can withstand the degradation by nature and mankind. Bridges are very much prone to natural hazards like earthquake and flooding. They are also exposed to modern methods of transportation (e.g., ships, trains, truck overloading). The regular movement of vehicles, heavy machinery, and supported loads causes bridges to be tested on the stage of time. By designing and having proper regular maintenance, accident levels can be reduced which will further affect the durability of bridges. At the initial stage, the methods of hazard identification and risk assessment are essential for bridge design and construction. The given paper is an extension of NDT on the concept of resistive crack measurement system and IoT which will be used to sense the initial faults and presumed deformation that will be helpful in averting the disaster.

Keywords Hazard identification · Risk assessment · NDT · IoT
Resistive crack measurement

A. Agarwal (✉) · V. Sharma · V. Shukla · B. P. Yadav · R. Singh
University of Petroleum and Energy Studies, Bidholi, Dehradun 248007, India
e-mail: adityaa235@gmail.com

V. Sharma
e-mail: vishalsharma7r@gmail.com

V. Shukla
e-mail: shuklavishakha52@gmail.com

B. P. Yadav
e-mail: bpyadav@ddn.upes.ac.in

R. Singh
e-mail: rsingh@ddn.upes.ac.in

N. A. Siddiqui et al. (eds.), *Advances in Fire and Process Safety*,
Springer Transactions in Civil and Environmental Engineering,
https://doi.org/10.1007/978-981-10-7281-9_25

1 Introduction

The Kadalundi River rail bridge failure, India, on July 21, 2001, was considered as a high-profile failure of a 139-year-old rail bridge which took life of 57 people, who all were drowned in the Kadalundi River [1]. The latest collapse of Vivekananda Flyover Bridge in Kolkata, India, on March 31, 2016, was a steel girder flyover bridge which killed 24 and injured 100 people [2]. All over the world, every year there are noticeable accidents in bridges which may be old, having design failure, or are prey to natural disasters or sabotage.

The hazard and risk identification which is done through visual inspection in order to maintain bridge management was categorized by Chiaramonte and Gattulli [3] which is represented in Fig. 1. The four different stimulation models were proposed by them which were as follows: bridge inventory, computer-aided visual inspection design system, defect catalog, and priority-ranking procedure. This was essential for bridge assessment. This categorization is helpful in predicting the failure mode of the bridge which is the part of hazard and risk identification. From this data, we can design the form of instrument to mitigate any risk or hazard.

It has been analyzed and noticed that there are five reasons for bridge failure and collapse: low-grade material, poor maintenance, defective design, construction failure, and natural calamities like earthquakes. While making bridges, use of low-grade material or techniques can be the cause of its failure [4]. The silver bridge breakdown on the Ohio River occurred in the year 1967 [5]. Poor maintenance and neglecting the reasons for decreased life of a structure are the essential points which are in need of research. With proper maintenance, most accidents can be prevented. In order to get a prolonged bridge durability and longevity, regular inspection and maintenance routines are required. In order to prevent bridge from collapsing, a

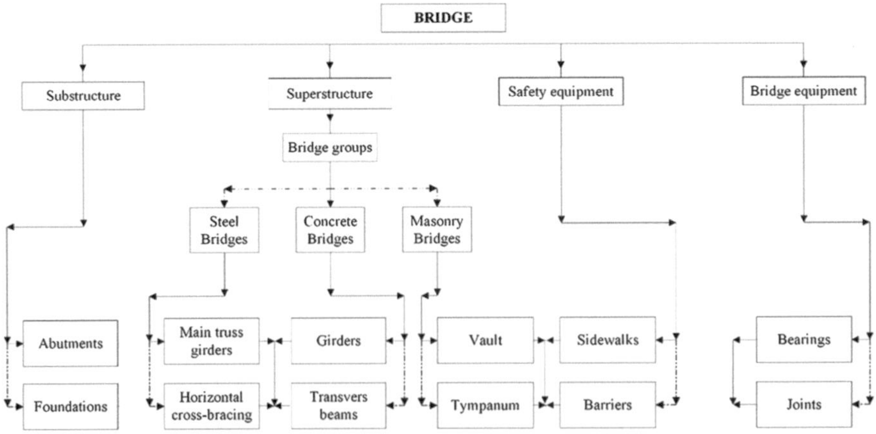

Fig. 1 Description of structure of bridge and its inventory module by Chiaramonte and Gattulli [3]

proper maintenance is required likewise say replacement of rusted metals, new coats of paints, addition of support, and also accumulation of draining water. The bridge could collapse even if the design is great but its maintenance is poor. Defective design is the other reason which is purely related to human error. If the design is not perfect, the construction no matter how much better it is will be hardly of any use [6]. This would lead to accidents and property damage if the bridge is not designed properly. Construction failure is one of the reasons which affects the longevity of the bridge. A great amount of manpower and labor are required for the construction of the bridge. In case the bridge collapses, the workers assigned for the construction will be in appalling position. This means construction workers are in appalling position in case the bridge will collapse [7]. The full design is not yet realized because the bridge construction is still in progress. The accidents could occur due to engineering miscalculation. At least 80 lives were lost when the Quebec Bridge was collapsed [8]. Natural calamities (earthquake and flooding) and manmade disasters (fire and vehicle traffic load) account for failure of the bridge which is required to be reported and checked in the periodic way. Between 1989 and 2000, five hundred bridges failed in the USA. The main reasons were flooding and scour due to which 53% of the failures took place [7]. To avoid such risks and identify failures at the initial level, there is a need of methodology and instrument which can detect initial failures and help in studying the changes brought up in the structure of bridge which could be helpful in averting any failure.

2 Non-Destructive Testing (NDT)

Our designed instrument is the part of non-destructive testing (NDT) and will give the results in the pre-disaster scenario. Any type of initial failure which can then become a disaster situation can be mitigated by the use of the proposed system, which can be helpful in emptying the bridge by not using it for transportation, hence saving lives and property. The NDT will be the part of risk and hazard identification and control that will be helpful for the civil and construction professionals to have frequent data by analyzing it through this system [9] (Table 1).

3 Hardware

3.1 Internal Crack Measurement

The proposed system consists of geosensor, microcontroller (Arduino UNO), power supply, resistive crack measurement system, and IoT modem. The geosensor will sense the magnitude of the frequency generated by any shock (earthquake, load by vehicles) which will be then compared with the recorded bridge frequency. The structure's natural frequency is calculated during the design and implementation

Table 1 Types of NDT test [10]

Test type	Applications	Limitations
Impact echo	(a) Finds defect in concrete	(a) Detection is feasible for viscous material
	(b) Finds delamination PCC and RC elements	(b) Concrete overlay detectable
	(c) Finds surface opening cracks	(c) On limited dimensions, it is observed that interference of boundary on signal is more prominent and result oriented, like girders and piers
	(d) Finds ducts, voids	
	(e) Overlay debonding, grouting characterization	
Ultrasonic pulse velocity	(a) Depth measurement, assess faults	(a) Time-consuming method
	(b) Finds delamination, shallow cracking	(b) Close grid spacing is required
	(c) Finds grounding defects	(c) Since the lower frequencies are used, some defects might remain undetected
	(d) Finds interfaces between concrete and air and concrete and steel [11–15]	(d) Small defects remain undetected
Ground-penetrating radars	(a) Finds buried objects	(a) Frozen environment remains undetectable
	(b) Find voids and abnormalities	(b) It is an expensive method
	(c) Mcasures concrete thickness	(c) Unable to estimate stress, strength properties
	(d) Helpful in estimating concrete properties	(d) Extreme cold conditions affect the GPR data
	(e) Sensitive to corrosiveness	
Electrical resistivity	(a) Finds moisture	(a) It provides raw data and interpretations
	(b) Finds the presence of cracks	(b) There is a need to always wet the surface
	(c) It can be used for corrosion activity	(c) Expensive technique
Infrared thermography	(a) Finds delamination in structures	(a) It is not applicable to surfaces which are vertical
	(b) Gives information for further testing	(b) Depends upon the hearing skills of the operator
	(c) Gives couture maps for the sound surface	(c) Detects initial delamination

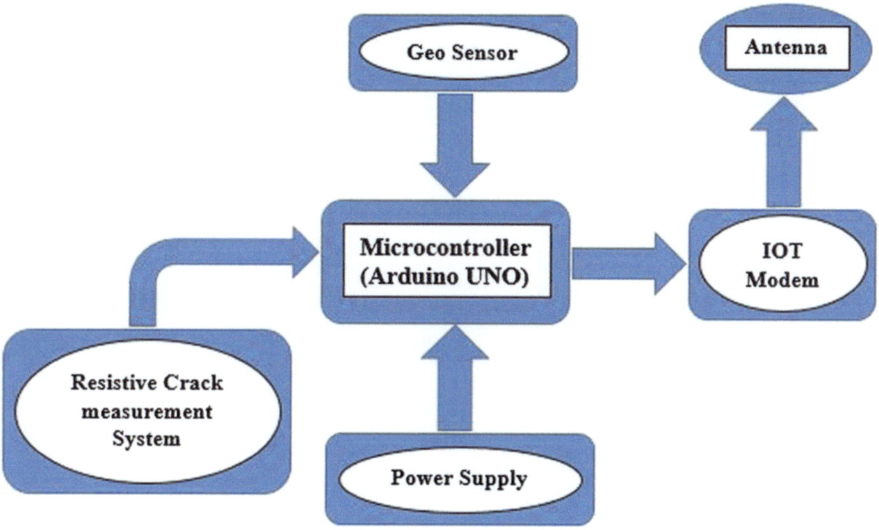

Fig. 2 Block diagram of internal crack measurement

stage. Henceforth, the natural frequency will be recorded in the microcontroller which will then evaluate the frequency and shock observed.

The working of resistive crack measurement system is based on the process of current flowing in different loops. Here the steel bars present in the bridge will act as a current carrying conductor. A continuous DC (direct current) supply of current is given to the steel bars as AC (alternating current) supply is not uniformly distributed through the steel bar thickness, and it tends to be concentrated near the surface of the bar and will flow as a current loop [16]. As a result, when there is any crack in the steel bar, then the current loop will show a break point, and from that we get the information about the internal cracks and their exact location in the bridge structure. With the help of IoT modem, the data will be transmitted to the managing or controlling section (Fig. 2).

3.2 External Crack Measurement

The external crack measurement system consists of a rotatory system consisting of an IR sensor, microcontroller (Arduino UNO), IoT modem, and power supply. On the bridge system, the system will be placed externally where human evaluation is not possible, i.e., below the bridge. The IR sensor will emit rays which will flow in a rotating way throughout the concrete section of the bridge. The transmitted rays will then be reflected from the surface defects and will be received. The result will be send to microcontroller (Arduino UNO), and then with the help of IoT modem, it will be transmitted to the managing or controlling section. The system will be

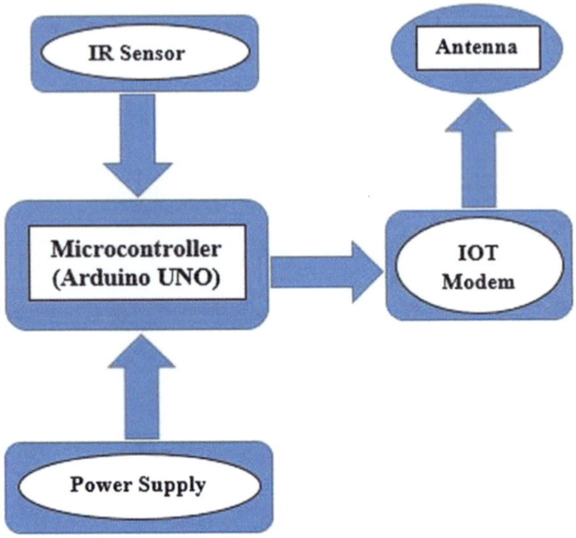

Fig. 3 Block diagram of external crack measurement

having DC power supply which will be helpful in case of power failure. The external crack measurement system is also an NDT design which will be taking care of the external surfaces (Fig. 3).

3.3 Receiver Section

The data collected by internal and external crack measurement systems gives the output data through the use of IoT modem to the required receiver section (mobile, laptops). As the information is collected on the periodic basis, it is easy to evaluate the system and its performance which is helpful in predicting any type of incident (Fig. 4).

4 Software

4.1 Internal Crack Measurement

In this, the main controlling unit is Arduino UNO in which the TxD and RxD pins of IoT modem are connected to RxD and TxD pins of Arduino UNO, respectively, that is used to send the data on periodic basis. Geosensor is connected to the single pin PD7 of Arduino UNO which can sense the magnitude of frequency generated by heavy loads or earthquakes. Also a resistive crack measurement system is connected to the PB2 pin of Arduino UNO which can find internal cracks (Fig. 5).

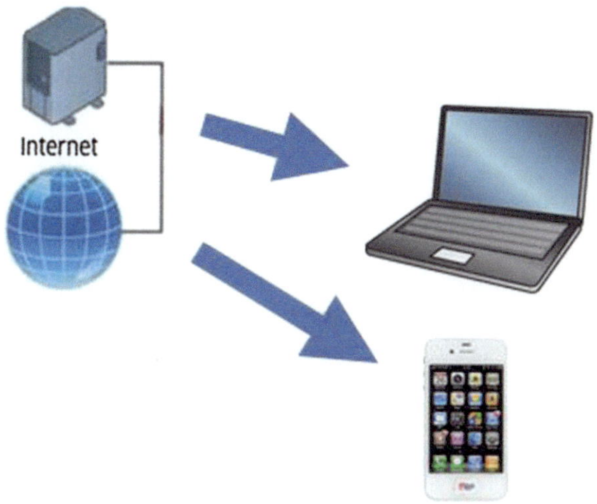

Fig. 4 Diagram of receiver section

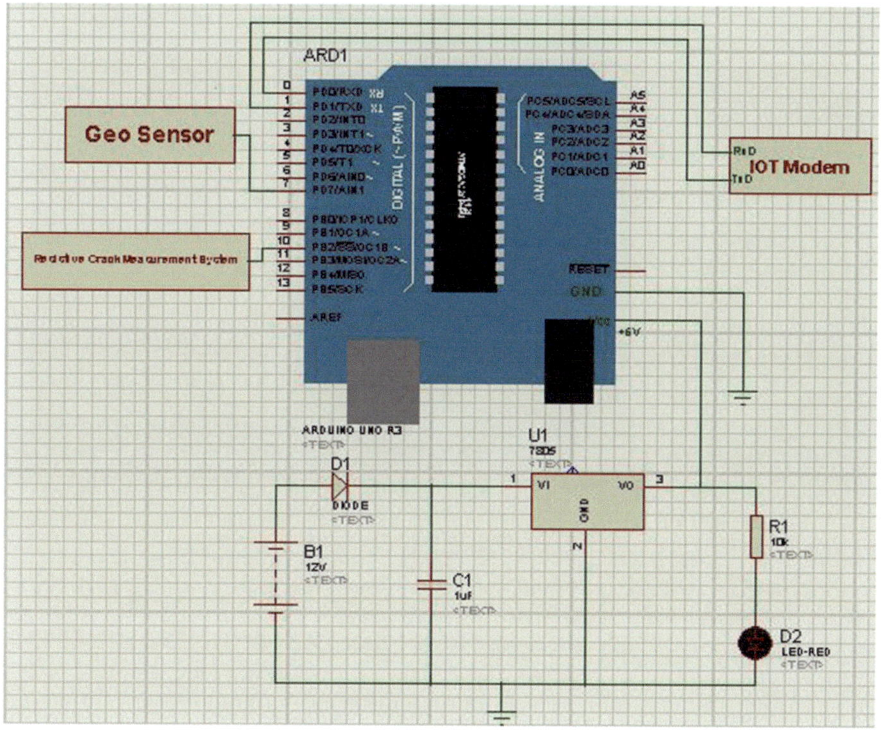

Fig. 5 Circuit diagram of internal crack measurement

4.2 *External Crack Measurement*

The main controlling unit is Arduino UNO. IoT modem is then connected to TxD and RxD pins of Arduino UNO that is used to receive and send the data on periodic basis. In Arduino UNO, a power supply is connected and also to the PD6 pin, an IR sensor is connected which is used to measure the external cracks of the bridge (Fig. 6).

5 Component Description

See Table 2.

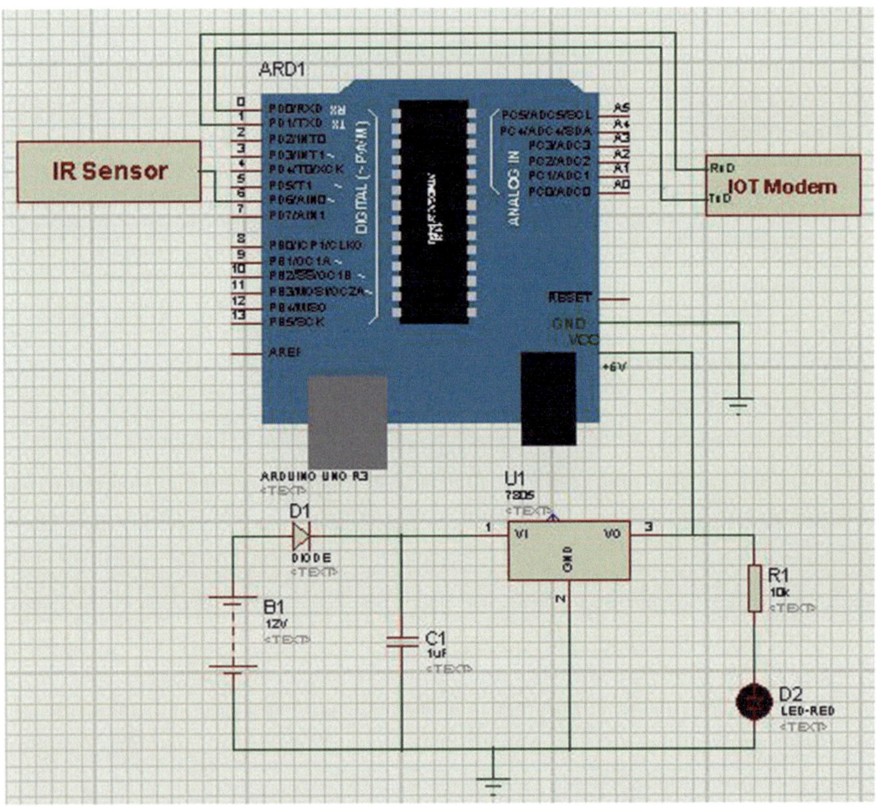

Fig. 6 Circuit diagram of external crack measurement

Table 2 Instruments used

S. No.	Device/Module	Specification and working
1	Arduino UNO	It is a microcontroller board based on the ATmega328P
2	IoT modem	It sends the data on the server
3	IR sensor	It is a light-emitting diode (LED) which emits infrared radiations
4	Geosensor	It senses the magnitude of frequency
5	Resistive crack measurement system	It is based on the process of current flowing in the loop
6	Power supply	12 V/1 A lithium-ion battery, 12 V/1 A power supply

6 Result and Discussion

The proposed instrument based on infrared, IoT, and resistive crack measurement will be helpful in the bridge structure to mitigate disaster. The instrument will give required details of the bridge response against environment and traffic loads which can then be evaluated by output data. If any structure or specific place requires any maintenance or support, then it will be easy to trace by the use of this system. Placing this system on bridge structures can be helpful in various ways for the civil engineers as it gives them data by analyzing the structure without destructing any part. Hence, this instrument which is in the category of NDT will be helpful in the generation of data and will make the evaluation quite simple and progressive.

7 Future Scope

The given system has an ability to detect natural disasters (e.g., earthquakes, floods) which while occurring hit the natural frequency of the bridge. As a consequence, there is loss of life and property. The proposed instrument will be placed on bridge and analyzing data on a regular basis which is helpful for managing officials. Hence, any form of disaster or accident can be averted by responding at the initial stage. The loss of life and property is the major concern, and this instrument can be beneficial to minimize the disastrous or accidental outcomes.

References

1. Sridhar, V. 2001. *Was it the bridge, or the bogies*? Vol. 18, Issue 15, 21 July–03 Aug 03 2001. Available at: http://www.frontline.in/static/html/fl1815/18151250.htm.
2. *Kolkata flyover collapse: Five officials detained, forensic team collects sample from accident site*. 1 Apr 2016. Available at: http://indianexpress.com/article/india/india-news-india/kolkata-under-construction-bridge-collapses-several-feared-dead-latest-updates/.
3. Gattulli, V., and L. Chiaramonte. 2005. Condition assessment by visual inspection for a bridge management system. *Computer-Aided Civil and Infrastructure Engineering* 20 (2): 95–107.
4. Morton Karrie. 22 November 2013. *5 reasons bridges fail*. Available at: http://www.equipmentworld.com/guest-blog-5-reasons-bridges-fail/.
5. *West Virginia historical society quarterly*, Vol XV, No. 4 October, 2001. The collapse of the Silver Bridge by Chis LeRose.
6. Khursheed, S., B.M. Al-Hashimi, S.M. Reddy, and P. Harrod. 2009. Diagnosis of multiple-voltage design with bridge defect. *IEEE Transactions on Computer-Aided Design of Integrated Circuits and Systems* 28 (3): 406–416.
7. Wardhana, K., and F.C. Hadipriono. 2003. Analysis of recent bridge failures in the United States. *Journal of Performance of Constructed Facilities* 17 (3): 144–150.
8. Pearson, C., and N. Delatte. 2006. Collapse of the Quebec Bridge, 1907. *Journal of Performance of Constructed Facilities* 20 (1): 84–91.
9. Colombo, M., and R. Felicetti. 2007. New NDT techniques for the assessment of fire-damaged concrete structures. *Fire Safety Journal* 42 (6): 461–472.
10. Rehman, S.K.U., Z. Ibrahim, S.A. Memon, and M. Jameel. 2016. Nondestructive test methods for concrete bridges: A review. *Construction and Building Materials* 107: 58–86.
11. Srikanth, B., et al. 2017. Recent advancements in supporting materials for immobilised photocatalytic applications in waste water treatment. *Journal of Environmental Management* 200: 60–78.
12. Afshari, A., D. Frazer, and R. Creese. 1996. Ultrasonic techniques for the bonding of rebar in concrete structures, In *Structural materials technology, an NDT conference*, 3–8. San Diego, USA.
13. Krause, M., B. Milmann, F. Mielentz, D. Streicher, B. Redmer, K. Mayer, et al. 2008. Ultrasonic imaging methods for investigation of post-tensioned concrete structures: A study of interfaces at artificial grouting faults and its verification. *Journal of Nondestructive Evaluation* 27 (1–3): 67–82.
14. Taffe, A., and H. Wiggenhauser. 2016. Validation for thickness measurement in civil engineering with ultrasonic echo, In *Proceedings of the 9th European conference on NDT*, 2006.
15. Hevin, G., O. Abraham, H. Pedersen, and M. Campillo. 1998. Characterization of surface cracks with Rayleigh waves: A numerical model. *NDT and E International* 31 (4): 289–297.
16. Wei, R. P., and R. L. Brazill. 1981. An assessment of AC and DC potential systems for monitoring fatigue crack growth. In *Fatigue crack growth measurement and data analysis*. ASTM International.

Performance Evaluation of Rectangular Fins by Modeling and Simulations

Akshay Maan, Praveen Pitta and Jitendra Yadav

Abstract Rectangular fins are widely used in many engineering applications, especially in IC engines, where the temperature of gases is around 2500 °C. A universal explanation of the performance of the rectangular fins is always a challenging task for researchers. Present work is intended to identify the effect of influencing parameters, and thereby checking the rectangular fins performance for motorbike by computer simulation. The input parameters selected for variation are the length of fin, thickness of fin, width of the fin and the performance is obtained in terms of efficiency, effectiveness, temperature distribution between fins. It is observed that if there is an increase in thickness, there is an increase in efficiency and in the temperature of the fin at the tip, but there is decrease in effectiveness, whereas while increasing the length of the fin, it experiences a decrease in the temperature at the tip, and efficiency decreases but there will be an increase in effectiveness. Present work provides guidelines for selecting the influencing parameters to ensure the optimum performance of rectangular fins.

Keywords Rectangular fins · Efficiency · Effectiveness · Heat flux
Modeling and simulation

1 Introduction

Selection of a particular type of fin is very important for the proper heat dissipation from the hot surfaces to endure failure-free operation. In IC engines, fins are considered as the extended surfaces employed to remove the unnecessary heat from

A. Maan (✉) · P. Pitta · J. Yadav
University of Petroleum and Energy Studies, Dehradun 248007, India
e-mail: akshaymaan93@gmail.com

P. Pitta
e-mail: praveen8423@yahoo.in

J. Yadav
e-mail: jyadav@ddn.upes.ac.in

N. A. Siddiqui et al. (eds.), *Advances in Fire and Process Safety*,
Springer Transactions in Civil and Environmental Engineering,
https://doi.org/10.1007/978-981-10-7281-9_26

the system during exhaust so as to maintain the temperature of the system. If fins are not used in the heat transfer system, then the temperature of the system goes on increasing and the system will burst out. Although there are many aspects through which we can reduce the temperature of the system like using coolant for cooling the system but it may lead to the cost of construction. So use of extended surfaces is always an alternative solution to optimize the cost of cooling.

The increase use of product model explores the research work to find out the fin performance by modeling and simulation. That will lead to analyze the system before actual fabrication takes place. A lot of research work is reported in literature to ascertain the performance of fins by modeling and simulation. Bodoia and Ostrele [1], investigated the developing flow numerically for the channels, their approach was numerical, to study the flow which is developing in a channel, the transfer of heat among equally heated, constant temperature plate in a determination to expect the total channel length that is required to be achieved for the fully developed flow, as the function of wall temperature and channel width. Ofi and Hetherington [2] used a finite element analysis method for the study of convective heat transfer which is occurring naturally from vertical channels which are open. Culham et al. [3] used a mathematical code for the simulation of heat transfer in free convective mode for the vertical fin array. Several attempts have been made to optimize the finned enclosures and heat sinks [4–8].

Although a lot of works were done to estimate the optimized performance of finned structure but still a lot remain in terms of optimized modeling with desired efficiency and effectiveness. Present work deals with the modeling of fins for the motorbike to get its insight into terms of effect of different influencing parameters on efficiency and effectiveness.

2 Modeling of the System

The system selected for the analysis is the rectangular fins for IC engines of motorbike. The fin is finite, and the parameters selected for the engine and fin are shown in Table 1.

The modeling of system is done by CAD modeling software, and the model is shown in Fig. 1.

The different mathematical expression used for the simulation for different performing parameters is shown by Eqs. 1–7.

2.1 Heat Dissipation

$$Q_{\text{fin}} = n \times \sqrt{hPKA} \times (t_0 - t_a) \times \left\{ \frac{\tanh(ml) + \frac{h}{km}}{1 + \tanh(ml) \times \frac{h}{km}} \right\} \quad (1)$$

Table 1 System parameters

S. No	Parameters	Symbol	Value	Unit
1	Length of the fin	L	15, 20, 25	mm
2	Base thickness	T	2, 4, 6	mm
3	Width	B	103, 123, 133	mm
4	Temperature of the fin base	t_o	200	°C
5	Temperature of ambient	t_a	30	°C
6	Heat transfer coefficient	h	40	W/m^2 °C
7	Thermal conductivity	k	237.5	W/m °C
8	No. of fins	N	6	
9	Displacement	V	160	Cc
10	Bore	D	63	mm
11	Stroke	L	65	mm

Fig. 1 CAD model of fins for motorbike

where

$$m = \sqrt{\frac{hP}{KA_{cs}}} \tag{2}$$

$$Q_{without-fin} = hA_{cs}(t_0 - t_a) \tag{3}$$

2.2 Fin Effectiveness

Fin effectiveness can be said as the ratio of heat transfer that is happened due to the presence of fin to the heat that is transferred if suppose the fin was not present. In brief, this tells us the quantity of extra heat that will be transferred by the fin.

$$\varepsilon = \frac{Q(\text{with fin})}{Q(\text{without fin})} \tag{4}$$

The desire is to have this ratio as large as possible while keeping the additional cost of adding the fins as low as possible.

2.3 Temperature at the End of the Fin

$$\frac{t - t_{\mathrm{a}}}{t_0 - t_{\mathrm{a}}} = \left[\frac{\cosh\{m(l - x)\} + \frac{h}{km}\sinh\{m(l - x)\}}{\cosh(ml) + \frac{h}{km}\sinh(ml)}\right] \tag{5}$$

At $x = 1$ (at tip)

$$\frac{t - t_a}{t_0 - t_a} = \frac{1}{\cosh(ml) + \frac{h}{km}\sinh(ml)} \tag{6}$$

2.4 Fin Efficiency

Fin efficiency can be defined as the ratio of heat that is transferred from the actual fin to the heat that is transferred by an imaginary fin of the exact same geometry conditions but with an infinite conductivity. (In other words, the whole fin surface was at a temperature equal to that of the fin base.)

$$\eta = \frac{\tanh\left\{\left(\sqrt{\frac{2h}{kt}}\right) \times \left(l + \frac{t}{2}\right)\right\}}{\left\{\sqrt{\frac{2h}{kt}} \times \left(1 + \frac{t}{2}\right)\right\}} \tag{7}$$

This ratio will always be smaller than one.

3 Result and Discussion

The model shown in Fig. 1 is imported to ANSYS and the thermal analysis has been performed and the sample results in the form of temperature distribution and heat flux analysis are done and shown in Figs. 2 and 3.

These were the results obtained after the primary analysis of the fins under steady state thermal conditions.

Temperature Distribution: 200–197.98 °C.

Total Heat Flux: 17913–161.05 W/m^2.

The similar analysis is done for varying the length, width, and thickness as shown in Table 1 and the different findings are tabulated from Tables 2, 3, and 4.

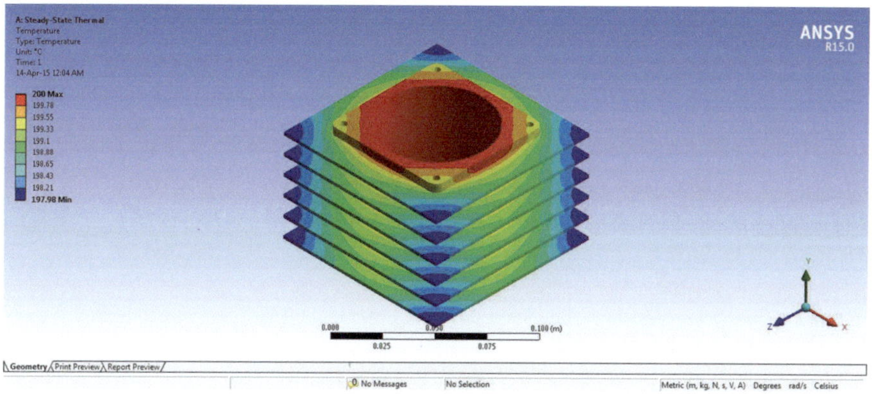

Fig. 2 Temperature distribution analysis for $T = 2$ mm, $B = 103$ mm, $L = 15$ mm

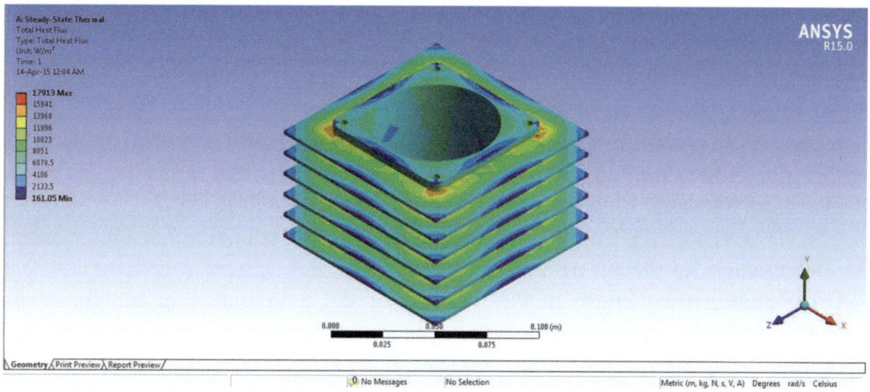

Fig. 3 Total heat flux analysis for $T = 2$ mm, $B = 103$ mm, $L = 15$ mm

Table 2 Parameters tabulated by varying thickness

Thickness (mm)	Efficiency	Effectiveness	Total heat flux (W/m^2)	Temperature distribution (°C)
2	98.6	96.3	17913–161.05	200 –197.98
4	99.1	55.9	12076–258.73	200–198.99
6	99.1	37.4	10738–713.65	200–199.16

Table 3 Parameters tabulated by varying width

Width (mm)	Efficiency	Effectiveness	Total heat flux (W/m^2)	Temperature distribution (°C)
103	98.6	96.3	17913–161.05	200–197.98
123	98.4	95.99	32441–292.27	200–196.79
133	98.4	96.2	40052–429.55	200–195.66

Table 4 Parameters tabulated by varying length

Length (mm)	Efficiency	Effectiveness	Total heat flux (W/m^2)	Temperature distribution (°C)
15	98.6	96.3	17913–161.05	200–197.98
20	97.6	96.8	24021–174.09	200–197.74
25	96.3	118.9	32441–292.27	200–196.79

4 Discussion of Result

Bar charts are drawn for the data from above tables, and they are shown in Figs. 4, 5, and 6.

On changing the thickness, it was found that the temperature at the tip of fin keeps increasing with the increase in thickness. The difference of total heat flux and directional heat flux decreases.

The efficiency of the fin also increases but not considerably, whereas the effectiveness of the fin decreases.

It can be observed from Fig. 5 that on changing the width it was found that the temperature at the tip of fin decreases with the increase in width, even the difference of total heat flux and directional heat flux increases.

The efficiency of the fin remains same since it is independent of width, whereas the effectiveness of the fin decreases.

On changing the length, it was found that the temperature at the tip of fin decreases with the increase in length, even the difference of total heat flux and directional heat flux increases.

The efficiency of the fin decreases gradually, whereas the effectiveness of the fin increases.

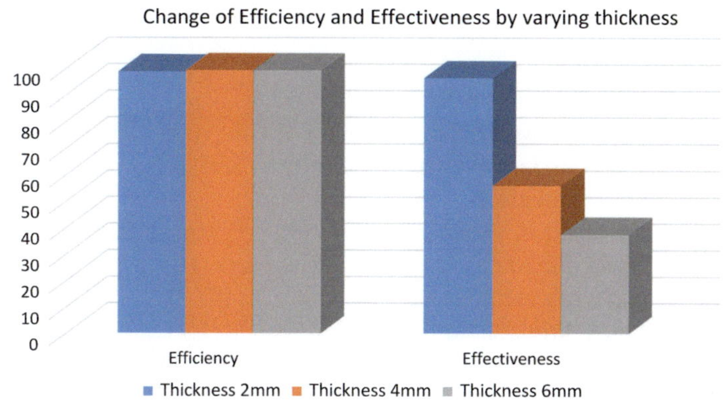

Fig. 4 Change of efficiency and effectiveness by varying thickness

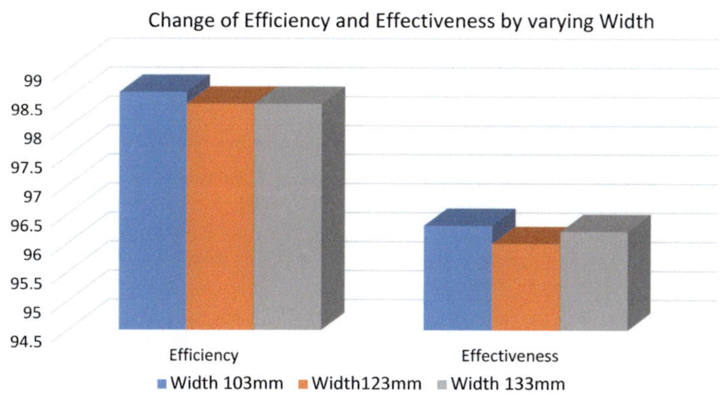

Fig. 5 Change of efficiency and effectiveness by varying width

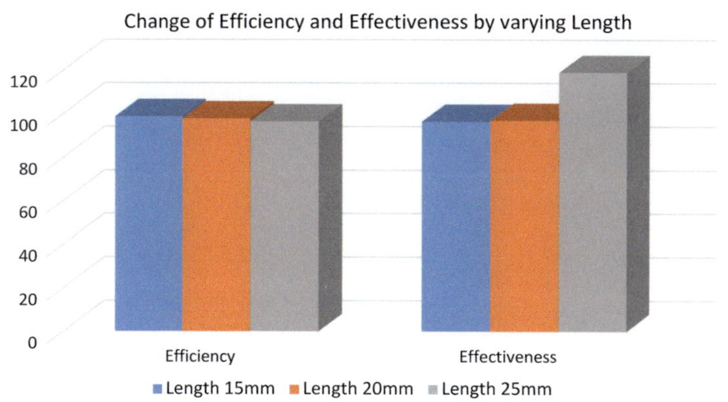

Fig. 6 Change of efficiency and effectiveness by varying length

5 Conclusion

Fins are an effective medium in heat transfer mechanism. They are very efficient modes of heat transfer as there happen to be no moving parts. Though fins can only be used where there is some movement of air nearby and so their use is only limited to motorcycles and aeroplanes.

Though fins can result in increased weight due to extra surface and cover more area, the effect caused due to this is more counteracting. The heat that the fins loose helps engine to prevent wear and tear caused due to the internal heat produced. It omits the use of any kind of liquid cooling which would result in less use of complex equipment making the engine overall lighter and making the cooling process cheaper and more effective.

Thus, this data give us sufficient understanding about the fins, make us realize about the various parameters that are crucial in the heat transfer of the fins, give scope for the researchers for doing more work in this area.

References

1. Bodoia, J.R., and J.F. Ostrele. 1962. The development of free convection between heated vertical plates. *ASME Journal of Heat Transfer* 84: 40–44.
2. Ofi, O., and H.J. Hetherington. 1977. Application of the finite element method to natural convection heat transfer from the open vertical channel. *International Journal Heat Mass Transfer* 20 (11): 1195–1204.
3. Culham, J.R., M.M. Yovanovich, and S. Lee. 1995. Thermal modeling of isothermal cuboids and rectangular heat sinks cooled by natural convection. *IEEE Transactions on Components, Packaging, and Manufacturing Technology* 18: 559–566.
4. Yuncu, Anbar. 1998. An experimental investigation on performance of fins on a horizontal base in free convection heat transfer. *Journal of Heat and Mass Transfer* 33: 507–514.
5. Yuncu, H., and B. Yazicioglu. 2007. Optimum fin spacing of rectangular fins on a vertical base in free convection heat transfer. *Journal of Heat and Mass Transfer* 44: 11–21.
6. Yuncu, H., and A. Guvenc. 2001. An experimental investigation on performance of rectangular fins on a vertical base in free convection heat transfer. *Journal of Heat and Mass Transfer* 37: 409–416.
7. Mobedi, M., and H. Yuncu. 2003. A three dimensional numerical study on natural convection heat transfer from short horizontal rectangular fin array. *J. Heat and Mass Transfer* 39 (2003): 267–275.
8. Harahap, F., and H.N. McManus. 1967. Natural convection heat transfer from horizontal rectangular fin array. *J. Heat Transfer* 89: 32–38.

Design of Trajectory and Perturbation Analysis for Satellite Orbital Parameters

M. Raja, Saurabh Pandey, Rishabh Kumar and Aman Dalmia

Abstract The aviation and space industry is advancing at a fast pace. In light of the launch and reentry accidents that have happened in the past, it is essential to have accurate analysis of the parameters required for error-free launch and placement of the launched body in correct trajectory. In this paper, we discuss various orbital parameters required for placing a satellite, launched from earth, in correct orbit and the designing of its trajectory using patched conic approximation (PCA) method. Also presented are various perturbations factors like lunar gravity and atmospheric drag. The effect of these factors is considered and compared with the ideal cases.

Keywords Keplerian elements · Perturbation · Patched conic approximation method · Atmospheric drag

1 Introduction

The Student Space Exploration and Technology Initiative (SSETI) paper started to create and build a microsatellite. Also it should be completed with the development of a Moon Rover in the third mission. The launch described in this paper is the microsatellite. The goal of this launch is to make the ESMO satellite orbit into Moon. One of the teams works on control of the attitude and the orbit of the ESMO satellite. To add control to the orbit of the satellite, forces acting on the satellite need to be described. The number of celestial bodies has to be decided by making the problem a two-, three-, or four-body problem depending on the number of celestial bodies included. These will form the largest forces, and other forces such as atmospheric drag and solar radiation pressure can also be included. Also, there are many possibilities from among possible trajectories to get to the Moon. Some are more fuel-efficient than others, but these often use longer time. But no matter

M. Raja (✉) · S. Pandey · R. Kumar · A. Dalmia
Department of Aerospace Engineering, University of Petroleum and Energy Studies, Dehradun 248007, Uttarakhand, India
e-mail: mraja@ddn.upes.ac.in

N. A. Siddiqui et al. (eds.), *Advances in Fire and Process Safety*,
Springer Transactions in Civil and Environmental Engineering,
https://doi.org/10.1007/978-981-10-7281-9_27

which is used, there will always be perturbations and combinations to consider. Safety is the probability of causing injury or loss of life. Unreliable launchers are not necessarily unsafe, whereas reliable launchers are usually, but not invariably safe. Apart from catastrophic failure of the launch vehicle itself other safety hazards and Van Allen radiation belts which preclude orbits which spend long periods within them. Trajectory optimization is the process of designing a trajectory that minimizes or maximizes some measure of performance within prescribed constraint boundaries. While not exactly the same, the goal of solving a trajectory optimization problem is essentially the same as solving an optimal control problem. This problem was first studied by Robert H. Goddard and is also known as the Goddard problem.

a. The two-body problem

The simplest of the n-body problems is the two-body problem; only two masses are considered at a time. Let the masses be denoted by m_1 and m_2.

$$\ddot{r}_2 - \ddot{r}_1 = -G(m_1 + m_2)\frac{r_1 - r_2}{r_3}$$

This is the equation of motion for the two-body problem.

2 Dynamics of Orbits

When orbital mechanics is to be described, there are many different types of coordinate systems to choose from. It is quite easily expressed in polar coordinates. The plane polar coordinates are (r, θ), and the unit vectors are shown in Fig. 1 [1]. The velocity vector is

$$\vec{v} = \dot{r}\vec{e_r} + r\theta\vec{e_\theta},$$

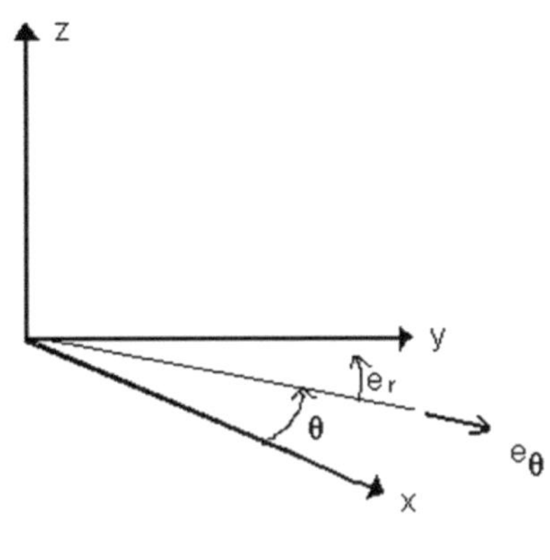

Fig. 1 Polar coordinates

and acceleration vector is

$$\vec{a} = \left(\ddot{r} - r\dot{\theta}^2\right)\overrightarrow{e_r} + \left(r\ddot{\theta} + 2\dot{r}\dot{\theta}\right)\overrightarrow{e_\theta}$$

The equations of motion can be divided up into radial and transverse directions. In the radial direction, the equation of motion is

$$\ddot{r} - r\dot{\theta}^2 = -\frac{\mu}{r^2}$$

where $\mu = Gm$ (G—gravitational constant, m—mass of spacecraft) and the whole expression on the right-hand side is gravity. This is the only acceleration that works in radial direction.

$$r(\theta) = \frac{\frac{h^2}{\mu}}{1 + \frac{Ah^2}{\mu}\cos(\theta - \theta_0)}$$

where A and θ_0 are constants. These are polar coordinates of an ellipse equation.

2.1 Orbit Geometry

The simplest orbits follow basic geometry of conic sections. Conic sections are different intersections of a plane and a cone. The circle intersects the cone horizontally, and the ellipse intersects the cone with a tilt; see Fig. 2. Both are closed curves. The hyperbola intersects the cone resulting in an open curve. There is yet another basic conic section; the parabola [2]. The parabola is the single curve which divides the closed ellipse from the open hyperbola. Here the plane is parallel to the side of the cone. There are two points of particular interest on the orbits; the pericenter and the apocenter. The pericenter is the point where *a* spacecraft will be closest to the object it is orbiting, and the epicenter is the point furthest away *b* semi-minor axis, a semi-major axis, *ae* distance from the center to the focal point. The distance is determined by the conic section of eccentricity *e* (Fig. 3).

2.2 Elliptical Orbits

The orbit period can be calculated from the equation for the area of an ellipse, and the definition of an orbit period results in

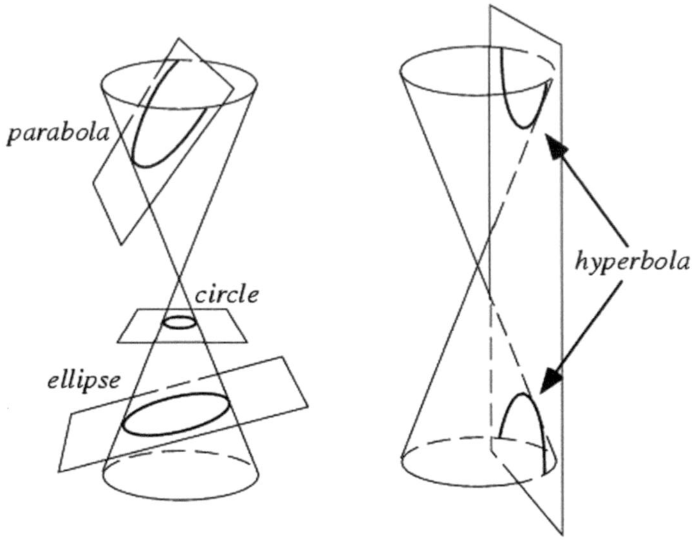

Fig. 2 Conic sections

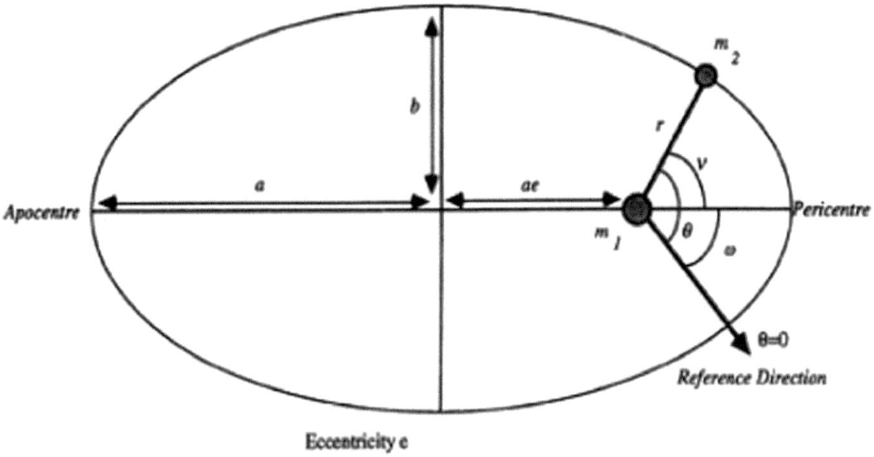

Fig. 3 Orbital parameter

$$T = 2\pi\sqrt{\frac{a^3}{\mu}}$$

2.3 Circular, Parabolic, and Hyperbolic Orbits

In circular orbit, the eccentricity is zero, which means that the radius is constant, R. This results in the following velocity and orbit period equations

$$v = \sqrt{\frac{\mu}{R}}$$

$$T = 2\pi\sqrt{\frac{R^3}{\mu}}$$

In a parabolic orbit, the eccentricity is one. This results in the velocity equation

$$v = \sqrt{\frac{2\mu}{r}}$$

The orbit period $T \to \infty$ since $a \to \infty$.

In hyperbolic orbit, the eccentricity is greater than one [3]. The velocity equation is then

$$v^2 = 2\frac{\mu}{r} + V_\infty^2$$

where $V1$ is the hyperbolic excess speed expressed as

$$V_\infty = \sqrt{\frac{\mu}{a}}$$

3 Perturbations of Orbits

3.1 The Flattening of the Earth

Earth is in everyday life thought of as being a perfect sphere. But this is not entirely true. Earth is slightly flattened at top and bottom (Fig. 4).

Besides being flat at top and bottom, Earth has a bulge on equator. It is not important to take this effect into account for low Earth orbits (LEOs) as it will average out after many revolutions, but it should be taken into account when

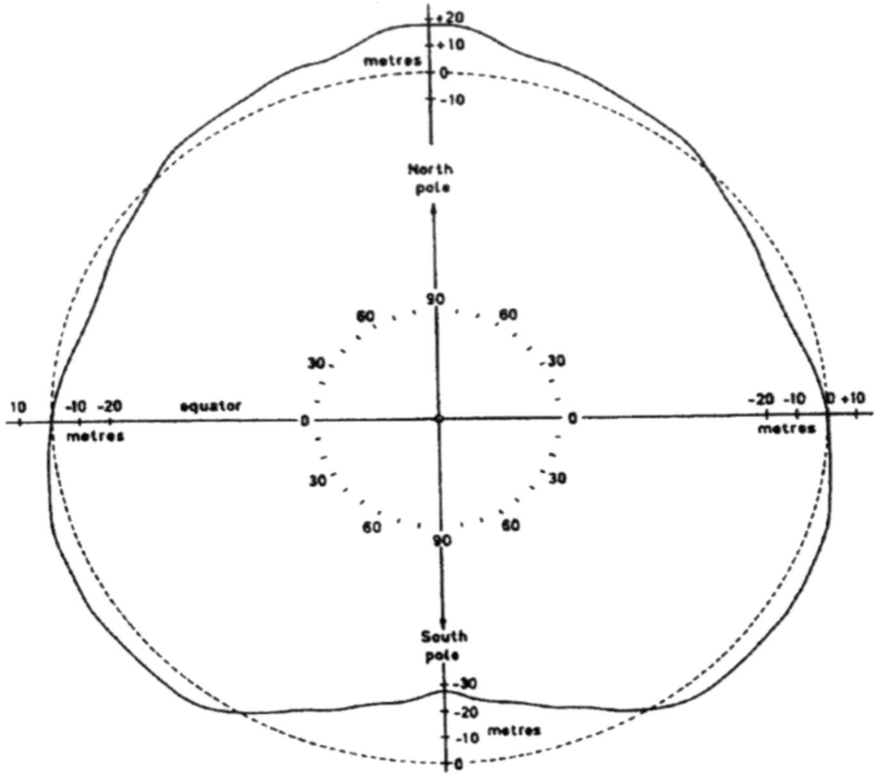

Fig. 4 Flattening of the Earth

determining orbits for geosynchronous Earth orbits (GEOs). As the ESMO satellite will keep a high altitude orbit around Earth before being launched into Moon orbit, it is relevant [4].

3.2 Atmospheric Drag

As we are considering LEO satellite, atmospheric drag dissipates energy from the satellite in orbit. The orbital height of the satellite will reduce slightly. It is inversely proportional to air density. Air density decreases with rise in altitude [5]. The ESMO satellite will be in an orbit where atmospheric drag is relevant and possibly in the start of the transfer orbit.

The drag force FD on a body acts in the opposite direction of the velocity vector and is given by the equation

$$\mathrm{AD_{drag}} = \frac{1}{2}\rho v^2 Acd_{co}$$

where

$\mathrm{AD_{drag}}$ is the drag force acting on the satellite,
ρ is the density of atmosphere at that level,
v is velocity of satellite,
A is frontal area of satellite,
Cd_{co} is the drag coefficient.

3.3 *Solar and Lunar Gravity Perturbation*

This solar and lunar perturbation causes tidal forces that perturb the satellite from its orbit [6]. The formulae for the perturbation calculation due to solar and lunar gravity are given as follows:

$$\Omega\mathrm{moon} = -\frac{0.00338\cos(i)}{n}$$

$$\Omega\mathrm{sun} = -\frac{0.00154\cos(i)}{n}$$

$$\omega\mathrm{moon} = \frac{0.00169(4 - 5\sin^2(i))}{n}$$

$$\omega\mathrm{sun} = \frac{0.00074(4 - 5\sin^2(i))}{n}$$

where

i orbit inclination,
n number of orbit revolutions per day,
Ω and ω degrees per day.

4 Trajectories

4.1 *Hohmann Transfer*

The Hohmann transfer is the traditional way for constructing a satellite transfer to the Moon. It uses two-body dynamics and is constructed by determining an elliptic transfer of orbit from an Earth parking orbit to the Moon's orbit [7]. It is an expensive approach when the ratio of the two radii of the orbits is large as it requires a large velocity. This subsection will therefore only describe it briefly (Fig. 5).

4.2 *Patched Conic Approximation (PCA) Method*

PCA is a well-known method. When used on a transfer between Earth and the Moon, it is also referred to as the lunar patched conic [8]. It is a good way to make an approximation of a simulation of a lunar transfer orbit. Still it is restricted to the two-body problem, but more than one, two-body problem are used, hence the name of the method (Fig. 6).

Well beyond the orbit of the Moon, so the patched conic method is a rough approximation.

1. Earth departure; Earth's gravitational pull dominates
2. Arrival at the Moon; Moon's gravitational pull dominates

This trajectory is a Hohmann-transfer ellipse around the Sun. The Hohmann transfer was described in Sect. 3.2.1. In the second region, motions are relative to Earth. This is really the first part of the trajectory. Here, the satellite escapes Earth and arrives at the SOI with the required velocity to enter into the heliocentric transfer orbit of region one. The satellite needs to increase its velocity in the parking orbit by a certain amount. In the third region, motions are relative to the Moon. Here, the satellite needs to be slowed down. If not, it will only swing by the Moon on a hyperbolic trajectory and depart the SOI on the other side.

Fig. 5 Hohmann transfer

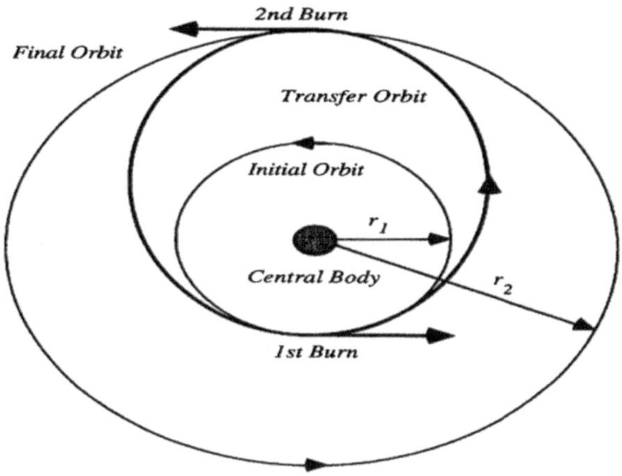

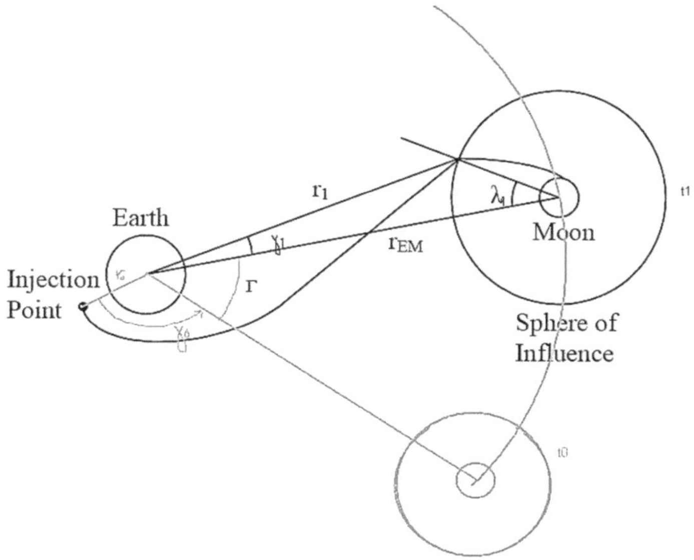

Fig. 6 Patched conic approximation method

5 Results

Satellite solar radiation impact

- Area of satellite facing the Sun decreases to a certain limit (threshold limit) with respect to the decrease in deceleration due to solar radiation, and after the threshold limit it increases (Fig. 7).
- Mass of satellite facing the Sun decreases to a certain limit (threshold limit) with respect to the decrease in deceleration due to solar radiation, and after the threshold limit it increases.

Effect of atmospheric drag (Fig. 8: changes in satellite semi-major axis)

This graph is hyperbolic in nature. Ballistic constant increases with the decrease in acceleration due to atmospheric drag.

- Change in velocity per revolution due to atmospheric drag versus ballistic constant
  This graph has a logarithmic decrement kind of nature. Ballistic constant increases with the decrease in change in velocity per revolution due to atmospheric drag.
- Lifetime of satellite in seconds versus deceleration due to atmospheric drag
  The lifetime of satellites decreases with the decrease in deceleration due to atmospheric drag
- Change in revolution period due to atmospheric versus ballistic constant

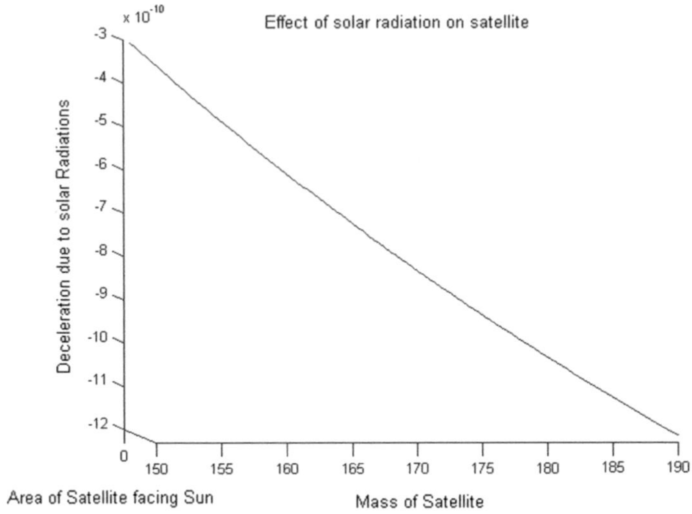

Fig. 7 Solar radiation on satellite mass versus deceleration

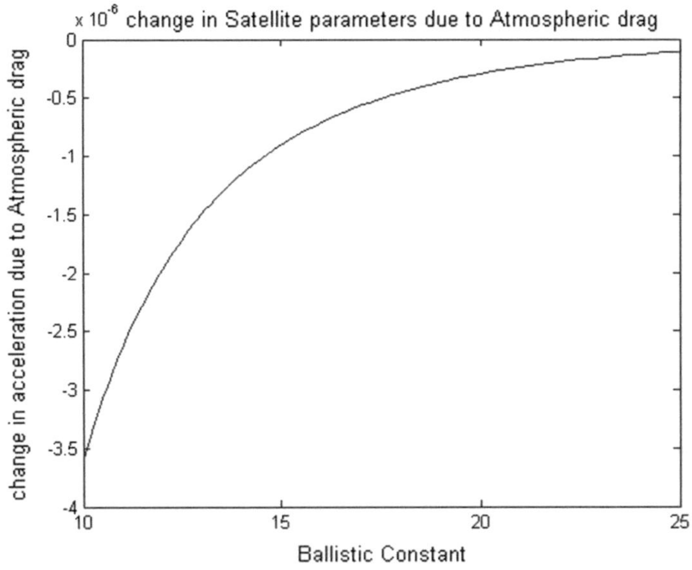

Fig. 8 Changes in satellite semi-major axis

This graph is hyperbolic in nature. Ballistic constant increases with decrease in change in revolution period due to atmospheric drag (Figs. 9 and 10).

Perturbation due to earth shape

Variation in longitude of ascending node with semi-major axis and angle of inclination.

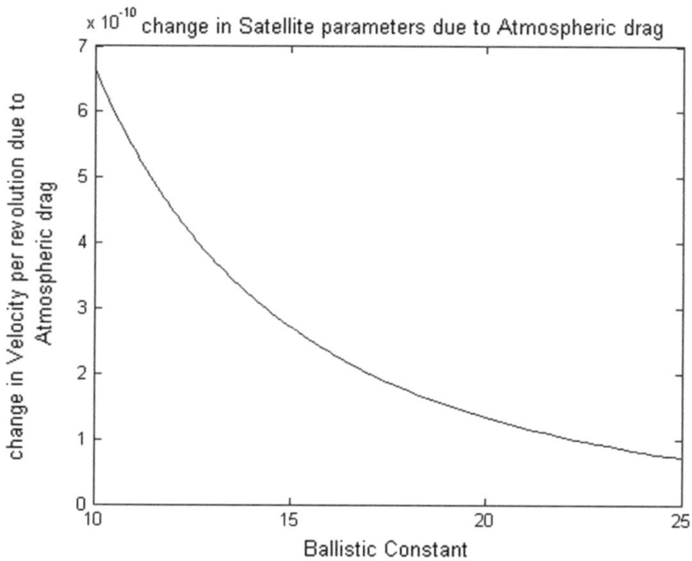

Fig. 9 Change in satellite time period

Semi-major axis increases with increase in longitude of ascending node to a maximum point after which it tends to remain constant and same goes for angle of inclination (Fig. 11).

- Variation in argument of perigee with semi-major axis and angle of inclination.

Semi-major axis decreases with the increase in argument of perigee to a minimum value after which it tends to remain constant, and angle of inclination increases with the increase in argument of perigee to a minimum value after which it tends to remain constant (Fig. 12).

Perturbation due to lunar and solar gravity

- Variation in longitude of ascending node with semi-major axis angle of inclination due to moon
  Semi-major axis decreases with the decrease in longitude of ascending node while angle of inclination increases (Fig. 13).
- Variation in longitude of ascending node with semi-major axis angle of inclination due to Sun
- Semi-major axis decreases with the decrease in longitude of ascending node while angle of inclination increases (Fig. 14).
- Variation in argument of perigee with semi-major axis angle of inclination due to moon
  Argument of perigee has no dominant effect on semi-major axis while angle of inclination slightly increases with the increase in argument of perigee (Fig. 15).

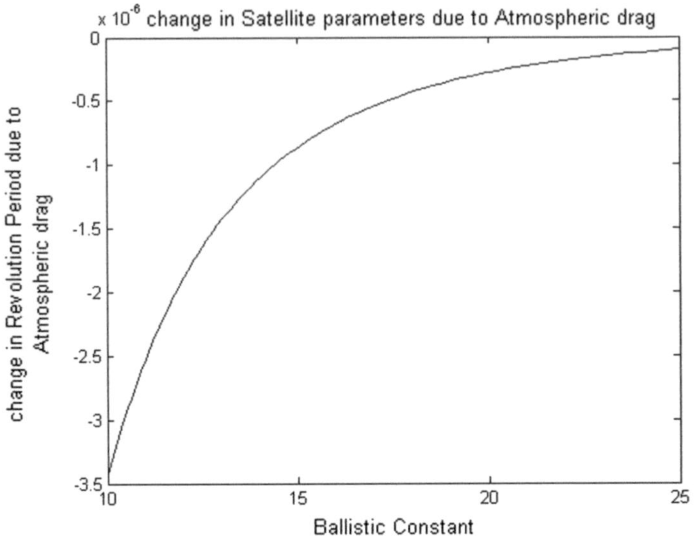

Fig. 10 Change in satellite velocity

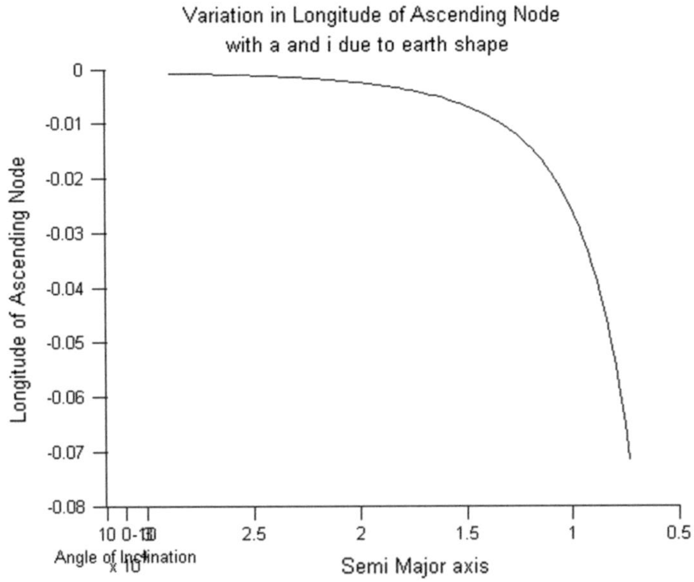

Fig. 11 Variation in ascending node for shape of Earth

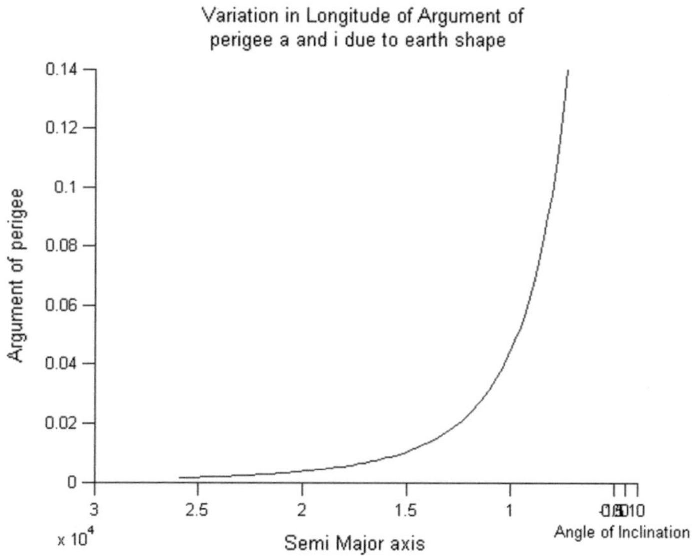

Fig. 12 Variation in longitude of ascending node

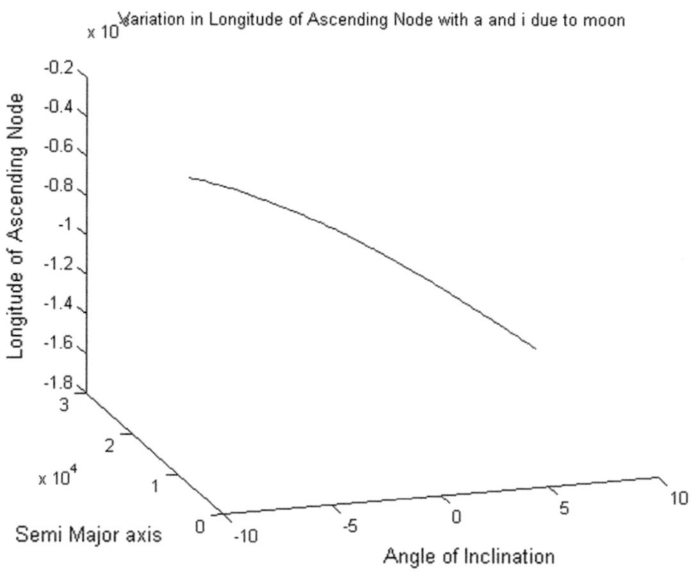

Fig. 13 Variation in LAAN due to Moon

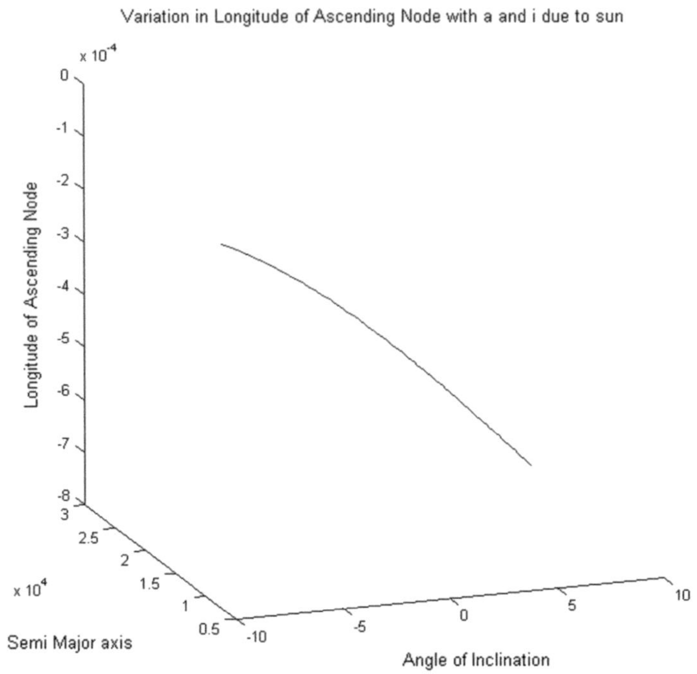

Fig. 14 Variation in LAAN due to Sun

- Variation in argument of perigee with semi-major axis angle of inclination due to Sun
  Argument of perigee has no dominant effect on semi-major axis while angle of inclination slightly increases with the increase in argument of perigee (Fig. 16).

Plotting of trajectories

- **Trajectory analysis**
  As per the requirements, platform for the testing of satellite is chosen as geosynchronous orbit. As in this orbit, the satellite remains constant over a particular point, so the data accumulated is more accurate and precise. Steps for reaching this orbit are:

Hohmann transfer

- First the satellite is launched in a highly elliptical orbit.
- When the satellite reaches apogee of the orbit, velocity injection is done to send the satellite in a circular orbit with firing retro rockets for velocity reduction.
- When the satellite reaches the perigee position of the circular orbit, rockets fired put the satellite in the geosynchronous satellite (Fig. 17).

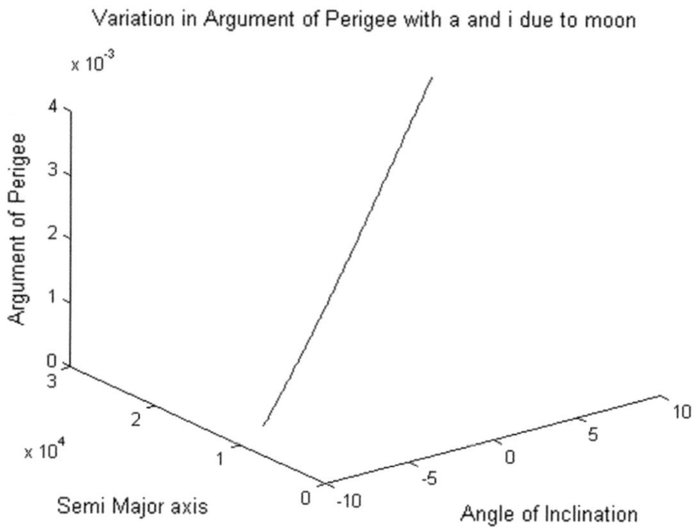

Fig. 15 Variation in argument of perigee due to Moon

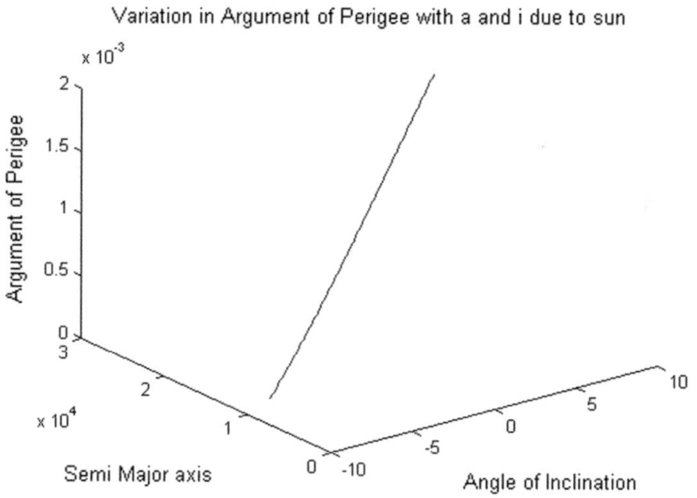

Fig. 16 Variation in argument of perigee due to Sun

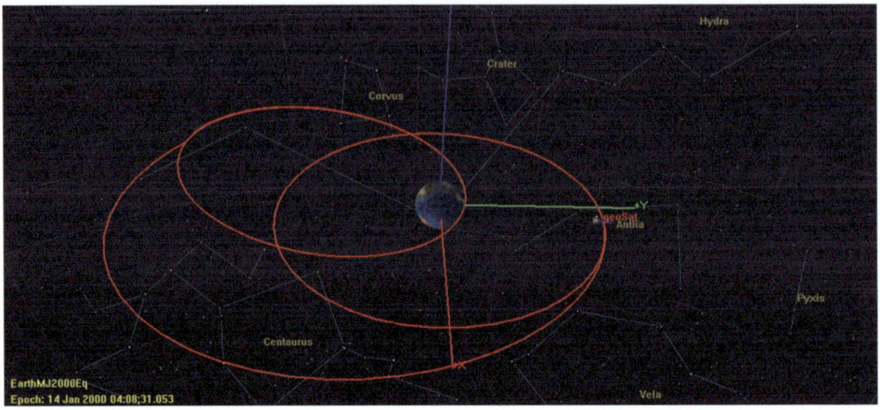

Fig. 17 Hohmann transfer

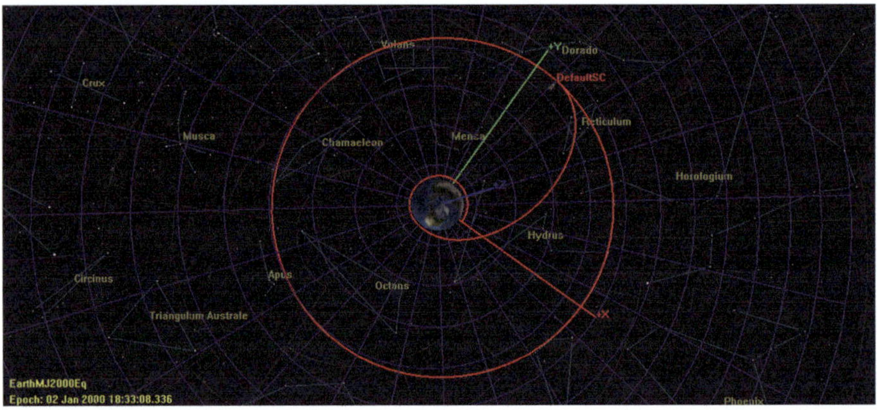

Fig. 18 One-tangent burn

One-tangent burn (Fig. 18)
Moon's influence on satellite

- Rockets are fired from the perigee of the geosynchronous orbit and hence it comes in the sphere of influence of moon.
- Due to the firing of rockets for short interval, spiral transfer is actuated in the trajectory of the satellite.
- The trajectory as seen from Earth is shown in (Fig. 19):

Trajectory as viewed from Moon (Fig. 20)

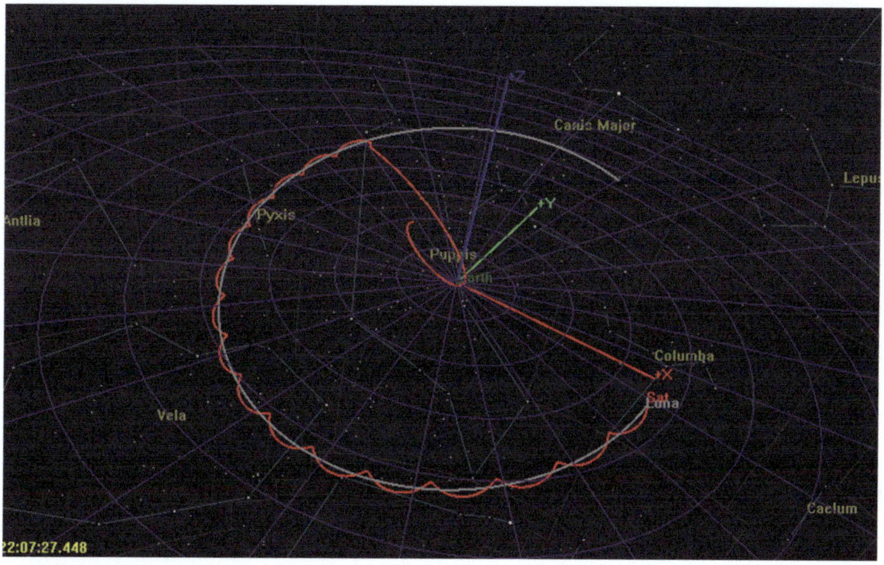

Fig. 19 Lunar transfer

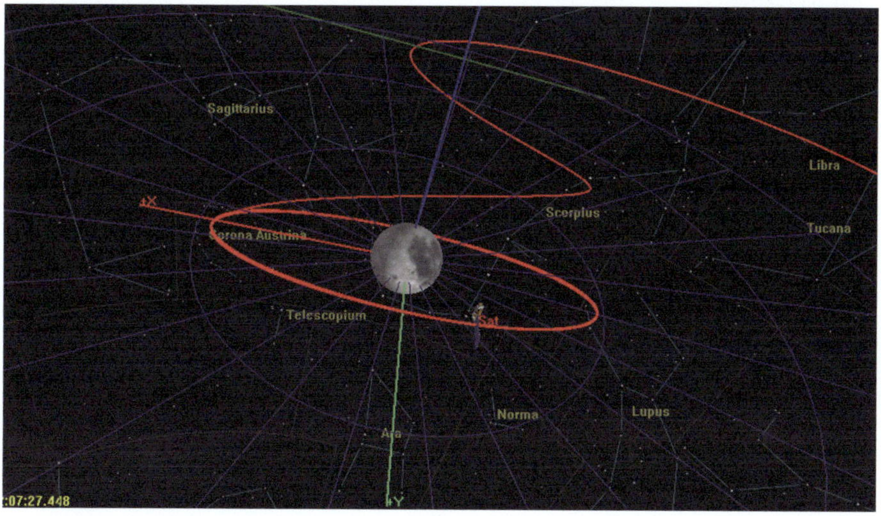

Fig. 20 Lunar transfer from Moon

6 Conclusion

This paper is mainly focused how to maintain the orientation of satellite in low Earth orbit (LEO) from perturbation and also reduce the **orbital decay** and increase the **lifetime** of the satellite and various perturbations like lunar gravity and atmospheric drag will be considered and will be compared with the ideal cases. This perturbation force is affecting the satellite from the original orbit. Then will affect the Keplerian elements. This variation is called as secular variation might be less than orbital period or greater than orbital period. This study states for low Earth orbit satellite have more aerodynamic drag and gravitational attraction due to Earth, and high Earth orbit causes more force due to moon attraction and magnetic effect and flattering of earth.

References

1. Anderson, John D. 2004. Introduction to flight. 5th ed. McGraw-Hill, ISBN 0-07-282569-3.
2. Wertz, J.R. 1978. Spacecraft attitude determination and control. Kluwer.
3. Bate, Mueller, White. 1971. Fundamentals of astrodynamics. Dover.
4. Perry, W.R. 1967. Orbital mechanics. In *Marks' Standard Handbook for Mechanical Engineers,* 7th ed, ed. Theodore Baumeister, 11:151–52. New York: McGraw Hill. ISBN 0-07-142867-4.
5. Thomson, W.T. 1961. Introduction to space dynamics. Dover.
6. Fellenz, D.W. 1967. Atmospheric entry. In *Marks' Standard Handbook for Mechanical Engineers,* 7th ed, ed. Theodore Baumeister, 11:155–58. New York: McGraw Hill. ISBN 0-07-142867-4.
7. Kromis, A.J. 1967. Powered-Flight-trajectory analysis. In *Marks' standard handbook for mechanical engineers,* 7th ed, ed. Theodore Baumeister, 11:154–55. New York: McGraw Hill. ISBN 0-07-142867-4.
8. Russell, J.W. 1967. Lunar and interplanetary flight mechanics. *In Marks' Standard Handbook for Mechanical Engineers,* 7th ed, ed. Theodore Baumeister 11:152–54. New York: McGraw Hill. ISBN 0-07-142867-4.

MIX
Papier aus verantwortungsvollen Quellen
Paper from responsible sources
FSC® C105338

If you have any concerns about our products,
you can contact us on
ProductSafety@springernature.com

In case Publisher is established outside the EU,
the EU authorized representative is:
Springer Nature Customer Service Center GmbH
Europaplatz 3, 69115 Heidelberg, Germany

Printed by Libri Plureos GmbH
in Hamburg, Germany